An Introduction to Mathematical Statistics

An Introduction to Mathematical Statistics

Fetsje Bijma, Marianne Jonker, Aad van der Vaart

Amsterdam University Press

Original publication: Fetsje Bijma, Marianne Jonker, Aad van der Vaart, *Inleiding in de statistiek*. Epsilon Uitgaven, 2016 [ISBN 978-90-5041-135-6]
© Fetsje Bijma, Marianne Jonker, Aad van der Vaart, 2016

Translated by: Reinie Erné

Cover design: V3-Services, Baarn

Amsterdam University Press English-language titles are distributed in the US and Canada by the University of Chicago Press.

ISBN	978 94 6298 510 0
e-ISBN	978 90 4853 611 5 (pdf)
NUR	916
DOI	10.5117/9789462985100

© Fetsje Bijma, Marianne Jonker, Aad van der Vaart / Amsterdam University Press B.V., Amsterdam 2017

PREFACE

This book gives an introduction into mathematical statistics. It was written for bachelor students in (business) mathematics, econometrics, or any other subject with a solid mathematical component. We assume that the student already has solid knowledge of probability theory to the extent of a semester course at the same level.

In Chapter 1, we give the definition and several examples of a statistical model, the foundation of every statistical procedure. Some techniques from descriptive statistics that can assist in setting up and validating statistical models are discussed in Chapter 2. The following chapters discuss the three main topics in mathematical statistics: estimating, testing, and constructing confidence regions. These subjects are discussed in Chapters 3, 4, and 5, respectively. Next, Chapter 6 provides deeper theoretical insight, in particular into the question under what circumstances and in what sense certain statistical models are mathematically optimal. In Chapter 7, we describe several regression models that are commonly used in practice. The theory from the previous chapters is applied to estimate and test unknown model parameters and give confidence regions for them. Finally, in Chapter 8, we discuss model selection. In that chapter, various criteria are presented that can be used to find the best-fitting model from a collection of (regression) models. Sections and examples marked with a * are more difficult and do not belong to the basic subject matter of mathematical statistics. Every chapter concludes with a summary.

In Appendix A, we recall elements from probability theory that are relevant for understanding the subject matter of this book. In Appendix B, we discuss properties of the multivariate normal distribution, which is used in several sections. Appendix C contains tables with values of distribution and quantile functions of several distributions to which we refer in the text. These are meant to be used at home or during problem sessions. In "real life," these tables are no longer used: the computer is faster, more accurate, and easier to use. The statistical package R, for example, contains standard functions for the distribution function, the density function, and the quantile function of all standard distributions.

The mathematical style of this book is more informal than that of many mathematics books. Theorems and lemmas are not always proved or may be formulated in an informal manner. The reason is that a pure mathematical treatment is only possible using measure theory, of which we do not assume any knowledge. On the other hand, the relevance and motivation of the theorems are also clear without going into all the details.

Each chapter concludes with a case study. It often contains a statistical problem that is answered as well as possible based on the collected data, using the statistical techniques and methods available at that point in the book. The R-code and data of these applications, as well as the data of several case studies described in the book, are available and can be downloaded from the book's webpage at http://www.aup.nl.

Though this book includes examples, practice is indispensable to gain insight into the subject matter. The exercises at the end of each chapter include both theoretical and more practically oriented problems. Appendix D contains short answers to most exercises. Solutions that consist of a proof are not included.

The book has taken form over a period of 20 years. It was originally written in Dutch and used yearly for the course "Algemene Statistiek" (General Statistics) for (business) mathematics and econometrics students given by the mathematics department of VU University Amsterdam. The various lecturers of the course contributed to the book to a greater or lesser extent. One of them is Bas Kleijn. We want to thank him for his contribution to the appendix on probability theory. More than 2000 students have studied the book. Their questions on the subject and advice on the presentation have helped give the book its present form. They have our thanks. The starting point of the book was the syllabus "Algemene Statistiek" (General Statistics) of J. Oosterhoff, professor of mathematical statistics at VU University Amsterdam until the mid-'90s. We dedicate this book to him.

In 2013, the first edition of this book was published in Dutch, and three years later, in 2016, the second Dutch edition came out. This second edition has been translated into English, with some minor changes. We thank Reinie Erné for translation.

Amsterdam and Leiden, March 2017

FURTHER READING

Reference [1] is an introduction to many aspects of statistics, somewhat comparable to *An Introduction to Mathematical Statistics*. References [3] and [4] are standard books that focus more on mathematical theory, and estimation and tests, respectively. Reference [6] describes the use of asymptotic methods in statistics, on a higher mathematical level, and gives several proofs left out in *An Introduction to Mathematical Statistics*. Reference [5] is a good starting point for whoever wants to delve further into the Bayesian thought process, and reference [7] provides the same for nonparametric methods, which are mentioned in *An Introduction to Mathematical Statistics* but perhaps less prominently than in current practice. Reference [2] elaborates on the relevance of modeling using regression models, for example to draw causal conclusions in economic or social sciences.

[1] Davison, A.C., (2003). *Statistical models*. Cambridge University Press.
[2] Freedman, D., (2005). *Statistical models: theory and applications*. Cambridge University Press.
[3] Lehmann, E.L. and Casella, G., (1998). *Theory of point estimation*. Springer.
[4] Lehmann, E.L. and Romano, J.P., (2005). *Testing statistical hypotheses*. Springer.
[5] Robert, C.P., (2001). *The Bayesian choice*. Springer-Verlag.
[6] van der Vaart, A.W., (1998). *Asymptotic statistics*. Cambridge University Press.
[7] Wasserman, L., (2005). *All of nonparametric statistics*. Springer.

TABLE OF CONTENTS

1 Introduction

1.1 What Is Statistics?

Statistics is the art of modeling (describing mathematically) situations in which probability plays a role and drawing conclusions based on data observed in such situations.

Here are some typical research questions that can be answered using statistics:
 (i) What is the probability that the river the Meuse will overflow its banks this year?
 (ii) Is the new medical treatment significantly better than the old one?
(iii) What is the margin of uncertainty in the prediction of the number of representatives for political party A?

Answering such questions is not easy. The three questions above correspond to the three basic concepts in mathematical statistics: *estimation*, *testing*, and *confidence regions*, which we will deal with extensively in this book. Mathematical statistics develops and studies methods for analyzing observations based on probability models, with the aim to answer research questions as above. We discuss a few more examples of research questions, observed data, and corresponding statistical models in Section 1.2.

In contrast to mathematical statistics, *descriptive statistics* is concerned with summarizing data in an insightful manner by averaging, tabulating, making graphical representations, and processing them in other ways. Descriptive methods are only discussed briefly in this book, as are methods for collecting data and the modeling of data.

1

1.2 Statistical Models

In a sense, the direction of statistics is precisely the opposite of that of probability theory. In probability theory, we use a given probability distribution to compute the probabilities of certain events. In contrast, in statistics, we observe the results of an experiment, but the underlying probability distribution is (partly) unknown and must be derived from the results. Of course, the experimental situation is not entirely unknown. All known information is used to construct the best possible statistical model. A formal definition of a "statistical model" is as follows.

Definition 1.1 Statistical model

A statistical model is a collection of probability distribution on a given sample space.

The interpretation of a statistical model is: the collection of all possible probability distributions of the observation X. Usually, this observation is made up of "subobservations," and $X = (X_1, \ldots, X_n)$ is a random vector. When the variables $X_1, \ldots, X_n$ correspond to independent replicates of the same experiment, we speak of a *sample*. The variables $X_1, \ldots, X_n$ are then independent, identically distributed, and their joint distribution is entirely determined by the marginal distribution, which is the same for all X_i. In that case, the statistical model for $X = (X_1, \ldots, X_n)$ can be described by a collection of (marginal) probability densities for the subobservations $X_1, \ldots, X_n$.

The concept of "statistical model" only truly becomes clear through examples. As simply as the mathematical notion of "statistical model" is expressed in the definition above, so complicated is the process of the statistical modeling of a given practical situation. The result of a statistical study depends on the construction of a good model.

Example 1.2 Sample

In a large population consisting of N persons, a proportion p has a certain characteristic A; we want to "estimate" this proportion p. It is too much work to examine everyone in the population for characteristic A. Instead, we randomly choose n persons from the population, with replacement. We observe (a realization of) the random variables $X_1, \ldots, X_n$, where

$$X_i = \begin{cases} 0 & \text{if the } i\text{th person does not have } A, \\ 1 & \text{if the } i\text{th person has } A. \end{cases}$$

Because of the set-up of the experiment (sampling with replacement), we know beforehand that $X_1, \ldots, X_n$ are independent and Bernoulli-distributed. The latter means that

$$P(X_i = 1) = 1 - P(X_i = 0) = p$$

for $i = 1, \ldots, n$. There is no prior knowledge concerning the parameter p, other then $0 \leq p \leq 1$. The observation is the vector $X = (X_1, \ldots, X_n)$. The statistical model for X consists of all possible (joint) probability distributions of X whose coordinates $X_1, \ldots, X_n$ are independent and have a Bernoulli distribution. For every possible value of p, the statistical model contains exactly one probability distribution for X.

It seems natural to "estimate" the unknown p by the proportion of the persons with property A, that is, by $n^{-1}\sum_{i=1}^{n} x_i$, where x_i is equal to 1 or 0 according to whether the person has property A or not. In Chapter 3, we give a more precise definition of "estimating." In Chapter 5, we use the model we just described to quantify the difference between this estimator and p, using a "confidence region." The population and sample proportions will almost never be exactly equal. A confidence region gives a precise meaning to the "margin of errors" that is often mentioned with the results of an opinion poll. We will also determine how large that margin is when we, for example, study 1000 persons from a population, a common number in polls under the Dutch population.

Example 1.3 Measurement errors

If a physicist uses an experiment to determine the value of a constant μ repeatedly, he will not always find the same value. See, for example, Figure 1.1, which shows the 23 determinations of the speed of light by Michelson in 1882. The question is how to "estimate" the unknown constant μ from the observations, a sequence of numbers $x_1, \ldots, x_n$. For the observations in Figure 1.1, this estimate will lie in the range 700–900, but we do not know where. A statistical model provides support for answering this question. Probability models were first applied in this context at the end of the 18th century, and the normal distribution was "discovered" by Gauss around 1810 for the exact purpose of obtaining insight into the situation described here.

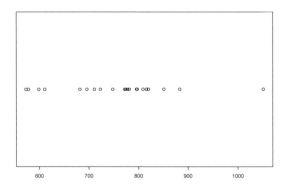

Figure 1.1. The results of the 23 measurements of the speed of light by Michelson in 1882. The scale along the horizontal axis gives the measured speed of light (in km/s) minus 299000 km/s.

3

If the measurements are all carried out under the same circumstances, independently of the past, then it is reasonable to include in the model that these numbers
are realizations of independent, identically distributed random variables $X_1, \ldots, X_n$.
The measurement errors $e_i = X_i - \mu$ are then also random variables. A common
assumption is that the expected measurement error is equal to 0, in other words, $Ee_i =
0$, in which case $EX_i = E(e_i + \mu) = \mu$. Since we have assumed that $X_1, \ldots, X_n$
are independent random variables and all have the same probability distribution, the
model for $X = (X_1, \ldots, X_n)$ is fixed by the choice of a statistical model for X_i.
For X_i, we propose the following model: all probability distributions with finite
expectation μ. The statistical model for X is then: all possible probability distributions
of $X = (X_1, \ldots, X_n)$ such that the coordinates $X_1, \ldots, X_n$ are independent and
identically distributed with expectation μ.

Physicists often believe that they have more prior information and make more
assumptions on the model. For example, they assume that the measurement errors
are normally distributed with expectation 0 and variance σ^2, in other words, that the
observations $X_1, \ldots, X_n$ are normally distributed with expectation μ and variance σ^2.
The statistical model is then: all probability distributions of $X = (X_1, \ldots, X_n)$ such
that the coordinates are independent and $N(\mu, \sigma^2)$-distributed.

The final goal is to say something about μ. In the second model, we know more,
so we should be able to say something about μ with more "certainty." On the other
hand, there is a higher "probability" that the second model is incorrect, in which case
the gain in certainty is an illusory one. In practice, measurement errors are often, but
not always, approximately normally distributed. Such normality can be justified using
the central limit theorem (see Theorem A.28) if a measurement error can be viewed
as the sum of a large number of small independent measurement errors (with finite
variances), but cannot be proved theoretically. In Chapter 2, we discuss methods to
study normality on the data itself.

The importance of a precisely described model is, among other things, that
it allows us to determine what is a meaningful way to "estimate" μ from the
observations. An obvious choice is to take the average of $x_1, \ldots, x_n$. In Chapter 6,
we will see that this is the best choice (according to a particular criterion) if the
measurement errors indeed have a normal distribution with expectation 0. If, on the
other hand, the measurement errors are Cauchy-distributed, then taking the average
is disastrous. This can be seen in Figure 1.2. It shows the average $n^{-1}\sum_{i=1}^{n} x_i$, for
$n = 1, 2, \ldots, 1000$, of the first n realizations $x_1, \ldots, x_{1000}$ of a sample from a
standard Cauchy distribution. The behavior of the averages is very chaotic, and they
do not converge to 0. This can be explained by the remarkable theoretic result that the
average $n^{-1}\sum_{i=1}^{n} X_i$ of independent standard Cauchy-distributed random variables
$X_1, \ldots, X_n$ also has a standard Cauchy distribution. So taking the averages changes
nothing!

Example 1.4 Poisson stocks

A certain product is sold in numbers that vary for different retailers and fluctuate over
time. To estimate the total number of items needed, the central distribution center

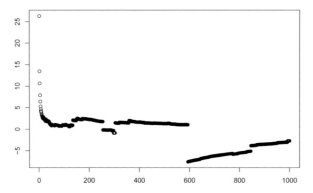

Figure 1.2. Cumulative averages (vertical axis) of $n = 1, 2, \ldots, 1000$ (horizontal axis) realizations from the standard Cauchy distribution.

registers the total number of items sold per week and retailer for several weeks. They observe $x = (x_{1,1}, x_{1,2}, \ldots, x_{I,J})$, where $x_{i,j}$ is the number of items sold by retailer i in week j. The observation is therefore a vector of length the product IJ of the number of retailers and the number of weeks, with integral coordinates. The observations can be seen as realizations of the random vector $X = (X_{1,1}, X_{1,2}, \ldots, X_{I,J})$. Many different statistical models for X are possible and meaningful in given situations. A common (because often reasonably fitting) model states:

- Every $X_{i,j}$ is Poisson-distributed with unknown parameter $\mu_{i,j}$.
- The $X_{1,1}, \ldots, X_{I,J}$ are independent.

This fixes the probability distribution of X up to the expectations $\mu_{i,j} = EX_{i,j}$. It is these expectations that the distribution center is interested in. The total expected demand in week j, for example, is $\sum_i \mu_{i,j}$. Using the Poisson-character of the demand $\sum_i X_{i,j}$, the distribution center can choose a stock size that gives a certain (high) probability that there is sufficient stock.

The goal of the statistical analysis is to deduce $\mu_{i,j}$ from the data. Up to now, we have left the $\mu_{i,j}$ completely "free." This makes it difficult to estimate them from the data, because only one observation, $x_{i,j}$, is available for each $\mu_{i,j}$. It seems reasonable to reduce the statistical model by including prior assumptions on $\mu_{i,j}$. We could, for example, postulate that $\mu_{i,j} = \mu_i$ does not depend on j. The expected number of items sold then depends on the retailer but is constant over time. We are then left with I unknowns, which can be "estimated" reasonable well from the data provided that the number of weeks J is sufficiently large. More flexible, alternative models are $\mu_{i,j} = \mu_i + \beta_i j$ and $\mu_{i,j} = \mu_i + \beta \mu_i j$, with, respectively, $2I$ and $I+1$ parameters. Both models correspond to a linear dependence of the expected demand on time. ▭

Example 1.5 Regression

Tall parents in general have tall children, and short parents, short children. The heights of the parents have a high predictive value for the final (adult) length of their children, their heights once they stop growing. More factors influence it. The gender of the child, of course, plays an important role. Environmental factors such as healthy eating habits and hygiene are also important. Through improved nutrition and increased hygiene in the past 150 years, factors that hinder growth like infectious diseases and malnutrition have decreased in most Western countries. Consequently, the average height has increased, and each generation of children is taller.

The target height of a child is the height that can be expected based on the heights of the parents, the gender of the child, and the increase of height over generations. The question is how the target height depends on these factors.

Let Y be the height a child will reach, let x_1 and x_2 be the heights of the biological father and mother, respectively, and let x_3 be an indicator for the gender ($x_3 = -1$ for a girl and $x_3 = 1$ for a boy). The target height EY is modeled using a so-called linear regression model

$$EY = \beta_0 + \beta_1 x_1 + \beta_2 x_2 + \beta_3 x_3,$$

where β_0 is the increase in average height per generation, β_1 and β_2 are the extent to which the heights of the parents influence the target height of their offspring, and β_3 is the deviation of the target height from the average final height that is caused by the gender of the child. Since men are, on average, taller than women, β_3 will be positive.

The model described above does not say anything about individual heights, only about the heights of the offspring of parents of a certain height. Two brothers have the same target height, since they have the same biological parents, the same gender, and belong to the same generation. The actual final height Y can be described as

$$Y = \beta_0 + \beta_1 x_1 + \beta_2 x_2 + \beta_3 x_3 + e,$$

where $e = Y - EY$ is the deviation of the actual final height Y from the target height EY. The observation Y is also called the dependent variable, and the variables x_1, x_2, and x_3 the independent or predictor variables. The deviation e is commonly assumed to have a normal distribution with expectation 0 and unknown variance σ^2. The final height Y then has a normal distribution with expectation $\beta_0 + \beta_1 x_1 + \beta_2 x_2 + \beta_3 x_3$ and variance σ^2.

In the Netherlands, the increase in the height of youth is periodically recorded. In 1997, the Fourth National Growth Study took place. Part of the study was to determine the correlation between the final height of the children and the heights of their parents. To determine this correlation, data were collected on adolescents and their parents. This resulted in the following observations: $(y_1, x_{1,1}, x_{1,2}, x_{1,3})$, $\ldots, (y_n, x_{n,1}, x_{n,2}, x_{n,3})$, where y_i is the height of the ith adolescent, $x_{i,1}$ and $x_{i,2}$ are the heights of the biological parents, and $x_{i,3}$ is an indicator for the gender of the ith adolescent. Suppose that the observations are independent replicates of linear regression model given above; in other words, given $x_{i,1}, x_{i,2}$, and $x_{i,3}$, the variable Y_i has expectation $\beta_0 + \beta_1 x_{i,1} + \beta_2 x_{i,2} + \beta_3 x_{i,3}$ and variance σ^2. The parameters $(\beta_0, \beta_1, \beta_2, \beta_3)$ are unknown and can be estimated from the observations. For a simple interpretation of the model, we choose $\beta_1 = \beta_2 = 1/2$, so that the target height is equal to the average height of the parents corrected for the gender of the child and the influence of time. The parameters β_0 and β_3 are equal to the increase in height in the previous generation and half the average height difference between men and women. These parameters are estimated using the least-squares method (see Example 3.44). The parameter β_0 is estimated to be 4.5 centimeters, and β_3 is estimated to be 6.5 centimeters.[†] The estimated regression model is then equal to

$$(1.1) \qquad Y = 4.5 + \frac{1}{2}(x_1 + x_2) + 6.5x_3 + e.$$

Figure 1.3 shows the heights of 44 young men (on the left) and 67 young women (on the right) set out against the average heights of their parents.[‡] The line is the estimated regression line found in the Fourth National Growth Study.

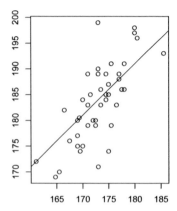 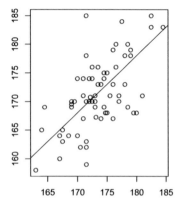

Figure 1.3. Heights (in cm) of sons (left) and daughters (right) set out against the average height of their parents. The line is the regression line found in the Fourth National Growth Study.

We can use the estimated regression model found in the Fourth National Growth Study to predict the final heights of children born now. We must then assume that the height increase in the next generation is again 4.5 centimeters and that the average height difference between men and women remains 13 centimeters. Based on the model presented above, the target heights of sons and daughters of a man of height 180 cm ($\approx$ 71 in or 5'9") and a woman of height 172 cm are $4.5 + (180 + 172)/2 + 6.5 = 187$ cm and $4.5 + (180 + 172)/2 - 6.5 = 174$ cm, respectively.

Other European countries use other models. In Switzerland, for example, the target height is

$$EY = 51.1 + 0.718 \frac{x_1 + x_2}{2} + 6.5x_3.$$

[†] An inch is approximately 2.54 cm, so 4.5 cm corresponds to $4.5/2.54 \approx 1.8$ in and 6.5 cm ≈ 2.6 in.

[‡] Source: The data were gathered by the department of Biological Psychology of VU University Amsterdam during a study on health, lifestyle, and personality. The data can be found on the book's webpage at http://www.aup.nl under heightdata.

The target heights of sons and daughters of parents of the same heights as above are now 184 and 171 centimeters, respectively.

In the example above, there is a linear correlation between the response Y and the unknown parameters $\beta_0, \ldots, \beta_3$. In that case, we speak of a linear regression model. The simplest linear regression model is that where there is only one predictor variable:

$$Y = \beta_0 + \beta_1 x + e;$$

this is called a simple linear regression model (in contrast to the multiple linear regression model when there are more predictor variables).

In general, we speak of a regression model when there is a specific correlation between the response Y and the observations $x_1, \ldots, x_p$:

$$Y = f_\theta(x_1, \ldots, x_p) + e,$$

where f_θ describes the correlation between the observations $x_1, \ldots, x_p$ and the response Y, and the random variable e is an unobservable measurement error with expectation 0 and variance σ^2. If the function f_θ is known up to the finite-dimensional parameter θ, we speak of a parameterized model. The linear regression model is an example of this; in this model, we have $\theta = (\beta_0, \ldots, \beta_p) \in \mathbb{R}^{p+1}$ and $f_\theta(x_1, \ldots, x_p) = \beta_0 + \beta_1 x_1 + \ldots + \beta_p x_p$. The regression model is then fixed if we know the values of θ and σ^2. The function f_θ can, however, also be known up to the finite-dimensional parameter θ and an infinite-dimensional parameter. We then speak of a semiparametric model. An example of a semiparametric model is the Cox regression model. This model is described at the end of this chapter, after the exercises. In Chapter 7, we discuss several regression models in detail, including the linear regression model and the Cox regression model.

Example 1.6 Water levels

In the 20th century (between 1910 and 2000), extreme water levels were measured 70 times in the river the Meuse near the town of Borgharen (Netherlands). Here, "extreme" is defined by Rijkswaterstaat (the Dutch government agency responsible for the management of waterways) as "more than 1250 m³/s." The maximal water flows during those 70 periods are shown in chronological order in Figure 1.4.[b] The problem is predicting the future. Rijkswaterstaat is particularly interested in how high the dikes must be to experience flooding at most once every 10 000 years. We can use a hydraulic model to compute the height of the water from the water flow.

---

[b] The data can be found on the book's webpage at http://www.aup.nl under maxflows and flows1965.

Since the maximal water flows $x_1, \ldots, x_{70}$ were measured in (mostly) different years, and the water level of the Meuse depends mainly on the weather in the Ardennes and further upstream, it is not unreasonable to view these numbers as realizations of independent random variables $X_1, \ldots, X_{70}$. The assumption that these parameters are also identically distributed is somewhat questionable because the course of the Meuse (and also the climate) has gradually changed during of the last century, but this assumption is usually made anyway. We can then view $X_1, \ldots, X_{70}$ as independent copies of one variable X and use the measured values $x_1, \ldots, x_{70}$ to answer the question.

Let E be the event that flooding takes place in an (arbitrary) year. The probability of event E is approximately equal to the expected number EN of extreme periods in a year, times the probability that there is a flood in an extreme period, that is, $P(E) \approx EN \; P(X > h)$ for X a maximal water flow in a period of extreme water flow, h the maximal water flow so that there is no flood, and N the number of times we have extremely high water levels in an arbitrary year. For this computation, we use that the probability of flooding in an extreme period $P(X > h)$ is small. The probability distribution of N is unknown, but it is reasonable to assume that the expectation of N is approximately equal to the average number of periods of extreme water flow per year in the past 90 years, so $EN \approx 70/90$. The question is now: for which number h do we have $P(X > h) = 1/10000 \cdot 90/70 = 0.00013$?

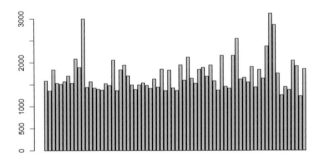

Figure 1.4. Maximal water flows in m^3/s (vertical axis) in the Meuse near Borgharen in the 20th century in chronological order (horizontal axis).

This question cannot easily be answered. If the observed maxima for a period of $100\,000$ years (or more) were available, then we could determine h with a reasonable accuracy, for example as the value of the 10%th highest measured water level ($10\% = 10\,000/100\,000$). Unfortunately, we dispose over only 70 observations, and must therefore extrapolate far into the future to a (probably) much more extreme situation than ever measured. If we can determine a good model for the distribution of X, then this is not a problem. If we, for example, knew that X has the standard exponential distribution, then we could determine h from the equation $0.00013 = P(X > h) = e^{-h}$. This is not, however, a realistic assumption.

An alternative is given by fitting an *extreme value distribution* to the data. These are probability distributions that are commonly used for modeling variables X that can be viewed as a maximum $X = \max(Y_1, \ldots, Y_m)$ of a large number of independent variables $Y_1, \ldots, Y_m$. Given the interpretation of X as a maximal water flow in a period, such distributions seem reasonable. Of the three types of extreme value distributions, one type proves to fit the data reasonably well. This is the *Fréchet family*, where the distribution function is given by

$$F(x) = \begin{cases} e^{-((x-a)/b)^{-\alpha}} & \text{if } x \geq a, \\ 0 & \text{if } x < a. \end{cases}$$

The Fréchet family has three parameters: $a \in \mathbb{R}$, $b > 0$, and $\alpha > 0$. If we are convinced of the usefulness of the resulting model, we can estimate these parameters from the 70 data points and then answer the question through a simple computation. In Chapter 3, we discuss suitable estimation methods and in the application after Chapter 6, we further work out the data of the water flows.

Example 1.7 Survival analysis

In survival analysis, we study the probability distribution of time spans. You can think of the life span of a light bulb, but also of the time before the next bug occurs in a computer program ("reliability analysis") and, in particular, of the remaining time until death or until the occurrence of a disease in medical statistics. Below is an example.

In persons with a leaking heart valve, the heart valve is often replaced by a biological or mechanical heart valve. A disadvantage of the biological over the mechanical heart valve is the relatively short life span (10 to 15 years). To study the distribution function F of the life span of a biological heart valve, n persons with such a valve are followed from the operation up to the moment that the valve must be replaced. At the end of the study, we have measured the life spans $t_1, \ldots, t_n$ of all of the n heart valves. We view these numbers as realizations of independent random variables $T_1, \ldots, T_n$ with distribution function F. The probability $F(t)$ that a biological heart valve must be replaced within t years can be estimated by the proportion of heart valves in the sample that is replaced within t years.

A special aspect of survival analysis is that, often, not all life spans are observed. At the moment that we want to draw conclusions from the data, for example, not all heart valves have needed replacement or a patient may have died with a heart valve that was still good. In those cases, too, we only observe a lower bound for the life spans, the time until the end of the study or until the death of the patient. We know that the heart valve still worked when the study was ended or the patient died. We then speak of *censored data*.

Long life spans are more frequently censored than short ones because the probability that the patient dies is greater during a long period of time than during a short one (and the same holds for the study ending). It would therefore be wrong to ignore censored data and estimate the distribution function F based on the uncensored data. This would lead to an overestimate of the distribution function of the life span and an underestimate of the expected life span because relatively longer life spans would be ignored. A correct approach is to use a statistical model for all observations, both censored and uncensored.

The statistical model becomes even more complex if we suspect that there are factors that could influence the life span of the heart valve, for example the age, weight, or gender of the patient. In such a case, the life span can be modeled using, for example, the Cox regression model. This model is studied at the end of this chapter (after the exercises) and in Chapter 7.

Example 1.8 Selection bias

To correctly answer a research question, it is important that this question, the collected data, and the statistical model are correctly aligned. This is illustrated below.

The Dutch Railways (Nederlandse Spoorwegen or NS for short) regularly receive complaints about crowding in the trains during rush hour. A study is set up to investigate whether these complaints are justified. There are two research questions. The first is what percentage of the passengers does not have a seat during rush hour. The second is what percentage of rush hour trains is too crowded. Note that these are two fundamentally different questions. The first question concerns people, a percentage of passengers, while the second question concerns trains. A passenger is probably only interested in the first research question, while the NS also attach importance to the answer of the second. They have to identify on which trains there are problems, and where measures must be taken.

To answer the first research question, a sample of size 50 is taken from train passengers that have just got off. Each person is asked whether they could sit. We observe the sequence $x_1, \ldots, x_{50}$, where x_i equals 1 if the ith person did not have a seat and x_i is equal to 0 if the ith person did have a seat. Then $x_1, \ldots, x_{50}$ are realizations of independent random variables $X_1, \ldots, X_{50}$ with a Bernoulli distribution with parameter p, where $p = P(X_i = 1)$ is the proportion of passengers that could not be seated. As in Example 1.2, we can estimate the proportion p using the sample mean $50^{-1} \sum_{i=1}^{50} x_i$. This is a correct way to answer the research question.

Answering the second research question is more difficult, because it concerns trains and not persons. To carry out this study, during rush hour, 50 head conductors are randomly chosen and asked whether the train they were just on was overcrowded. We observe the sequence $y_1, \ldots, y_{50}$, where y_i is equal to 1 if the ith head conductor indicates that the train was overcrowded and y_i is equal to 0 if this was not the case. We can again view $y_1, \ldots, y_{50}$ as realizations of $Y_1, \ldots, Y_{50}$, which are independent Bernoulli variables with probability $q = P(Y_i = 1)$. If we assume that there is only one head conductor on each train, the probability q equals the proportion of rush hour trains that were overcrowded. We can see $Y_1, \ldots, Y_{50}$ as a sample from the trains that just pulled in. The proportion q can be estimated using the sample mean $50^{-1} \sum_{i=1}^{50} y_i$.

It is simpler to also ask the sample of travelers we gathered to answer the first research question whether the train they were in was overcrowded. In that case, we observe a sequence of realization of the independent Bernoulli variables $Z_1, \ldots, Z_{50}$ with $r = P(Z_i = 1)$. Here, Z_i is defined analogously to Y_i. Since a train carries more than one passenger, not every train passenger will correspond to a unique train. Since there are more persons in crowded trains than in quiet ones, the percentage "people from crowded trains" in the population of train passengers will be much higher than the percentage of "crowded trains" in the population of trains. In other words, r will be greater than q. It is difficult to give a correlation between r and q without making additional assumptions. That is why the second research question could not easily be answered based on a sample from the passengers, while the first research question could.

In most of the examples given above, the statistical model is *parameterized* by a parameter, for example p, (μ, σ^2), $(\beta_0, \beta_1, \beta_2, \beta_3)$, or (a, b, α). Many statistical models are known up to a parameter. In this book, we often denote that parameter by θ ("theta"). The statistical model can then be denoted by $\{P_\theta : \theta \in \Theta\}$, where P_θ is the probability distribution of the observation X and Θ is the set of possible parameters. There is a tacit assumption that exactly one of the parameter values (or exactly one element of the model) gives the "true" distribution of X. The purpose of statistics is to find that value. What makes statistics difficult, is that we never fully succeed and that statements about the true parameter value always contain a certain element of uncertainty (by definition).

Exercises

1. Suppose that n persons are chosen randomly from a population and asked their political affiliation. Denote by X the number of persons from the sample whose affiliation is with political party A. The proportion of individuals in the population affiliated with party A is the unknown probability p. Describe a corresponding statistical model. Give an intuitively reasonable "estimate" of p.

2. Suppose that $m + n$ patients with high blood pressure are chosen randomly and divided arbitrarily into two groups of sizes m and n. The first group, the "treatment group," is given a particular blood-pressure-lowering drug; the second group, the "control group," is given a placebo. The blood pressure of each patient is measured before and one week after administering the drug or placebo, and the difference in blood pressure is determined. This gives observations $x_1, \ldots, x_m$ and $y_1, \ldots, y_n$.
   (i) Formulate a suitable statistical model.
   (ii) Give an intuitively reasonable "estimate" of the effect of the drug on the height of the blood pressure, based on the observations (several answers are possible!).

3. We want to estimate the number of fish, say N, in a pond. We proceed as follows. We catch r fish and mark them. We then set them free. After some time, we catch n fish (without putting them back). Of these, X are marked. Consider r and n as constants we choose ourselves, and let X be the observation.
   (i) Formulate a suitable statistical model.
   (ii) Give an intuitively reasonable "estimate" of N based on the observation.
   (iii) Answer the previous questions if, the second time we catch fish, they are put back directly after catching them (sampling with replacement).

4. When assessing a batch of goods, we continue until 3 items are rejected.
   (i) Formulate a suitable statistical model.
   (ii) The third rejected item is the 50th we assess. Give an estimate of the percentage of defect items in the batch. Justify your choice.

5. The number of customers in the post office seems to depend on the day of the week (weekday or Saturday) and half-day (morning or afternoon). On workdays, the post office is open in the morning and in the afternoon, and on Saturday, is it open only in the morning. To determine how many employees are required to provide prompt service, the number of customers is registered over a period of ten weeks. Every day, the number of customers in the post office in the morning (on weekdays and Saturdays) and in the afternoon (on weekdays only) is noted.
   (i) Formulate a suitable statistical model.
   (ii) Give an intuitively reasonable "estimate" of the number of clients on a Monday afternoon. Justify your choice.
   (iii) The biggest difference in numbers of customers is between the half-days during the workweek (Monday through Friday, mornings and afternoons) and the Saturday morning. It was therefore decided to only take into account this difference in the staff planning. Reformulate the statistical model and give a new estimate.

6. The yearly demand for water in the African city of Masvingo is greater than the amount that can be recovered from the precipitation in one year. Therefore, water is supplied from a nearby lake according to the need. The amount of water that needs to be supplied per year depends on the precipitation in that year and on the size of the population of Masvingo. Moreover, rich people use more water than poor people. Describe a linear regression model with "amount of water to be supplied" as dependent variable and "population size," "precipitation," and "average income" as predictor variables. Indicate for each of the parameters whether you expect them to be positive or negative.

7. A linear correlation is suspected between the income of a person and their age and level of education (low, middle, high).
   (i) Describe a linear regression model with "income" as dependent variable and "age" and "education" as predictor variables. Think carefully about how to include the variable "education" in the model.
   (ii) We want to study whether the gender of a person has an influence on the income. Adapt the linear regression model so that this can be studied.

8. We want to estimate the average length of wool fibers in a large bin. The bin is first shaken well, after which we take a predefined number of fibers from the bin, one by one and with closed eyes. We estimate the average length of the wool fibers in the bin to be the average length of the wool fibers in the sample. Is the estimated length systematically too long, systematically too short, or just right?

9. At a call center, we want to estimate how long a customer must wait before being helped. For one day, we register how long each customer must wait. If the customer looses patience and hangs up, their waiting time up to that moment is noted. Afterward, we calculate the average waiting time by taking the average of the noted times. This average is used as an estimate of the waiting time of a new customer. What do you think of this method?

In survival analysis, we are interested in the distribution function of the time span before the occurrence of a particular event, for example, the time before dying after a serious operation, the time before a certain device breaks down, or the time before an ex-convict commits a new crime. Several factors can influence this distribution function. For example, a young woman will presumably have a lower probability of dying after a serious operation than an older woman, and, hopefully, more time will pass before an ex-convict commits a new crime if he receives financial support than if he does not. It is important to gain insight in how and how much these factors influence the "life span," so that we can determine a more person-specific risk and take measures to reduce risks. If, for example, ex-convicts are more likely to commit a new crime if they are in financial difficulty after returning to society, then financial support or help in finding a job may help these people stay on the right track. In this application, we will use a sample to delve more deeply into survival analysis.

Ex-convicts often fall back into their old habits and come back into contact with police and justice. Suppose that we want to study the distribution function of the time span between release and recidivism and whether financial support after release has a positive effect on the time before an ex-convict comes back into contact with police. To begin with, we assume that there are no other factors.

Suppose that 100 ex-convicts are followed during one year. We know of each of them whether they commit a new crime within a year and if so, how many weeks after their release. We want to use these data to research which percentage of ex-convicts commit a new crime within t weeks (with $t \in [0, 52]$). We first set up a statistical model for these data. Define Y_i^t for $i = 1, \ldots, 100$ as the indicator that tells us whether the ith ex-convict has committed a new crime within t weeks; $y_i^t = 0$ if they have not, and $y_i^t = 1$ if they have. Then $Y_1^t, \ldots, Y_{100}^t$ are Bernoulli-distributed with unknown parameter $p_t = P(Y_i^t = 1)$, the probability of recidivism within t weeks. Under the assumption that the variables $Y_1^t, \ldots, Y_{100}^t$ are, moreover, independent, the statistical model is fixed. We could "estimate" the probability p_t using the fraction $\sum_{i=1}^{100} y_i^t / 100$. If the number of ex-convicts we follow, in our case 100, is large, then the proportion we find in the sample will lie close to the actual proportion p_t; this follows from the law of large numbers.

Often, studies are set up in a different way. Instead of following all ex-convicts for a year, we choose to restrict the length of the study to one year. We follow the convicts released during that year until they commit a new crime (if they do) or until the study ends. We have followed a total of 432 convicts in such a study. Figure 1.5 shows the observed time spans of 5 ex-convicts. Along the x-axis, the image on the left has the time from the beginning of the study (the vertical line at time 0) until the end of the study (the vertical line at week 52). The numbers along the y-axis are the personal numbers of the ex-convicts. The first person was released 10 weeks after the study began and arrested 31 weeks after the beginning. This person was free for $31 - 10 = 21$ weeks. The second individual was released 27 weeks after the study began and had not committed a new crime before the end of the study. We do not know *whether this second person committed a new crime after the end of the study. The first $52 - 27 = 25$ weeks, he did not. For the first person, the measurements are complete, while for the second we only have a lower bound for the time span until a new arrest. We call this data right-censored.*

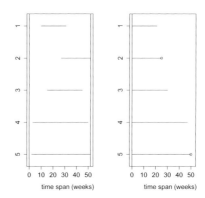

Figure 1.5. Left: The time between release and new arrest ("time span") of 5 ex-convicts. The x-axis indicates the time from the beginning of the study. Right: The same data, where the x-axis indicates the times from release to new arrest or censoring (end of the study or death). A circle indicates that the data for that person is right-censored.

The image on the right in Figure 1.5 shows the same information, a different way. The x-axis now shows the time from release to a new arrest. For individuals who were not arrested again during the study, we only know a lower bound for the time span; this is indicated by a small circle.

Suppose that, based on the observed data, we want to estimate the percentage of ex-convicts who are arrested again within 26 weeks (a half year). For person 2 in Figure 1.5, for example, we know that he was still free after 25 weeks, but we do not know anything about the period between 25 and 26 weeks; this person is right-censored. A natural solution is to remove all right-censored persons from the data set and use the same estimation method as described before. This proves to be a terrible choice. The longer an ex-convict is free, the higher the probability that he will be right-censored and therefore removed from the data set. An ex-convict who is rearrested 51 weeks after his release will only be included in the data set if he was released in the first week of the study. If he is released in the second week, then his data will be censored by the end of the study. By ignoring right-censored individuals, relatively many long "life spans" will be removed, and the proportion of ex-convicts who are rearrested within t weeks in the stripped set will be (much) too high. For example, p_{51} now has probability close to 1. How should we estimate p_t? To do this correctly, we first describe a statistical model for the observed data.

For an arbitrary ex-convict, we define T as the time span between release and recidivism. We view T as a random variable with distribution function $t \mapsto F(t) = P(T \leq t)$ and density function f. The survival function S is defined as $t \mapsto S(t) = 1 - F(t) = P(T > t)$ and describes the probability of not having been arrested again *after t weeks. We could assume that the distribution function or the survival function has a particular form (for example, that the distribution function corresponds to the exponential or normal distribution), but if there is no prior knowledge of the form of the distribution, it is better not to make any assumptions. An incorrect assumption can lead to incorrect conclusions.*

For some ex-convicts, the time span T is not observed during the study; at the end of the study, they had not been rearrested (or the person had died). We therefore also define, for each individual, the censoring time C as the time span between release and the end of the study or death. If $T \leq C$, then we will observe that the individual in questions commits a new crime, and if $T > C$, then we will observe not T but C. We therefore define $\tilde{T} = \min\{T, C\}$, so that $\tilde{T}$ is observed for every individual in the study. We, moreover, define Δ as the indicator function $\Delta = 1_{\{T \leq C\}}$; that is, $\Delta = 1$ if $T \leq C$ or, equivalently, $\tilde{T} = T$, and $\Delta = 0$ if $T > C$ or, equivalently, $\tilde{T} = C$. We observe the pair $(\tilde{T}, \Delta)$ for every ex-convict in the data set. The data set consists of the values $(\tilde{t}_i, \delta_i)$ for $i = 1, \ldots, 432$ for the 432 ex-convicts, where $\tilde{t}_i$ and δ_i are the observed values of $\tilde{T}_i$ and Δ_i.[♯]

Suppose that we want to estimate the probability of an arbitrary ex-convict not having committed a new crime within 26 weeks (half a year); in other words, we want to estimate $S(26)$. To explain how to do this, we assume, for now, that we only have data on 5 persons, shown in Figure 1.5. We have $\tilde{t}_1 < \tilde{t}_2 < 26 < \tilde{t}_3 < \tilde{t}_4 < \tilde{t}_5$ (see Figure 1.5). Individual 1 is the only one whose rearrest within half a year has been observed. We do not know anything about individual 2; he was censored within half a year. To estimate $S(26)$, we rewrite $S(26) = P(T > 26)$ as

$$P(T > 26) = P(T > 26|\tilde{T} > \tilde{t}_1)P(\tilde{T} > \tilde{t}_1)$$
$$= P(T > 26|\tilde{T} > \tilde{t}_2, \tilde{T} > \tilde{t}_1)P(\tilde{T} > \tilde{t}_2|\tilde{T} > \tilde{t}_1)P(\tilde{T} > \tilde{t}_1)$$
$$= P(T > 26|\tilde{T} > \tilde{t}_2)P(\tilde{T} > \tilde{t}_2|\tilde{T} > \tilde{t}_1)P(\tilde{T} > \tilde{t}_1).$$

Instead of estimating $S(26)$ directly, we estimate the three factors on the last line separately. We begin at the end, with the probability $P(\tilde{T} > \tilde{t}_1)$. Individual 1 committed a crime in week 21. Of the five individuals, there are four for whom $\tilde{T} > \tilde{t}_1$. We therefore estimate the probability $P(\tilde{T} > \tilde{t}_1)$ to be $4/5$. To estimate the second factor, $P(T > \tilde{t}_2|\tilde{T} > \tilde{t}_1)$, we use only the data on the individuals who satisfy the condition $\tilde{T} > \tilde{t}_1$; these are individuals 2 through 5. Individual 2 was censored at time $\tilde{t}_2$ (end of the study), so of the four ex-convicts that are left, there are only three for whom $\tilde{T} > \tilde{t}_2$. We estimate the probability $P(\tilde{T} > \tilde{t}_2|\tilde{T} > \tilde{t}_1)$ to be $3/4$. We estimate the last probability, $P(T > 26|\tilde{T} > \tilde{t}_2)$ analogously. Of the three individuals with $\tilde{T} > \tilde{t}_2$, all have a T-value greater than 26; we therefore estimate this probability to be $3/3 = 1$. Multiplying the three estimates gives 0.6.

[♯] The data come from a study described in P.H. Rossi, R.A. Berk, and K.J. Lenihan, *Money, Work and Crime: Some Experimental Results* (1980), Academic Press, New York. In our example, we have modified the data (censored times randomly) to illustrate the concept of censoring. The modified data can be found on the book's webpage at http://www.aup.nl under convicts.

We can estimate S at another time than 26 or based on another data set analogously. When we estimate the function S this way in values between 0 and 52 weeks, we find a step function that only jumps (down) in points t_i where $\delta_i = 1$. Figure 1.6 shows the estimated survival curve S based on the full data set. If the number of observations increases, the intervals where the curve is constant will become shorter, as will the sizes of the jumps. We can prove that the estimate of S at a time t converges (in probability) to the true value $S(t)$ if the number of observations goes to infinity. If we had assumed that F is a known continuous function, for example the distribution function corresponding to the exponential distribution, then we would estimate F not with the method described above but with the (parametric) maximum likelihood method described in Chapter 3. The survival curve would then be estimated by a continuous decreasing curve. If, however, the assumption about the form of the curve was wrong, then the estimate of S in an arbitrary point t would not converge (in probability) to $S(t)$.

Figure 1.6 also shows the estimated curve that would have been found if all censored data was removed; for estimating $S(t)$, all individuals censored before time t are removed. The difference between the two curves increases with t. This is because the more t increases, the more values are removed, and the greater the error that is made. This approach leads to an underestimate of the survival curve.

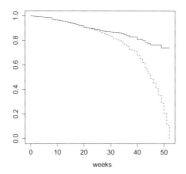

Figure 1.6. Estimated survival curve based on all data (solid) and based on the data set from which the censored observations have been removed (dashed).

Another way to represent the distribution of T is by using a so-called risk or hazard function. The hazard function associated with a probability density f and distribution function F is defined as

$$t \mapsto \lambda(t) = \frac{f(t)}{1 - F(t)} = \frac{f(t)}{S(t)}.$$

If we view $f(t)\,dt$ as the probability that T lies in the interval $[t, t + dt)$, then $\lambda(t)\,dt$ has the interpretation

$$\lambda(t)\,dt \approx \frac{P(t \leq T < t + dt)}{P(T > t)} = P(t \leq T < t + dt | T > t).$$

The value $\lambda(t)$ can therefore be viewed as the conditional probability of committing a new crime right after time t given that at time t, the ex-convict had not been rearrested. It is because of this interpretation as an "instantaneous probability" that the hazard function is often used for modeling survival data. The hazard function is the derivative of $t \mapsto -\log(1 - F(t))$ with respect to t, and given the hazard function λ, we can recover the distribution function F using the formula $F(t) = 1 - e^{-\Lambda(t)}$, where Λ is the cumulative hazard function, that is, $\Lambda(t) = \int_0^t \lambda(s)ds$ if $F(0) = 0$. The density f is then equal to $f(t) = \lambda(t)e^{-\Lambda(t)}$. As for the survival curve, we could now assume that the hazard function takes on a particular form. For example, if we assume that the hazard function is constant, $\lambda(t) \equiv \nu$, then the corresponding density is $f(t) = \lambda(t)e^{-\Lambda(t)} = \nu e^{-\nu t}$; in other words, T has an exponential distribution. We can also make no assumption at all on the form. To obtain an estimate of the distribution function of T, we can also turn to the hazard function and the formulas given above.

If factors such as age, gender, and education possibly influence the time span, then it is wise to include these in the model. Often, a so-called Cox model is chosen. In this model, the hazard function for the ith ex-convict with observed variables $X_i = x_i$ is of the form

$$\lambda(t|X_i = x_i) = e^{\beta_1 x_{i1} + \beta_2 x_{i2} + \ldots + \beta_K x_{iK}} \lambda_0(t),$$

where x_{ik} is the value of the kth variable for the ith individual and K is the number of variables in the model. The function $t \mapsto \lambda_0(t)$ is called the baseline hazard function and is equal to the hazard function when all predictor variables are equal to 0. According to the Cox model, the hazard functions of two ex-convicts with predictor variables x_i and x_j are proportional; this means that

$$\frac{\lambda(t|X_i = x_i)}{\lambda(t|X_j = x_j)} = e^{\beta^T (x_i - x_j)},$$

does not depend on t.

This gives a simple interpretation for the parameter β: it determines the size of the relative risks attached to certain predictor variables. For example, suppose that two ex-convicts score the same for all predictor variables, with the exception of financial support. Person i receives financial support ($x_{i1} = 1$), while person j does not ($x_{j1} = 0$). The ratio of the hazard functions then reduces to

$$\frac{\lambda(t|X_i = x_i)}{\lambda(t|X_j = x_j)} = e^{\beta_1 (x_{i1} - x_{j1})} = e^{\beta_1}.$$

If β_1 takes on the value -0.400, then the risk of being arrested again is $e^{-0.400} = 0.670$ times as great for the ith ex-convict as it is for the jth. The relative risk is therefore independent of the time since the release of the ex-convicts.

In our example, the Cox model includes the following predictor variables: financial support or not, age, race, marital status, number of prior convictions. Based on the data of the 432 ex-convicts, we can estimate the regression parameter $\beta = (\beta_1, \ldots, \beta_5)$ and the unknown baseline hazard function λ_0. A suitable method is elaborated in Chapter 7. Here, we will only give the results. Table 1 shows the estimates of the regression parameters $\beta_1, \ldots, \beta_5$.

	β_1	β_2	β_3	β_4	β_5
estimate	-0.400	-0.0425	0.282	-0.590	0.0977
exp(estimate)	0.670	0.958	1.326	0.554	1.103

Table 1.1. Estimates of $\beta_1, \ldots, \beta_5$; the parameters correspond to, respectively, financial support or not (0: no support, 1: support), age, race (0: other; 1: black), marital status (0: not married, 1: married), and number of prior convictions.

The estimate of the regression parameter β_1 is negative, financial support after release therefore has a positive effect. However, it seems that being married has a stronger positive effect than receiving financial help. If this effect is causal (see the application after Chapter 7), then it would be wiser to help an ex-convict find a partner than to help him find a job.

In the above, we have made several assumptions for our model. For example, we have assumed that the hazard functions of two ex-convicts are proportional and that the predictor variables are additive. These assumptions must, of course, be verified. We can do so, for example, by plotting suitable graphs. We refer to the literature for more information.

2 Descriptive Statistics

2.1 Introduction

A statistical model is an expression of our prior knowledge of the probability experiment that led to the observed data. The model postulates that the observation X is generated from one of the probability distributions in the model. How do we find a good model? In some cases, the model is clear from the way the experiment was set up. For example, if in a poll, the sample has been taken randomly and with replacement from a well-defined population, then the binomial distribution is inevitable. If the observations concern numbers of emitted radioactive particles, then the Poisson distribution is the right choice because of the physical theory of radioactivity. It is also possible that the experiment strongly resembles past experiments, and that a particular model is suggested by experience. The choice of a statistical model is certainly not always uncontroversial. At the very least, the chosen model must be validated. In some cases, this is done before estimating the model parameters, and in other cases after. These methods are not only applied to the data itself but often also to "residuals," after the estimation of, for example, a regression model. In this chapter, we discuss some of these methods.

2.2 Univariate Samples

Suppose that the numbers $x_1, \ldots, x_n$ are the outcomes of a repeated experiment. From the manner in which the n experiments are carried out (with the same initial situation, without any "memory" of the previous experiments), we deduce that it is reasonable to view the n numbers as realization of a univariate sample $X_1, \ldots, X_n$; the random variables are independent and identically distributed. This already fixes the statistical model for a large part. The remaining question is: which collection of (marginal) distributions do we use? In this section, we discuss a number of numerical and graphical methods that can help.

Two important numerical properties of a distribution are location and dispersion. The expectation and median are often used for the location of a distribution; they are equal to each other when the distribution is symmetric. When the distribution has a tail to the right (respectively, to the left), the expectation is greater (respectively, less) than the median. To obtain an idea of the location of the underlying distribution based on observations $x_1, \ldots, x_n$, we can use the sample mean or sample median. The *sample median* is the middle value in a sequence of sorted observations.

Definition 2.1 Sample mean

The sample mean of a sample $X_1, \ldots, X_n$ is the random variable

$$\overline{X} = \frac{1}{n} \sum_{i=1}^{n} X_i.$$

The dispersion of a distribution can be represented by the variance (or standard deviation) and the interquartile range. The *interquartile range* is the distance between the upper and lower quartiles of the distribution. Using the observations $x_1, \ldots, x_n$, we can compute the sample interquartile range and the sample variance to obtain an idea of the dispersion. The sample interquartile range is the distance between the upper and lower quartiles of the data.

Definition 2.2 Sample variance

The sample variance of a sample $X_1, \ldots, X_n$ is the random variable

$$S_X^2 = \frac{1}{n-1} \sum_{i=1}^{n} (X_i - \overline{X})^2.$$

In practice, the observed sample mean and observed sample standard deviation are often given when the distribution appears to be symmetric. For asymmetric distributions, the observed sample median and observed sample interquartile range are preferred. The best way to determine whether a distribution is symmetric is to use a graphical method. In the next section, we present three graphical methods: histograms, boxplots, and QQ-plots.

2.2.1 Histograms

A simple technique to obtain an idea of the probability density giving the data $x_1, \ldots, x_n$ is the *histogram*. For a partition $a_0 < a_1 < \cdots < a_m$ that covers the range of the data $x_1, \ldots, x_n$, this is the function that, on each interval $(a_{j-1}, a_j]$, takes on the value equal to the number of sample points x_i in that interval divided by the length of the interval. If the intervals $(a_{j-1}, a_j]$ all have the same length, then the histogram is sometimes also defined without dividing by the interval length. In that case, the heights of the bars of the histogram are equal to the numbers of observations in the various intervals.

To obtain an idea of the probability density giving certain data, it is useful to represent the histogram and possible densities in a single diagram. This can be done by scaling the histogram by $1/n$, where n is the total number of data points. The area under the histogram is then equal to 1, as it is for a probability density. In $x \in (a_{j-1}, a_j]$, the scaled histogram is given by

$$h_n(x) = \frac{\#(1 \leq i \leq n \colon x_i \in (a_{j-1}, a_j])}{n(a_j - a_{j-1})} = \frac{1}{n(a_j - a_{j-1})} \sum_{i=1}^{n} 1_{a_{j-1} < x_i \leq a_j},$$

where the *indicator function* $1_{a_{j-1} < x_i \leq a_j}$ is equal to 1 for $a_{j-1} < x_i \leq a_j$ and 0 elsewhere. Another way to write this indicator function is $1_{(a_{j-1}, a_j]}(x_i)$.

A scaled histogram provides a good impression of the density giving the data $x_1, \ldots, x_n$, provided that the partition $a_0 < a_1 < \cdots < a_m$ has been chosen well and that the number of sample points n is not too small. To see this, we view $x_1, \ldots, x_n$ as realizations of random variables with density f and compute the expected value of the scaled histogram h_n in terms of $X_1, \ldots, X_n$ in an arbitrary point x where $f(x) > 0$. Suppose that for some $1 < j \leq m$, we have $a_{j-1} < x \leq a_j$; then this expected value is equal to

$$\mathrm{E}h_n(x) = \mathrm{E}\frac{1}{n(a_j - a_{j-1})} \sum_{i=1}^{n} 1_{a_{j-1} < X_i \leq a_j} = \frac{1}{a_j - a_{j-1}} \mathrm{E} 1_{a_{j-1} < X_1 \leq a_j}$$

$$= \frac{1}{a_j - a_{j-1}} \mathrm{P}(a_{j-1} < X_1 \leq a_j) = \frac{\int_{a_{j-1}}^{a_j} f(s) \, ds}{a_j - a_{j-1}}.$$

If f does not vary too much over the interval $(a_{j-1}, a_j]$, then the last expression is approximately equal to the value of f on this interval. The computation shows that the expected value of $h_n(x)$ is approximately equal to $f(x)$. By the law of large numbers (Theorem A.26), we moreover know that the value $h_n(x)$ converges in probability to this expected value.

23

A histogram therefore provides an impression of the distribution giving a sample. Unfortunately, we only obtain a good impression if the sample is sufficiently large (for example, $n = 100$ or even better $n = 500$) and the intervals have been chosen well. The choice of the intervals is a question of taste. If the chosen intervals are too short, then, in general, the histogram is too spiky to notice properties of the true density. If the intervals are too long, then all detail is lost, and little can be said about the true density based on the histogram. Hence, we may not expect more than a first impression from the histogram. Other, more complicated, techniques can give better results.

Example 2.3 Height

Figure 2.1 shows histograms for the heights (in cm) of 44 men (on the left) and 67 women (on the right).[†] The histograms have been scaled in such a way that the areas under the histograms are equal to 1. Both figures also show the density of a normal distribution. The expectation and variance of these normal distributions are equal, respectively, to the sample mean and sample variance of the corresponding data. Based on the forms of the histograms, there is some doubt whether the data can come from a normal distribution. The deviation from symmetry in the histogram on the left may be due to the small number of observations. Further research is certainly recommended.

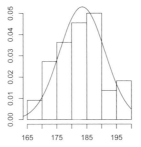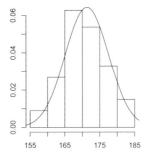

Figure 2.1. Histograms of the heights (in cm) of 44 men (on the left) and 67 women (on the right), together with the densities of the normal distributions with expectations equal to the sample means and variations equal to the sample variations of the data.

[†] Source: The data were gathered by the department of Biological Psychology of VU University Amsterdam during a study on health, lifestyle, and personality. The data can be found on the book's webpage at http://www.aup.nl under heightdata.

Example 2.4 Normal distribution

Figure 2.2 shows the density of the standard normal distribution, together with four realizations of the histogram, based on 30, 30, 100, and 100 observations, where the partitions were chosen by the statistical software package R. The figures at the top left and at the bottom right show a clear deviation from symmetry. Because the data come from the normal distribution, this is merely due to chance variations. ▭

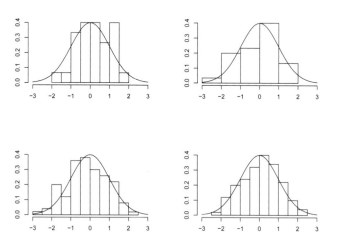

Figure 2.2. Histograms of samples with 30 (top row) and 100 (bottom row) observations from the standard normal density, shown together with the true density.

2.2.2 Boxplots

A *boxplot* is a graphical representation of the data that gives an idea of the location and dispersion of the data, of possible outliers in the observations, and of the symmetry of the distribution giving the observations. In the boxplot, the observations are set out along the vertical axis. The bottom of the "box" is drawn at the level of the lower quartile, and the top at the level of the upper quartile of the data. The lower (respectively, upper) quartile of the data is the value x for which one fourth of the data points are less (respectively, greater) than x. The width of the box is arbitrary. The box has a horizontal line at the level of the median of the data. The *median* is the middle value in a sorted row of data. At the top and bottom of the box, *whiskers* are drawn. The whisker at the top links the box to the greatest observation that lies within 1.5 times the interquartile range of the upper quartile. The *interquartile range* is the distance between the lower and upper quartiles, that is, the height of the box. The whisker at the bottom is drawn analogously. Observations that lie beyond the whiskers are indicated separately, for example by a star, a small circle, or a dash.

25

Example 2.5 Some common distributions

Figure 2.3 shows boxplots of samples from the exponential distribution with parameter 1, the standard normal distribution, and the standard Cauchy distribution. The samples from the exponential and Cauchy distributions have outliers, shown by the small circles beyond the whiskers. The boxplot in the middle shows that the data from the standard normal distribution are quite symmetric with respect to the median and do not contain any outliers.

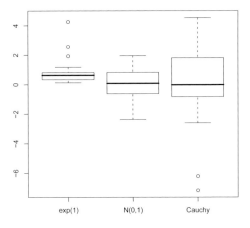

Figure 2.3. Boxplots of samples of size 20 from the standard exponential distribution (left), the standard normal distribution (middle), and the standard Cauchy distribution (right).

2.2.3 Location-Scale Family and QQ-plots

A third graphical method that is commonly used to find a suitable class of distributions (a so-called location-scale family) given a sample $x_1, \ldots, x_n$ is drawing QQ-plots. In this section, we first discuss location-scale families and then QQ-plots.

Definition 2.6 Location-scale family

If the random variable X has a distribution function F, then $Y = a + bX$ has the distribution function $F_{a,b}$ given by

$$F_{a,b}(y) = P(a + bX \leq y) = F\left(\frac{y - a}{b}\right), \qquad b > 0.$$

The family of distributions $\{F_{a,b} : a \in \mathbb{R}, b > 0\}$ is called the location-scale family associated with F (or "for X").

If F has probability density f, then $F_{a,b}$ has probability density $f_{a,b}$ given by

$$f_{a,b}(y) = \frac{d}{dy} F\left(\frac{y-a}{b}\right) = \frac{1}{b} f\left(\frac{y-a}{b}\right).$$

If $EX = 0$ and $\text{var } X = 1$, then a and b^2 are, respectively, the expected value and variance of Y, hence those corresponding to the distribution $F_{a,b}$.

To every (standard) distribution (normal, Cauchy, exponential, etc.) corresponds a location-scale family. We note that members of one location-scale family do not always all have the same name: the members of the location-scale family associated with the standard Cauchy distribution are not all Cauchy distributions. Conversely, distributions with the same name are not always members of the same location-scale family: for example, χ^2-distributions with different numbers of degrees of freedom are not in the same location-scale family.

Example 2.7 Normal distribution

Let X be a $N(0,1)$-distributed random variable. From probability theory, we know that $Y = a + bX$ with $b > 0$ has the $N(a, b^2)$-distribution. Hence, all members of the location-scale family associated with the $N(0,1)$-distribution are normally distributed. Conversely, if Y has the $N(a, b^2)$-distribution, then Y has the same distribution as $a + bX$, where X has the standard normal distribution, and therefore the $N(a, b^2)$-distribution is a member of the location-scale family associated with the standard normal distribution. In other words, all members of the location-scale family associated with the $N(0,1)$-distribution are normal distributions, and conversely, all normal distributions are in the location-scale family associated with the $N(0,1)$-distribution.

"QQ-plots" are a graphical tool for finding a suitable location-scale family for a given sample $x_1, \ldots, x_n$. They are based on quantile functions. If, for a given $\alpha \in (0,1)$, there exists exactly one number $x_\alpha \in \mathbb{R}$ such that $F(x_\alpha) = \alpha$, then x_α is called the α-*quantile* of F, denoted by $F^{-1}(\alpha)$. As suggested by the notation, the function $\alpha \mapsto F^{-1}(\alpha)$ is the *quantile function*, the inverse of F, provided that this is well defined. If F is strictly increasing and continuous, then $F(F^{-1}(\alpha)) = \alpha$ for all $\alpha \in (0,1)$ and $F^{-1}(F(x)) = x$ for all $x \in \mathbb{R}$.

Example 2.8 Exponential distribution

Let X be a random variable with an exponential distribution with parameter λ. The distribution function F of X is given by $F(x) = 1 - e^{-\lambda x}$ for $x \geq 0$, and the quantile function F^{-1} is given by $F^{-1}(\alpha) = -\log(1-\alpha)/\lambda$ for $\alpha \in (0,1)$.

Because a distribution function can exhibit both jumps and constant sections, in general the equation $F(x) = \alpha$ for a given α can have no solutions, exactly one solution, or infinitely many solutions (see Figure 2.4). To also be able to speak of an α-quantile in the first and last case, we define the quantile function of F as follows.

Definition 2.9 α-Quantile

The α-quantile (or α-point) of F is equal to

$$F^{-1}(\alpha) = \inf\{x \colon F(x) \geq \alpha\}, \qquad \alpha \in (0, 1).$$

In words, $F^{-1}(\alpha)$ is the smallest value x with $F(x) \geq \alpha$.

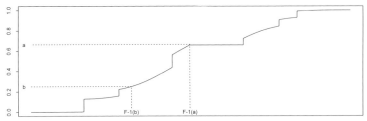

Figure 2.4. A distribution function and two quantiles.

There is a linear relationship between quantile functions of distributions within a given location-scale family:

$$F_{a,b}^{-1}(\alpha) = a + b\,F^{-1}(\alpha)$$

(see Exercise 2.2). In other words, the points $\{(F^{-1}(\alpha), F_{a,b}^{-1}(\alpha)) \colon \alpha \in (0, 1)\}$ are on the straight line $y = a + bx$. Figure 2.5 illustrates the fact that two normal distributions belong to the same location-scale family.

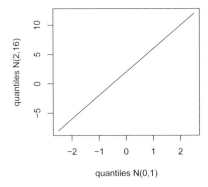

Figure 2.5. The quantiles of the $N(2, 4^2)$-distribution (y-axis) plotted against those of the $N(0, 1)$-distribution (x-axis).

Notation 2.10 Order statistics

The order statistics of a sample $X_1, \ldots, X_n$ are given by the sequence $X_{(1)}, \ldots,$ $X_{(n)}$, where the quantities have been placed in increasing order, $X_{(1)} \leq X_{(2)} \leq \ldots \leq X_{(n)}$. In particular, the first and last order statistics are equal to

$$X_{(1)} = \min_{1 \leq i \leq n} X_i \qquad \text{and} \qquad X_{(n)} = \max_{1 \leq i \leq n} X_i.$$

For the ith order statistic $X_{(i)}$ of a given sample $X_1, \ldots, X_n$ from a distribution F, we have $\mathrm{E}F(X_{(i)}) = i/(n+1)$ (See Exercise 2.8). We may therefore expect the points $\left\{ \left(i/(n+1), F(x_{(i)}) \right) : i = 1, \ldots, n \right\}$ in the xy-plane to lie approximately on the line $y = x$. The same must then hold for the points

$$\left\{ \left(F^{-1} \left(\frac{i}{n+1} \right), x_{(i)} \right) : i = 1, \ldots, n \right\}.$$

More generally, if the sample $x_1, \ldots, x_n$ comes from an element $F_{a,b}$ of the location-scale family associated with F, then we expect the points mentioned above to lie on the line $y = a + bx$; after all, we then have $x_{(i)} \approx F_{a,b}^{-1}(i/(n+1)) = a + bF^{-1}(i/(n+1))$.

Definition 2.11 QQ-plot

A QQ-plot of the data set $x_1, \ldots, x_n$ for a distribution function F is a plot of the points

$$\left\{ \left(F^{-1} \left(\frac{i}{n+1} \right), x_{(i)} \right) : i = 1, \ldots, n \right\}.$$

A QQ-plot provides a graphical method to verify whether a sample can come from the location-scale family associated with F. The Q stands for "quantile."

Example 2.12 Normal distribution

Figure 2.6 shows QQ-plots of six samples simulated from the $N(2, 4^2)$-distribution using a random number generator, plotted against the $N(0, 1)$-distribution. Because two normal distributions are in the same location-scale family, we can expect the points to lie more or less on a straight line. The top and bottom figures represent data sets of 10 and 50 observations, respectively. The points in the QQ-plots are not exactly on a straight line, but rather vary slightly around a straight line. In small samples, this variation is much greater than in larger samples.

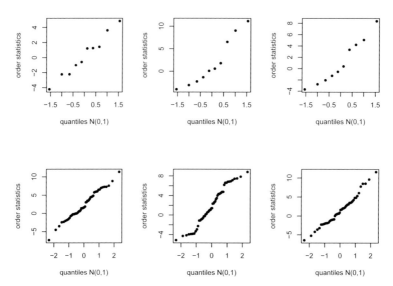

Figure 2.6. Six QQ-plots of 10 (top row) or 50 (bottom row) data points from the $N(2, 4^2)$-distribution plotted against the $N(0, 1)$-distribution.

If a QQ-plot of a sample $x_1, \ldots, x_n$ against the quantiles of F shows approximately the straight line $y = x$, then this is an indication that the data come from the distribution F. Deviations from the line $y = x$ give an indication of the deviation of the true distribution of the data from F. The simplest case is that the plot does show a straight line, but not the line $y = x$. This implies that the data come from another member of the location-scale family associated with F, as in Example 2.12. The values of a and b can then be approximated roughly by fitting the line $y = a + bx$ to the QQ-plot. In Chapter 3, we will see other methods to estimate the parameters. Curved lines are more difficult to evaluate. These mainly give an indication of the relative weight of the tails of the distribution of the data with respect to F. To illustrate the different types of deviations from linearity, Figure 2.7 shows some QQ-plots of "true" quantile functions. These are plots of the points $\{(F^{-1}(\alpha), G^{-1}(\alpha)): \alpha \in (0, 1)\}$ for various distribution functions F and G.

Example 2.13 Height

Based on the form of the histograms in Figure 2.1, there is some doubt whether the heights can come from a normal distribution. To study this further, QQ-plots have been drawn in Figure 2.8 that show the heights of men (on the left) and women (on the right) plotted against the standard normal distribution. To study whether the points lie on a straight line, a suitable line $y = a + bx$ has been drawn in both figures. For the men, this is the line $y = 183.5 + 7.5x$, and for the women, it is the line $y = 171.3 + 6.2x$. These lines have been determined by estimating a and b^2 using the maximum likelihood estimators for the expected value and variance (see Example 2.7

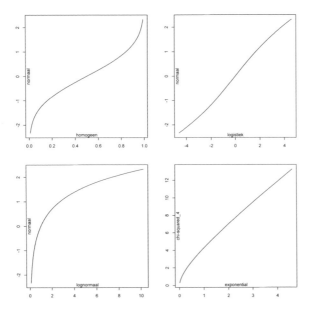

Figure 2.7. Plots of pairs of quantile functions: uniform-normal, logistic-normal, lognormal-normal, exponential-χ_4^2.

and Chapter 3). As the data follow these lines fairly well, we can conclude that the location-scale family associated with the standard normal distribution is a good fit for these two data sets. Since this family contains only normal distributions, this supports the assumption that the data come from normal distributions.

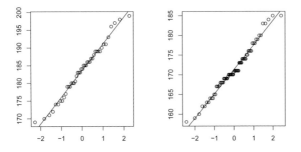

Figure 2.8. QQ-plots of the heights of 44 men (on the left) and 67 women (on the right) plotted against the quantiles of a standard normal distribution.

2.3 Correlation

In many cases, the observations x_i are not numbers but vectors $x_i = (x_{i,1}, \ldots, x_{i,d})$. We are then often interested in the correlation between the different coordinates. In this section, we will restrict ourselves to vectors with two coordinates and denote these by (x_i, y_i) (instead of $(x_{i,1}, x_{i,2})$).

Definition 2.14 Scatter plot

A scatter plot of a sample of 2-dimensional data points $(x_1, y_1), \ldots, (x_n, y_n)$ is a graph of these points in the xy-plane.

When there is a strong correlation between the x- and y-coordinates of the data in a scatter plot, this is clearly visible. For example, the variables in the image on the right in Figure 2.9 show a clear linear correlation, while no correlation is apparent in the image on the left.

The linear correlation in the image on the right in Figure 2.9 is unmistakable, but not perfect. The points do not lie exactly on a straight line but vary around an (imaginary) line.

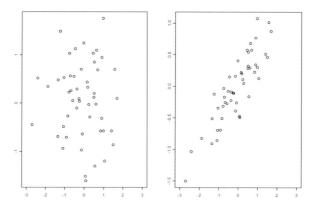

Figure 2.9. Scatter plots of two samples of 50 points: on the left with independent coordinates ($r_{x,y} = -0.05$) and on the right with coordinates with linear correlation ($r_{x,y} = 0.87$).

Definition 2.15 Sample correlation coefficient

The sample correlation coefficient of a sample consisting of pairs $(X_1, Y_1), \ldots, (X_n, Y_n)$ is

$$r_{X,Y} = \frac{\sum_{i=1}^{n}(X_i - \overline{X})(Y_i - \overline{Y})}{(n-1)\sqrt{S_X^2}\sqrt{S_Y^2}}.$$

The sample correlation coefficient $r_{x,y}$ of the observed pairs $(x_1, y_1), \ldots,$ (x_n, y_n) is a numerical measure for the strength of the linear correlation and lies between -1 and 1. The value can be interpreted as follows:

(i) If $r_{x,y} = 1$, then the n points in the scatter plot lie exactly on the line $y = \bar{y} + (s_y/s_x)(x - \bar{x})$ (total positive correlation).

(ii) If $r_{x,y} = -1$, then the n points in the scatter plot lie exactly on the line $y = \bar{y} - (s_y/s_x)(x - \bar{x})$ (total negative correlation).

(iii) If $X_1, \ldots, X_n$ and $Y_1, \ldots, Y_n$ are independent samples, then the resulting $r_{x,y}$ will take on values close to 0.

The first two statements and the inequality $|r_{x,y}| \leq 1$ follow from the Cauchy–Schwarz inequality from linear algebra.[‡] The third statement follows from the fact that independent random variables are uncorrelated, combined with the intuitively plausible fact that the sample correlation coefficient will approach the *population correlation coefficient*

$$\rho = \frac{\mathrm{cov}(X, Y)}{\sqrt{\mathrm{var}\, X}\sqrt{\mathrm{var}\, Y}} = \frac{\mathrm{E}(X - \mathrm{E}X)(Y - \mathrm{E}Y)}{\sqrt{\mathrm{E}(X - \mathrm{E}X)^2}\sqrt{\mathrm{E}(Y - \mathrm{E}Y)^2}}$$

when n is large. Since $\mathrm{cov}(X, Y) = \mathrm{E}(X - \mathrm{E}X)(Y - \mathrm{E}Y) = \mathrm{E}(XY) - \mathrm{E}X\mathrm{E}Y$, this coefficient ρ is equal to 0 for independent random variables X and Y: independent random variables are uncorrelated. We give a further interpretation of the sample correlation coefficient when we discuss linear regression in Chapter 7.

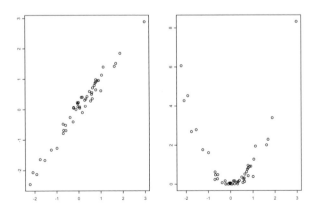

Figure 2.10. Scatter plots of two samples of 50 data points, with sample correlation coefficients 0.98 and -0.05, respectively. The image on the right gives the points (x_i, y_i^2) for the points (x_i, y_i) in the image on the left.

[‡] The inner product of vectors a and b in $\mathbb{R}^n$ satisfies $|\langle a, b \rangle| \leq \|a\|\,\|b\|$, where $\|\cdot\|$ is the Euclidean norm.

We may not invert statement (iii) to claim that a correlation close to 0 implies that the two coordinates are independent. This is illustrated in Figure 2.10. The image on the left shows a clear linear correlation, corresponding to a correlation coefficient of 0.98. The image on the right is a scatter plot of the points (x_i, y_i^2) for the points (x_i, y_i) in the image on the left. The quadratic correlation is clear. The "strength of the correlation" between the two coordinates in the image on the right is no less than the strength of the correlation in the image on the left. However, the sample correlation coefficient for the points in the image on the right is equal to -0.05. Apparently, this numerical quantity is blind for the quadratic relationship that is present.

Example 2.16 Twin data

Height is largely hereditary. We already saw this in Example 1.5, which models the correlation between the heights of parents and those of their children. This is also apparent in studies of twins. Because identical twins are genetically identical and fraternal twins in general share 50% of their genetic material, the correlation between the heights of identical twins will be greater than that for fraternal twins (of the same gender). In Figure 2.11, the heights of identical twins (men on the left, women on the right) have been plotted against each other.[b] Both scatter plots show a strong correlation. The sample correlation of the 46 male identical twins is equal to 0.87. For the 70 female identical twins, it is an impressive 0.96. We can do the same for fraternal twins of the same gender; see the scatter plots in Figure 2.12 (men on the left, women on the right). The figure clearly shows that the correlation is less for fraternal twins. The sample correlation between the heights of the 29 male fraternal twins is equal to 0.55, while that for the 56 female fraternal twins is 0.50. In the application given after the exercises in Chapter 3, we will come back in detail to genetic research based on data on twins.

2.3.1 Autocorrelations

Scatter plots can also be used to verify the common assumption that a sample $x_1, \ldots, x_n$ is a realization of independent variables. For example, we can plot the points (x_{2i-1}, x_{2i}) for $i = 1, \ldots, \lfloor n/2 \rfloor$ or the points (x_i, x_{i+1}) for $i = 1, \ldots, n-1$. If the assumption is correct, then these scatter plots should not show much structure.

The *sample autocorrelation coefficient* of order $h \in \mathbb{N}$ of an observed sample $x_1, \ldots, x_n$ is defined by

$$r_x(h) = \frac{\sum_{i=1}^{n-h} (x_{i+h} - \overline{x})(x_i - \overline{x})}{(n-h)s_x^2}.$$

---

[b] Source: The data used in this example were gathered by the department of Biological Psychology of VU University Amsterdam during a study on health, lifestyle, and personality. The data can be found on the book's webpage at http://www.aup.nl under twindata.

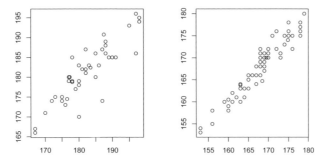

Figure 2.11. Scatter plots of the heights of 46 male (on the left) and 70 female (on the right) identical twins.

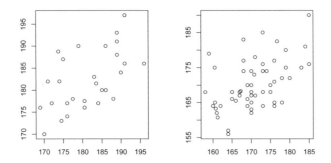

Figure 2.12. Scatter plots of the heights of 29 male (on the left) and 56 female (on the right) fraternal twins.

The sample correlation coefficient corresponding to the points (x_i, x_{i+1}) for $i = 1, \ldots, n - 1$ is (in essence) the sample autocorrelation coefficient of order 1. These coefficients are especially interesting when the index i of the data point x_i corresponds to a time parameter and the data are thought to exhibit a time dependence. We then measure the correlation between the variables X_i and X_{i-h} from h points of time earlier.

Example 2.17 Share prices

The top image of Figure 4.14 shows the value of a share of Hewlett Packard at the New York stock exchange plotted against the time, in the period 1984–1991. The values a_i of the share at closing time on consecutive exchange days ($i = 1, 2, \ldots, 2000$) are plotted; in the graph, these values have been interpolated linearly. Because share prices

35

generally form an exponentially increasing (or decreasing) sequence, it is common to analyze the "log returns," defined as

$$x_i = \log \frac{a_i}{a_{i-1}},$$

instead of the share prices themselves. These values are plotted in the bottom image of Figure 4.14.[♯] Since the index i of x_i corresponds to the ith exchange day, it would not be surprising if $x_1, \ldots, x_{2000}$ could not be modeled well as realizations of independent variables $X_1, \ldots, X_{2000}$. After all, a significant change on day i would greatly influence the change on day $i + 1$. Regardless, the converse assumption of independence, the "random walk hypothesis," has long been accepted in econometrics.

A first step to verify this hypothesis is computing the sample autocorrelations of the sequence $x_1, \ldots, x_{2000}$. These are shown graphically in the image on the left in Figure 2.13, where the values $h = 0, 1, 2, \ldots, 30$ have been set out along the horizontal axis, and the heights of the line segments give the corresponding sample autocorrelation coefficients of order h (the sample autocorrelation of order 0 is, of course, equal to 1). Almost all sample autocorrelation coefficients are small, which justifies the conclusion that the log returns show little linear correlation.

The image on the right gives the sample autocorrelation coefficients of the squares $x_1^2, \ldots, x_{2000}^2$ of the log returns. Although these coefficients are small, the conclusion that the quadratic log returns show little correlation is debatable: too many coefficients differ too much from 0. If the squares are not independent, then the log returns themselves are of course also not independent. It is therefore not a good assumption that $x_1, \ldots, x_{2000}$ can be modeled as realizations of independent variables: the time effect should be taken into account. Share prices do not form a random walk.

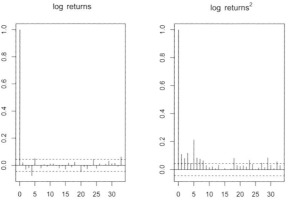

Figure 2.13. Sample autocorrelation function of the log returns of HP shares in the period 1984–1991 (on the left) and of the squares of the log returns (on the right). The dashed lines are at heights $\pm 1.96/\sqrt{2000}$ (see Example 4.40).

[♯] The data can be found on the book's webpage at http://www.aup.nl under hpprices (share prices) and hplogreturns (log returns).

In the above, we wrote that the coefficients in the image on the left in Figure 2.13 are "small," while they "differ from 0" in the image on the right. We can support these claims objectively using statistical tests such as those discussed in Chapter 4. The dashed horizontal lines in the two figures give critical values for the sample autocorrelations as sample variables for the null hypothesis that $x_1, \ldots, x_{2000}$ can be interpreted as a sample of independent variables (with a margin of error of 5%). Coefficients that do not land between the lines lead to rejecting the null hypothesis. We must take into account that when we assume the null hypothesis, about 1 in 20 coefficients will land outside the lines because of "random variations," because of the 5% margin of error (see Chapter 4). In the image on the right, too many values land outside the lines. ▭

2.4 Summary

Quantitative measures for a univariate sample $X_1, \ldots, X_n$ from an unknown distribution:

- The *sample mean* gives an idea of the location of the underlying distribution:

$$\overline{X} = \frac{1}{n} \sum_{i=1}^{n} X_i.$$

- The *sample variance* gives an idea of the dispersion of the underlying distribution:

$$S_X^2 = \frac{1}{n-1} \sum_{i=1}^{n} (X_i - \overline{X})^2.$$

  The *sample standard deviation* is S_X, the square root of S_X^2.
- Other commonly used measures are the sample median and the sample interquartile range.

Graphical methods for a univariate sample $X_1, \ldots, X_n$ from an unknown distribution:

- A *histogram* gives an idea of the form of the underlying distribution.
- A *boxplot* shows the median, the interquartile interval, and the outliers of the sample. It gives an idea of the location and scale of the underlying distribution, as well as the symmetry and thickness of the tails.
- A *QQ-plot* shows the sample quantiles plotted against the quantiles of a chosen distribution. If the chosen distribution and the underlying distribution are in the same location-scale family, then the points will lie near a straight line.

Graphical method and quantitative measure for a bivariate sample $(X_1, Y_1), \ldots, (X_n, Y_n)$ from an unknown distribution:

- A *scatter plot* gives a graphical representation of the correlation between the coordinates.
- The *sample correlation coefficient* is a quantitative measure for the linear correlation between the coordinates:

$$r_{X,Y} = \frac{\sum_{i=1}^{n} (X_i - \overline{X})(Y_i - \overline{Y})}{(n-1)\sqrt{S_X^2}\sqrt{S_Y^2}}.$$

Quantitative measure for a sequence of (possibly dependent) observations $X_1, \ldots, X_n$ from an unknown distribution:

- The *sample autocorrelation coefficient* of order h is used to find a possible (time) dependence among the observations:

$$r_X(h) = \frac{\sum_{i=1}^{n-h} (X_{i+h} - \overline{X})(X_i - \overline{X})}{(n-h)S_X^2}.$$

Exercises

1. Let h_n be the scaled histogram of a sample $X_1, \ldots, X_n$ from a distribution with density f. The partition of the histogram is given by $a_0 < a_1 < \ldots < a_m$. Prove that for $a_{j-1} < x \le a_j$, we have $h_n(x) \to (a_j - a_{j-1})^{-1} \int_{a_{j-1}}^{a_j} f(s)\, ds$ with probability 1 as $n \to \infty$.

2. Let X be a random variable with distribution function F and quantile function Q. Define x_α as the α-quantile of F and y_α as the α-quantile of the distribution of $Y = a + bX$.
   (i) Suppose that F is strictly increasing and continuous, so that the inverse of F exists and is equal to Q. Show that there is a linear correlation between $x_\alpha = F^{-1}(\alpha)$ and $y_\alpha = F_{a,b}^{-1}(\alpha)$ by using the inversibility of F.
   (ii) Show that the same linear correlation exists between x_α and y_α for a general distribution function F. Use the general definition of the α-quantile.

3. The standard exponential distribution has distribution function $x \mapsto 1 - e^{-x}$ on $[0, \infty)$.
   (i) Does the exponential distribution with parameter λ belong to the location-scale family associated with the standard exponential distribution?
   (ii) Express the parameters a and b in the location-scale family $F_{a,b}$ associated with the standard exponential distribution in terms of the expected value and variance of a random variable with distribution $F_{a,b}$.

4. Let X be a random variable with a uniform distribution on $[-3, 2]$.
   (i) Determine the distribution function F of X.
   (ii) Determine the quantile function F^{-1} of X.

5. Let X be a random variable with probability density
$$f(x) = \frac{2}{\theta^2} x \mathbb{1}_{[0,\theta]}(x),$$
where $\theta > 0$ is a constant.
   (i) Determine the distribution function F of X.
   (ii) Determine the quantile function F^{-1} of X.

6. Which line is plotted in Figure 2.5?

7. Let $X_1, \ldots, X_n$ be a sample from a continuous distribution with distribution function F and density f. Show that the probability density of the kth order statistic $X_{(k)}$ is equal to
$$f_{(k)}(x) = \frac{n!}{(k-1)!(n-k)!} F(x)^{k-1}(1 - F(x))^{n-k} f(x)$$
by first determining the distribution function of $X_{(k)}$. (Hint: We have $X_{(k)} \le x$ if and only if at least k observations X_i are less than or equal to x. The number of X_i that are less than or equal to x is binomially distributed with parameters n and $P(X_i \le x)$.)

8. Let $X_1, \ldots, X_n$ be a sample from a continuous distribution with distribution function F. In this exercise, we want to show that $EF(X_{(k)}) = k/(n+1)$. Define $U_i = F(X_i)$ for $i = 1, \ldots, n$.
   (i) Show that the random variables $U_1, \ldots, U_n$ form a sample from the uniform distribution on $[0, 1]$.
   (ii) Show that the distribution function $F_{(k)}$ of $U_{(k)}$ is given by
$$F_{(k)}(x) = \sum_{j=k}^{n} \binom{n}{j} x^j (1-x)^{n-j}.$$

39

(iii) Show that the density $f_{(k)}$ of $U_{(k)}$ is given by

$$f_{(k)}(x) = \frac{n!}{(k-1)!(n-k)!} x^{k-1}(1-x)^{n-k}.$$

(iv) Show that $EU_{(k)} = k/(n+1)$.

9. Draw a graph of the quantiles of the $N(2, 2^2)$-distribution plotted against the quantiles of the $N(0, 3^2)$-distribution. What line is this?

10. Let X be a standard normal random variable. Compute the correlation coefficient between the random variables X and $Y = X^2$.

11. Explain why it is plausible that the sample correlation $r_{X,Y}$ is approximately equal to the correlation coefficient ρ for large values of n.

12. Assume that X and Y are independent and that both have the standard normal distribution. Compute the correlation coefficient between X and $Z = X + Y$.

In 1938, the physicist Benford published an academic paper in which he claims that in a data set, the frequency of the leading digit of the numbers is greater the smaller the digit. In other words, in a data set, more numbers begin with a 1 than with a 2, more numbers begin with a 2 than with a 3, and so on. This pattern does not correspond to the general feeling that all leading digits, from 1 to 9, occur with about the same frequency. In his paper, Benford even states that the probability of an arbitrary number from a data set starting with the digit d is equal to $\log_{10}(1 + 1/d)$ for $d \in \{1,\ldots,9\}$ (where $\log_{10}$ is the base 10 logarithm). So according to Benford, the probability that an arbitrary number begins with a 1 is about 0.30, and the probability that it begins with a 9 has dropped to less than 0.05. Figure 2.14 shows the probabilities. The claim stated above is known as "Benford's law."

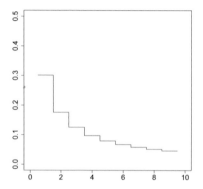

Figure 2.14. The probabilities of the different leading digits according to Benford's law.

Benford was not the first to discover the regularity mentioned above. More than fifty years earlier, in 1881, the American astronomer Newcomb published an academic paper with the same findings. Newcomb noted that the first pages of books with logarithmic tables were dirtier and showed more wear than the later pages. Since the books started with the numbers with low leading digits and ended with those with high leading digits, Newcomb concluded that logarithms of numbers with low leading digits were consulted more often than those of numbers with high leading digits.

Let us try it out. We compose a data set with the numbers of inhabitants of all countries in the world† using the CIA World Factbook (February 2006). Figure 2.15 shows a histogram (of area 1) with the leading digits of the numbers of inhabitants, together with the Benford frequencies. The frequencies of the leading digits seem to follow "Benford's law" fairly well.

---

† The data come from http://www.worldatlas.com/aatlas/populations/ctypopls.htm and can be found on the book's webpage at http://www.aup.nl under populationsize.

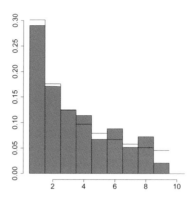

Figure 2.15. Histogram of the leading digits from 1 to 9 in the data set of the population sizes of all countries in the world. The step function in the figure gives the expected frequencies based on Benford's law.

Many data sets have been studied for the validity of Benford's law, from physical quantities measured in a laboratory to geographical information (like the lengths of rivers and population sizes of capitals), and from business accounting to currency conversion factors. In almost all cases, the law holds by approximation. Of course, not every data set is suitable. Purely random numbers (for example the outcomes of repeatedly casting a die) or numbers that are subject to restrictions, such as the ages of the inhabitants of the Netherlands or the phone numbers in a telephone directory, do not satisfy Benford's law.

Numbers that occur in financial statements, for example the accounts of rather large companies, often approximately satisfy Benford's law. This law can therefore be used to verify accounts and to investigate fraud and inconsistencies. An employee who commits fraud and tries to mask this will often fabricate or manipulate numbers in such a way that the leading digits occur in the same measure. If the employee manipulates or fabricates numbers regularly, then his actions will change the distribution of the leading digits, which will then deviate from the one predicted by Benford's law. If, for example, 9% of the numbers in the accounts begin with a 9, then the accounts will almost certainly be investigated because, by Benford's law, only 4.6% of the numbers should begin with a 9. However, a deviation from Benford's law does not automatically mean that there is fraud. In some cases, people prefer numbers that begin with a 9; for example, a product sells better if the price is 99 euros rather than 100.

Only structural fraud can be detected using Benford's law. If there is a single transfer of a large amount to a private account, this will not be noticed if one only looks at deviations from Benford's law. Figure 2.16 shows a histogram (of area 1) of the leading digits of 1.5 million numbers in the accounts of a large company, together with the frequencies one would expect based on Benford's law. The numbers in the accounts seem to follow Benford's law fairly well.

Despite much research into Benford's law, it is still not completely clear why one

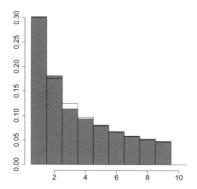

Figure 2.16. Histogram of the leading digits from 1 to 9 of figures in the bookkeeping. The staircase function in the histogram gives the expected frequencies based on Benford's law.

data set satisfies it, while another does not. An example where it is satisfied, is the case of exponential growth. Let us study this case more closely. Since we are only interested in the leading digit of a number, we write a number z as $z = x \times 10^n$ with $1 \leq x < 10$ and $n \in \mathbb{Z}$. This notation is possible for all positive numbers. We will call x the normed observation corresponding to $z = x \times 10^n$. The leading digit of z is equal to the leading digit of x. Let D be the random variable that gives the leading digit of an arbitrary (random) number $Z = X \times 10^n$ in a data set. Suppose that X is distributed according to a b^Y with $a, b > 0$, and that Y is uniformly distributed on the interval $[0, 1/\log_{10} b]$. Then

$$
\begin{aligned}
P(D = k) &= P(k \leq X < k + 1) \\
&= P(k \leq a\, b^Y < k + 1) \\
&= P(\log_{10}(k/a) \leq Y \log_{10} b < \log_{10}((k+1)/a)) \\
&= \log_{10}(k+1) - \log_{10} a - (\log_{10} k - \log_{10} a) \\
&= \log_{10}(1 + 1/k),
\end{aligned}
$$

where the fourth equality follows from the distribution of $Y \log_{10} b$, the uniform distribution on the interval $[0, 1]$. The probability that the leading digit D is equal to k is therefore exactly the probability according to Benford's law. If $b = 10$, then $\log_{10} b = 1$ and the assumption is that Y is uniformly distributed on $[0, 1]$.

Figure 2.17 shows a QQ-plot of the order statistics of $\log_{10}$ of the normed population sizes from Figure 2.15 plotted against the quantiles of the uniform distribution on $[0, 1]$. For this data set, the assumption apparently holds.

The assumption that X is distributed according to a b^Y, where $a, b > 0$ and Y is uniformly distributed on the interval $[0, 1/\log_{10} b]$ is not very insightful, and therefore seems unrealistic. The following example, however, shows that this impression is misleading. Suppose that a company has a market value of d million euros, which grows by $x\%$ each year. After t years, the market value of the company has increased to $d(1 + x/100)^t$ million euros. After $t = 1/\log_{10}(1 + x/100)$ years, we have

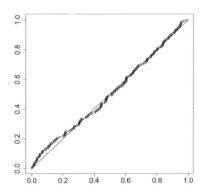

Figure 2.17. QQ-plot of $\log_{10}$ of the normed population sizes plotted against the quantiles of the uniform distribution on $[0, 1]$. The line in the figure is the line $y = x$.

$(1 + x/100)^t = 10$ and the initial value has increased tenfold. The leading digit is then equal to the leading digit at time $t = 0$. Since this time span does not depend on the initial amount d, the time 0 can be chosen arbitrarily, and we are only interested in the leading digit, it suffices to consider values of t in the interval $[0, 1/\log_{10}(1 + x/100)]$. Let T be a random variable that is uniformly distributed on the interval $[0, 1/\log_{10}(1 + x/100)]$. For an arbitrary company with market value d, the value at time T is then equal to $Z = d(1+x/100)^T = (d/10^n)(1+x/100)^T 10^n$, with $n \in \mathbb{N}$ such that $(d/10^n)(1 + x/100)^T \in [1, 10)$ with probability 1. We are now back in the situation of the previous example, with $Y = T$, $b = 1 + x/100$, and $a = d/10^n$. The probability that a company with market value d at time 0 has a market value beginning with a k at time T is equal to the Benford probability $\log_{10}(1 + 1/k)$.

Another argument that leads to the same answer is based on the assumption that the probability that an arbitrary company has a market value that begins with the digit k is directly proportional to the time span that the company has a market value that begins with the digit k. Let t_k be the time span (in years) during which the market value increases from k to $k + 1$ (million) euros; then $k(1 + x)^{t_k} = k + 1$, that is, $t_k = \log_{10}(1+1/k)/\log_{10}(1+x/100)$. The time span necessary to go from leading digit k (k million euros) to leading digit $k + 1$ ($k + 1$ million euros) is therefore proportional to $\log_{10}(1+1/k)$, the probability of having leading digit k according to Benford's law. This is, of course, independent of the chosen unit "millions of euros." We can again conclude that under our assumption, a fraction of approximately $\log_{10}(1+1/k)$ of all companies have a market value with leading digit k, exactly as predicted by Benford's law.

3 Estimators

3.1 Introduction

A statistical model consists of all probability distributions that, a priori, seem possible for the given data. Given a correctly set-up model, we assume that the data were generated from one of the distributions in the model. After setting up a suitable statistical model, the next step is determining which distribution within the model fits the data points best. If the model is described by a parameter, this is equivalent to determining the best-fitting value of the parameter, often called the "true" parameter. In statistics, this process is called "estimating." Other names are "fitting" and "learning."

Suppose that the distribution of X depends on an unknown parameter θ, so that the statistical model is of the form $\{P_\theta : \theta \in \Theta\}$, for P_θ the distribution of X if θ is the "true" parameter. Based on an observation x, we want to estimate the true value of θ, or perhaps the value of a function $g(\theta)$ of θ. Here, "estimating" means making a statement about θ or $g(\theta)$ of the form "I think that $g(\theta)$ is approximately equal to $T(x)$," for some value $T(x)$ that depends on the observed value x.

Definition 3.1 Estimator

An *estimator* or *statistic* is a random vector $T(X)$ that depends only on the observation X. The corresponding *estimate* for an observation x is $T(x)$.

By this definition, many objects are estimators. What matters is that $T(X)$ is a function of X that does not depend on the parameter θ: we must be able to compute $T(x)$ from the data x. Given the observed value x, the statistic T is realized as $t = T(x)$, which is used as an estimate of θ (or $g(\theta)$). We often shorten $T(X)$ to T. Mathematically, the word "statistic" has exactly the same definition as "estimator," but it is used in a different context.

Both estimators and estimates of θ are often indicated by $\hat{\theta}$. The hat indicates that $\hat{\theta}$ is a function of the observation, but this notation does not differentiate between the random vector and its observed value: $\hat{\theta}$ can mean both $\hat{\theta}(X)$ and $\hat{\theta}(x)$.

There are many estimation methods. In this chapter, we discuss several general principles, such as the maximum likelihood method, the method of moments, and the Bayes method. We begin, however, by setting up the framework necessary to compare the performance of the different estimators.

3.2 Mean Square Error

Although every function of the observation is an estimator, not every estimator is a good one. A good estimator of $g(\theta)$ is a function T of the observed data such that T is "close" to the estimand $g(\theta)$. The distance $\|T - g(\theta)\|$ is an unsatisfactory measure for two reasons:

- This measure depends on the unknown value θ.
- This measure is stochastic and cannot be computed before carrying out the experiment.

To avoid the second difficulty, we consider the *distribution* of the distance $\|T - g(\theta)\|$ *under the assumption that θ is the true value*. The best situation is that where this distribution is degenerate at 0, that is, if θ is the true value, then $\|T - g(\theta)\|$ has probability 1 to be equal to 0. This would mean that we do not make any estimation errors; the estimate $T(x)$ would be equal to the estimand with absolute certainty. Unfortunately, this is impossible in practice, and we must settle for the smallest possible (average) error. We are looking for an estimator whose distribution for the true value θ is concentrated as much as possible around $g(\theta)$ or, equivalently, for which the distribution of $\|T - g(\theta)\|$ is concentrated as much as possible in a neighborhood of 0.

Example 3.2 Uniform distribution

Let $X_1, \ldots, X_n$ be independent $U[0, \theta]$-distributed random variables. The observation is the vector $X = (X_1, \ldots, X_n)$, and we want to estimate the unknown θ. Since $E_\theta X_i = \frac{1}{2}\theta$, it is reasonable to estimate $\frac{1}{2}\theta$ using the sample mean $\overline{X}$ and θ using $2\overline{X}$; after all, by the law of large numbers (Theorem A.26), the sample mean converges (in probability) to $E_\theta X_i = \frac{1}{2}\theta$. Suppose that $n = 10$ and that the data have the following values: 3.03, 2.70, 7.00, 1.59, 5.04, 5.92, 9.82, 1.11, 4.26, 6.96, so that $2\overline{x} = 9.49$.

This estimate is certainly too small. Indeed, one of the observations is 9.82, so that we must have $\theta \geq 9.82$.

Can we think of a better estimator? We can avoid the problem we just mentioned by taking the maximum $X_{(n)}$ of the observations. However, the maximum is certainly also less than the true value, for all observations x_i lie in the interval $[0, \theta]$. An obvious solution is to add a small correction. We could, for example, take $(n+2)/(n+1)\,X_{(n)}$ as estimator.

So there are several candidates. Which estimator is the best? To gain insight into this question, we carried out the following simulation. We chose $n = 50$ and simulated 1000 independent samples of size 50 from the uniform distribution on $[0, 1]$. For each sample, we computed the estimators $2\overline{X}$ and $(n+2)/(n+1)X_{(n)}$. Figure 3.1 shows histograms of two sets with 1000 estimates each for the parameter θ. The image on the left uses the estimator $(n+2)/(n+1)X_{(n)}$, and the one on the right uses $2\overline{X}$.

These histograms can be viewed as approximations of the densities of the estimators. The density on the left is more concentrated around the true value $\theta = 1$ than the density on the right. We therefore prefer the estimator $(n+2)/(n+1)X_{(n)}$: "on average," it is closer to the true value. (The difference in the forms of the histograms is remarkable: the one on the left resembles an (inverse) exponential density, while the one on the right resembles a normal density. We can easily explain this theoretically. How?)

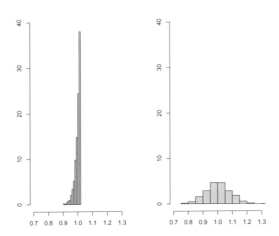

Figure 3.1. Histograms of 1000 realizations of the estimators $(n+2)/(n+1)X_{(n)}$ and $2\overline{X}$ for the parameter 1 of a uniform distribution, each based on $n = 50$ observations.

Note that it is not true that the estimator $(n + 2)/(n + 1)X_{(n)}$ gives the best estimate on each of the 1000 samples. This can be seen in Figure 3.2, where the difference $|(n + 2)/(n + 1)x_{(n)} - 1| - |2\overline{x} - 1|$ is set out along the vertical axis. In general, this difference is negative, but sometimes it is positive, in which case the estimator $2\overline{X}$ gives a value that is closer to the true value $\theta = 1$. Because in practice we do not know the true value, it is not possible to choose the "best of both worlds." We must use the estimator that is the best on average.

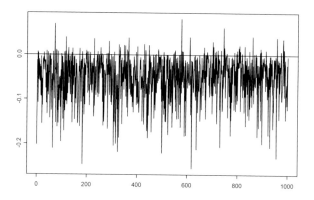

Figure 3.2. Differences $|(n+2)/(n+1)x_{(n)} - 1| - |2\bar{x} - 1|$ of the absolute distances from the estimates $(n+2)/(n+1)x_{(n)}$ and $2\bar{x}$ to the estimand 1 in Figure 3.1.

Our simulation experiment only shows that $(n+2)/(n+1)X_{(n)}$ is the better estimator if the true value of θ is equal to 1. To determine which estimator is better when θ has a different value, we would have to repeat the simulation experiment with simulated samples from the uniform distribution on $[0, \theta]$, for every θ. This is not something we want to do, of course, and that is one of the reasons to study estimation problems mathematically. Another reason is that instead of ordering pairs of estimators, we would like to find the overall best estimator.

Since a probability distribution is a complicated object, comparing the "concentration" is not well defined. It is therefore useful to express the concentration as a number, so that we only need to compare numbers. There are many ways to do this. One measure of concentration that is relatively simple to deal with mathematically is the mean square error or mean square deviation.

Definition 3.3 Mean square error

The mean square error or MSE of an estimator T for the value $g(\theta)$ is

$$\mathrm{MSE}(\theta; T) = \mathrm{E}_\theta \| T - g(\theta) \|^2.$$

The subscript θ in E_θ in the definition is essential: the MSE is the expected square deviation of T from $g(\theta)$ under the assumption that θ is the true value of the parameter (this sentence has the *same* θ twice). We view the mean square error as the function $\theta \mapsto \mathrm{MSE}(\theta; T)$ for a given statistic T. A more complete notation would be

$\text{MSE}(\theta; T, g)$, but since g is fixed in the context of the problem, we leave g out of the notation.

The first difficulty—that the measure of quality depends on θ—has not been solved yet; the mean square error is a function of θ. In principle, it suffices if $\text{MSE}(\theta; T)$ is as small as possible in the "true value" of θ. As we do not know that value, we try to keep the mean square error (relatively) small for all values of θ at once.

Convention 3.4

We prefer an estimator with small mean square error (MSE) for all values of the parameter θ at once.

If for two estimators T_1 and T_2, we have

$$\text{E}_\theta \|T_1 - g(\theta)\|^2 \leq \text{E}_\theta \|T_2 - g(\theta)\|^2 \qquad \text{for all } \theta \in \Theta,$$

with a strict inequality for at least one value of θ, then we prefer T_1. The estimator T_2 is then called *inadmissible*. However, the strict inequality may hold for some θ, while for other θ, the strict inverse inequality may hold. It is then not directly clear which estimator we should prefer. Because the true value of θ, say θ_0, is unknown, we do not know which of $\text{MSE}(\theta_0; T_1)$ and $\text{MSE}(\theta_0; T_2)$ is the smallest.

In Section 6.3, we discuss optimality criteria for estimators and how to find optimal estimators. In this chapter, we discuss several methods to find estimators of which it is intuitively clear that they are reasonable and compare mean square errors.

The mean square error of a real-valued estimator T can be decomposed in two terms:

$$\text{MSE}(\theta; T) = \text{var}_\theta T + \big(\text{E}_\theta T - g(\theta)\big)^2$$

(verify). Both terms in this decomposition are nonnegative. Hence, the mean square error can only be small if both terms are small. If the second term is 0, the estimator is called unbiased.

Definition 3.5 Unbiased estimator

An estimator T is called unbiased for the estimation of $g(\theta)$ if $\text{E}_\theta T = g(\theta)$ for all $\theta \in \Theta$. The bias is defined as $\text{E}_\theta T - g(\theta)$.

The second term in the decomposition of $\text{MSE}(\theta; T)$ is therefore the square of the bias. For an unbiased estimator, this term is identically 0. This seems very desirable, but is not always so. Namely, the condition that an estimator be unbiased can lead to the variance being very large, so that we amply loose in the first term what we would have gained in the second one. In general, a small variance leads to a large bias, and a small bias to a large variance. We must therefore balance the two terms against each other.

The standard deviation $\sigma_\theta(T) = \sqrt{\mathrm{var}_\theta\, T}$ of an estimator is also called the *standard error*. This should not be confused with the standard deviation of the observations. In principle, the standard error $\sigma_\theta(T)$ depends on the unknown parameter θ and is therefore itself an unknown. Because the bias of reasonable estimators is often small, the standard error often gives an idea of the quality of the estimator. An estimate of the standard error is often given along with the estimate itself. We will come back to this when we discuss confidence regions in Chapter 5.

We are thus looking for estimators with a small standard error and a small bias.

Example 3.6 Uniform distribution

Let $X_1, \ldots, X_n$ be independent, $\mathrm{U}[0, \theta]$-distributed random variables. The estimator $2\overline{X}$ is unbiased because for all $\theta > 0$,

$$\mathrm{E}_\theta(2\overline{X}) = \frac{2}{n}\sum_{i=1}^{n}\mathrm{E}_\theta X_i = \frac{2}{n}\sum_{i=1}^{n}\frac{\theta}{2} = \theta.$$

The mean square error of this estimator is

$$\mathrm{MSE}(\theta; 2\overline{X}) = 4\,\mathrm{var}_\theta\, \overline{X} = \frac{4}{n^2}\sum_{i=1}^{n}\mathrm{var}_\theta\, X_i = \frac{\theta^2}{3n}.$$

The estimator $X_{(n)}$ is biased because for all $\theta > 0$,

$$\mathrm{E}_\theta X_{(n)} = \int_0^\theta xnx^{n-1}\frac{1}{\theta^n}\,dx = \frac{n}{n+1}\theta$$

(see Exercise A.10 for the distribution of $X_{(n)}$). Nevertheless, (for n not too small) we prefer $X_{(n)}$ to $2\overline{X}$, because this estimator has a smaller mean square error:

$$\mathrm{MSE}(\theta; X_{(n)}) = \mathrm{var}_\theta\, X_{(n)} + \left(\mathrm{E}_\theta X_{(n)} - \theta\right)^2$$

$$= \theta^2\frac{n}{(n+2)(n+1)^2} + \theta^2\left(\frac{n}{n+1} - 1\right)^2 = \frac{2\theta^2}{(n+2)(n+1)}.$$

We can cancel out the bias of $X_{(n)}$ by multiplying by a constant: the estimator $(n+1)/n\, X_{(n)}$ is unbiased for θ. However, the biased estimator $(n+2)/(n+1)\, X_{(n)}$ is better than all estimators we have mentioned up to now, because

$$\mathrm{MSE}\left(\theta; \frac{n+2}{n+1}X_{(n)}\right) = \frac{\theta^2}{(n+1)^2}.$$

Figure 3.3 shows the mean square error of this last estimator, together with the mean square errors of $X_{(n)}$ and $2\overline{X}$ as functions of θ for $n = 50$. For values of θ close to 0, the differences between the mean square errors of $2\overline{X}$ and of the other two estimators are small, but they increase rapidly when θ increases.

On closer inspection, it turns out that for values of n that are not too small, the difference between the mean square errors of $(n+2)/(n+1)X_{(n)}$ and $X_{(n)}$ is small. The greater precision of $(n+2)/(n+1)X_{(n)}$ compared to $2\overline{X}$, however, rapidly becomes apparent when n increases, because the mean square error of the first is smaller by a factor of n.

We have already noted (see Figure 3.2) that the estimator $(n+2)/(n+1)\,X_{(n)}$ does not give a better result than the estimator $2\overline{X}$ on every sample. The fact that $\mathrm{MSE}\big(1;(n+2)/(n+1)\,X_{(n)}\big) < \mathrm{MSE}(1;2\overline{X})$ certainly does not exclude this, because the mean square error is an expected value and can be viewed as the average over a large number of realizations. An average can be negative without all terms being negative. On average, $(n+2)/(n+1)\,X_{(n)}$ is (much) better.

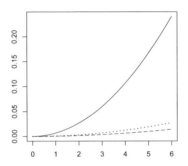

Figure 3.3. The mean square errors as functions of θ for the estimators $2\overline{X}$ (solid line), $X_{(n)}$ (dotted line), and $(n+2)/(n+1)X_{(n)}$ (dashed line) for the parameter in $\mathrm{U}[0,\theta]$, for $n=50$.

Example 3.7 Sample mean and sample variance

Let $X_1,\ldots,X_n$ be independent, identically distributed random variables, with an unknown marginal distribution. We want to estimate the expected value μ and variance σ^2 of the observations. Formally, we can take θ equal to an unknown distribution, the so-called "nonparametric model," which does not specify the underlying distribution. The "parameters" μ and σ^2 are functions of this underlying distribution.

The sample mean $\overline{X}$ and the sample variance S_X^2 are equal to (see Definitions 2.1 and 2.2)

$$\overline{X} = \frac{1}{n}\sum_{i=1}^{n}X_i, \qquad S_X^2 = \frac{1}{n-1}\sum_{i=1}^{n}(X_i - \overline{X})^2.$$

The sample mean is an unbiased estimator for μ since

$$\mathrm{E}_\theta\overline{X} = \frac{1}{n}\sum_{i=1}^{n}\mathrm{E}_\theta X_i = \mu.$$

51

The mean square error of this estimator is given by

$$\mathrm{MSE}(\theta; \overline{X}) = \mathrm{var}_\theta\, \overline{X} = \frac{1}{n^2}\sum_{i=1}^{n}\mathrm{var}_\theta\, X_i = \frac{\sigma^2}{n}.$$

The mean square error of $\overline{X}$ is therefore smaller by a factor of n than the mean square error of the estimator X_i based on a single observation, $\mathrm{MSE}(\theta, X_i) = \mathrm{var}_\theta\, X_i = \sigma^2$. Since the mean square error is an estimated *square* distance, we conclude that the quality of the estimator $\overline{X}$ increases by a factor of $\sqrt{n}$. So for an estimator that is twice as good, you need four times as many observations.

The sample variance is an unbiased estimator for σ^2 because

$$\mathrm{E}_\theta S_X^2 = \mathrm{E}_\theta \frac{1}{n-1}\sum_{i=1}^{n}((X_i - \mu) + (\mu - \overline{X}))^2$$

$$= \mathrm{E}_\theta \frac{1}{n-1}\sum_{i=1}^{n}\left[(X_i - \mu)^2 + (\mu - \overline{X})^2 + 2(\mu - \overline{X})(X_i - \mu)\right]$$

$$= \frac{1}{n-1}\sum_{i=1}^{n}\mathrm{E}_\theta(X_i - \mu)^2 - \frac{n}{n-1}\mathrm{E}_\theta(\overline{X} - \mu)^2 = \sigma^2,$$

where the last equality follows from $\mathrm{E}_\theta(X_i - \mu)^2 = \mathrm{var}_\theta\, X_i = \sigma^2$ and $\mathrm{E}_\theta(\overline{X} - \mu)^2 = \mathrm{var}_\theta\, \overline{X} = \sigma^2/n$. With a bit of work, the mean square error of S_X^2 can be expressed in the fourth sample moment of the observations, but we will not discuss this.

Suppose that we are looking for an unbiased estimator for μ^2. Since $\overline{X}$ is an unbiased estimator for μ, in first instance we take $\overline{X}^2$ as estimator for μ^2. However, this estimator is biased:

$$\mathrm{E}_\theta(\overline{X})^2 = \mathrm{var}_\theta\, \overline{X} + (\mathrm{E}_\theta\overline{X})^2 = \frac{\sigma^2}{n} + \mu^2.$$

It immediately follows that $\mathrm{E}_\theta(\overline{X}^2 - \sigma^2/n) = \mu^2$, but since σ^2 is an unknown parameter, $\overline{X}^2 - \sigma^2/n$ is not an estimator. If we replace σ^2 by its unbiased estimator S_X^2, then we see that $\overline{X}^2 - S_X^2/n$ is an unbiased estimator for μ^2. ⊏⊐

* **Example 3.8** Sample theory

Suppose that a proportion p of a population has a certain characteristic A. We will compare three methods to estimate p, based on a sample with replacement, a sample without replacement, and a stratified sample.

In the first method, we take a sample of size n from a population, with replacement, and estimate p using the fraction X/n, where X is the number of persons with characteristic A in the sample. Then X is $\mathrm{bin}(n, p)$-distributed and has expected value np and variance $np(1 - p)$. Since $\mathrm{E}_p(X/n) = p$ for all p, the estimator X/n is unbiased. The mean square error is

$$\mathrm{MSE}\left(p; \frac{X}{n}\right) = \mathrm{var}_p\left(\frac{X}{n}\right) = \frac{p(1 - p)}{n}.$$

It follows, among other things, that the estimator is better when $p \approx 0$ or $p \approx 1$, and the worst when $p = \frac{1}{2}$. The mean square error does not depend on the size of the population. By choosing n sufficiently large, for example $n \geq 1000$, we can obtain an estimator with a mean square error of at most $(1/4)/1000 = 1/4000$, regardless of whether the population consists of 800 or a trillion persons.

In the second method, we take a sample of size n from a population, without replacement, and estimate p using the faction Y/n, where Y is the number of persons with characteristic A in the sample. Then Y is $\mathrm{hyp}(N, pN, n)$-distributed and has expected value np and variance $np(1-p)(N-n)/(N-1)$. So the estimator Y/n is again unbiased; the mean square error is

$$\mathrm{MSE}\left(p; \frac{Y}{n}\right) = \mathrm{var}_p\left(\frac{Y}{n}\right) = \frac{p(1-p)}{n} \frac{N-n}{N-1}.$$

This is smaller than $\mathrm{MSE}(p; X/n)$, although the difference is negligible for $n \ll N$. This is not surprising: it is not useful to study persons that have already been studied again, but if $n \ll N$, the probability of this happening is negligible.

In the third method, we first divide the population into a number of subpopulations, called *strata*. This can be a classification by region, gender, age, income, profession, or some other background variable. Suppose that the entire population has size N, while the subpopulations have sizes $N_1, \ldots, N_m$. We now draw, for convenience without replacement, $(N_j/N)n$ persons from the jth population, a *stratified sample*, and estimate p using Z/n, where Z is the total number of persons with property A in our sample. So $Z = Z_1 + \cdots + Z_n$, where Z_j is the number of persons with characteristic A drawn from the jth population. Now, $Z_1, \ldots, Z_m$ are independent, $\mathrm{bin}((N_j/N)n, p_j)$-distributed variables, where p_j is the proportion of persons with characteristic A in the jth population. Then

$$\mathrm{E}_p\left(\frac{Z}{n}\right) = \frac{1}{n}\sum_{j=1}^{m}\mathrm{E}_p Z_j = \frac{1}{n}\sum_{j=1}^{m}\frac{N_j}{N}np_j = \frac{1}{N}\sum_{j=1}^{m}N_j p_j = p,$$

$$\mathrm{MSE}\left(p; \frac{Z}{n}\right) = \mathrm{var}_p\left(\frac{Z}{n}\right) = \frac{1}{n^2}\sum_{j=1}^{m}\mathrm{var}_p Z_j = \frac{1}{n^2}\sum_{j=1}^{m}\frac{N_j n}{N}p_j(1-p_j)$$

$$= \frac{p(1-p)}{n} - \frac{1}{n}\sum_{j=1}^{m}\frac{N_j}{N}(p_j - p)^2.$$

The estimator Z/n is therefore also unbiased, and its mean square error is less than or equal to the mean square error of X/n. The difference is mostly worth considering when the p_j differ greatly. Stratified sampling is therefore the preferred method in general, even though in practice, it can mean more work.

Similar results also hold for sampling without replacement, provided that the sizes of the strata and samples satisfy certain conditions. It is, however, not true that in this case stratification always leads to greater precision.

3.3 Maximum Likelihood Estimators

The "maximum likelihood estimation method" is the most common method to find estimators for an unknown parameter. Before presenting the method in general, we deduce the maximum likelihood estimator for the (simple) case of the binomial distribution.

Example 3.9 Binomial distribution

Suppose that we toss a biased coin 10 times. For this coin, the probability p of getting "head" is not necessarily $1/2$. Let X be the number of times we get "head" in the 10 tosses. The random variable X then has a binomial distribution with parameters 10 and the unknown $p \in [0, 1]$. Suppose that we get "head" 3 times. The probability of this outcome is equal to

$$P_p(X = 3) = \binom{10}{3} p^3 (1 - p)^7.$$

The probability p is unknown and must be estimated. What value for p is the most probable? Figure 3.4 shows the probability $P_p(X = 3)$ as a function of p. We see that there is exactly one value of p that maximizes this probability, namely 0.3. This value for p assigns the greatest possible value to the observation "3 times head." In this situation, the estimate $\hat{p} = 0.3$ turns out to be the maximum likelihood estimate.

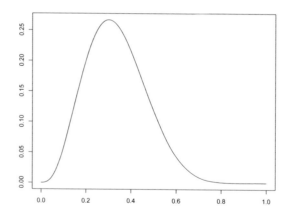

Figure 3.4. The probability $P_p(X = 3)$ as a function of p, where the random variable X is binomially distributed with parameters 10 and p.

The maximum likelihood method requires a likelihood function, which is deduced from the density of the observation. By a probability density p_θ of a random variable X, we mean the function $x \mapsto P_\theta(X = x)$ if X is discrete and the function p_θ such that $P_\theta(X \in B) = \int_B p_\theta(x) \, dx$ if X is continuous.

Definition 3.10 Likelihood function

Let X be a random vector with probability density p_θ that depends on a parameter $\theta \in \Theta$. For x fixed, the function

$$\theta \mapsto L(\theta; x) := p_\theta(x),$$

seen as a function of $\theta \in \Theta$ (where Θ is the parameter space), is called the likelihood function.

Often, $X = (X_1, \ldots, X_n)$ is a vector with independent, identically distributed coordinates X_i. The density of X in $(x_1, \ldots, x_n)$ is then equal to the product $\prod_{i=1}^{n} p_\theta(x_i)$ of the marginal probability densities of $X_1, \ldots, X_n$, and the likelihood function is equal to

$$\theta \mapsto L(\theta; x_1, \ldots, x_n) = \prod_{i=1}^{n} p_\theta(x_i),$$

where p_θ is now the (marginal) density of one X_i. However, the general definition of the likelihood function also holds for an observation vector whose elements are not independent or identically distributed. We therefore prefer to write the observation as x, rather than $(x_1, \ldots, x_n)$, and to write the likelihood function as $L(\theta; x) \equiv p_\theta(x)$.

Definition 3.11 Maximum likelihood estimate and estimator

The maximum likelihood estimate of θ is the value of $T(x) \in \Theta$ that maximizes the likelihood function $\theta \mapsto L(\theta; x)$. The maximum likelihood estimator is the corresponding estimator $T(X)$.

In the case of a discrete probability distribution, the maximum likelihood estimate can be described as *the value of the parameter that assigns the greatest probability to the observed value x.* Indeed, in that case, we maximize the probability density $p_\theta(x) = P_\theta(X = x)$ with respect to θ for fixed x (see Example 3.9). Intuitively, this is a reasonable principle for taking estimates. It also explains the name. This principle should, however, be seen only as a way to obtain estimates: maximum likelihood estimators are not necessarily the best estimators, regardless of their nice name. By a "best" estimator, we mean an estimator with the smallest possible mean square error.

For a given model, computing the maximum likelihood estimators is a matter of applying calculus. Often, we differentiate the likelihood function and set the derivatives equal to 0. A trick that limits the necessary calculations (especially with independent observations) is to first take the logarithm of the likelihood. Because the logarithm is a monotone function, the value $\hat{\theta}$ maximizes the function $\theta \mapsto L(\theta; x)$ if and only if this value maximizes the function $\theta \mapsto \log L(\theta, x)$. (Note that we are speaking only of the value where the maximum is reached, not of the value of the maximum!) For fixed x, the *log-likelihood function* is given by

$$\theta \mapsto \log L(\theta; x) = \log p_\theta(x).$$

If L is differentiable in $\theta \in \Theta \subset \mathbb{R}^k$ and takes on its maximum in an interior point of Θ, then

$$\frac{\partial}{\partial \theta_j} \log L(\theta; x)_{|\theta=\hat{\theta}} = 0, \qquad j = 1, \ldots, k.$$

This system is called the system of *likelihood equations* and cannot always be solved explicitly. If necessary, an iterative method is used to obtain an approximation of the solution.

Not only maxima, but also minima and inflection points are solutions of the likelihood equations. To verify whether a solution is indeed a maximum, we must consider the form of the (log-)likelihood function. One way to do this, is to determine the second derivative (or the Hessian matrix if the parameter has dimension greater than 1) of the log-likelihood function in the solution. If the function has a maximum in the solution, the second derivative in that point will be negative. For higher-dimensional parameters, all eigenvalues of the Hessian matrix must be negative.

If the observation $X = (X_1, \ldots, X_n)$ consists of independent, identically distributed subobservations X_i, then the likelihood $L(\theta; x)$ of the observation x is a product $L(\theta; x) = \prod_i p_\theta(x_i)$, where p_θ is the (marginal) density of one X_i. The log-likelihood is then

$$\theta \mapsto \log L(\theta; x_1, \ldots, x_n) = \log \prod_{i=1}^n p_\theta(x_i) = \sum_{i=1}^n \log p_\theta(x_i).$$

The derivative of $\log L$, the *score function*, is the sum of the score functions of the individual observations; see Definition 5.8. The likelihood equations are then of the form

$$\sum_{i=1}^n \frac{\partial}{\partial \theta_j} \log p_\theta(x_i) = 0, \qquad j = 1, \ldots, k.$$

Example 3.12 Exponential distribution

Let $X = (X_1, \ldots, X_n)$ be a sample from the exponential distribution with unknown parameter $\lambda > 0$. Then the log-likelihood function of a realization $x_1, \ldots, x_n$ is equal to

$$\lambda \mapsto \log L(\lambda; x_1, \ldots, x_n) = \log \prod_{i=1}^n \lambda e^{-\lambda x_i} = n \log \lambda - \lambda \sum_{i=1}^n x_i.$$

The parameter space for λ is $(0, \infty)$. Setting the derivative of the log-likelihood function with respect to λ equal to 0 gives

$$\frac{d}{d\lambda} \log L(\lambda; x_1, \ldots, x_n)_{|\lambda=\hat{\lambda}} = \frac{n}{\hat{\lambda}} - \sum_{i=1}^n x_i = 0,$$

with solution $\hat{\lambda} = 1/\bar{x}$. The second derivative of the log-likelihood function with respect to λ is

$$\frac{d^2}{d\lambda^2} \log L(\lambda; x_1, \ldots, x_n) = -\frac{n}{\lambda^2};$$

it is negative for all $\lambda > 0$, so the likelihood function indeed has a maximum in $\hat{\lambda}$. The maximum likelihood estimator for λ is equal to $\hat{\lambda} = 1/\overline{X}$.

Example 3.13 Binomial distribution, continued from Example 3.9

The variable X is defined as the number of heads when a coin is tossed 10 times. It is binomially distributed with parameter 10 and unknown probability p. The observed value is $x = 3$. The log-likelihood function is equal to the function

$$p \mapsto \log L(p; x = 3) = \log\left(\binom{10}{3} p^3 (1 - p)^7\right)$$

$$= \log\binom{10}{3} + 3\log p + 7\log(1 - p).$$

The maximum likelihood estimate of p is the value in $[0, 1]$ that maximizes this function with respect to p. This again gives the solution $\hat{p} = 0.3$.

In the general case of a binomially distributed quantity X with parameters n and p, the log-likelihood function is equal to

$$p \mapsto \log L(p; x) = \log\binom{n}{x} + x\log p + (n - x)\log(1 - p).$$

If $0 < x < n$, then $\log L(p; x) \to -\infty$ as $p \downarrow 0$ or $p \uparrow 1$, so that the log-likelihood function takes on its maximum in the interval $(0, 1)$. It follows that the likelihood function $L(p; x)$ also takes on its maximum in the interval $(0, 1)$. Setting the derivative with respect to p equal to 0 gives one solution, $\hat{p} = x/n$. This solution is therefore the maximum likelihood estimate, $\hat{p} = x/n$. Instead of looking at the form, we can also determine the second derivative of the (log-)likelihood function in $p = x/n$. If x is equal to 0 or n, then $L(p; x)$ has a local maximum in 0 or 1. In these cases, too, the maximum likelihood estimate can be written as $\hat{p} = x/n$. The maximum likelihood estimator is equal to $\hat{p} = X/n$.

Example 3.14 Normal distribution

The log-likelihood function for a sample $X = (X_1, \ldots, X_n)$ from the $N(\mu, \sigma^2)$-distribution is given by

$$(\mu, \sigma^2) \mapsto \log \prod_{i=1}^{n} \frac{1}{\sqrt{2\pi\sigma^2}} e^{-\frac{1}{2}(x_i - \mu)^2/\sigma^2}$$

$$= -\tfrac{1}{2}n\log 2\pi - \tfrac{1}{2}n\log\sigma^2 - \frac{1}{2\sigma^2}\sum_{i=1}^{n}(x_i - \mu)^2.$$

We take the natural parameter space for the parameter $\theta = (\mu, \sigma^2)$, namely $\Theta = \mathbb{R} \times (0, \infty)$. The partial derivatives of the log-likelihood with respect to μ and σ^2 are

$$\frac{\partial}{\partial \mu} \log L(\mu, \sigma^2; x_1, \ldots, x_n) = \frac{1}{\sigma^2} \sum_{i=1}^{n} (x_i - \mu)$$

$$\frac{\partial}{\partial \sigma^2} \log L(\mu, \sigma^2; x_1, \ldots, x_n) = -\frac{n}{2\sigma^2} + \frac{1}{2\sigma^4} \sum_{i=1}^{n} (x_i - \mu)^2.$$

Setting the first equation equal to 0 gives one solution: $\hat{\mu} = \bar{x}$. In this value for μ, the log-likelihood indeed has a global maximum for every $\sigma^2 > 0$, because the value of the log-likelihood goes to $-\infty$ as $\mu \to \pm\infty$. Next, we substitute $\mu = \hat{\mu}$ in the second partial derivative, set the latter equal to 0, and solve the likelihood equation for σ^2. This again gives one solution: $\hat{\sigma}^2 = n^{-1} \sum_{i=1}^{n} (x_i - \bar{x})^2$. For the same reason as before, the log-likelihood function has a maximum in this value. (Note that maximizing the log-likelihood function in σ instead of σ^2 gives the square root of $\hat{\sigma}^2$ as maximum likelihood estimator for σ.) To verify whether the (differentiable) log-likelihood function has a maximum in the solution of the likelihood equation that we found, we can also determine the Hessian matrix of the log-likelihood function in the point $(\hat{\mu}, \hat{\sigma}^2)$, which in this case is equal to

$$\frac{1}{\hat{\sigma}^4} \begin{pmatrix} -n\hat{\sigma}^2 & 0 \\ 0 & -n/2 \end{pmatrix}.$$

Both eigenvalues of this matrix are negative; consequently, the log-likelihood has a maximum in the point $(\hat{\mu}, \hat{\sigma}^2)$.

The resulting maximum likelihood estimator for (μ, σ^2) is equal to

$$\left(\overline{X}, \frac{1}{n} \sum_{i=1}^{n} (X_i - \overline{X})^2 \right) = \left(\overline{X}, \frac{n-1}{n} S_X^2 \right)$$

with

$$S_X^2 = \frac{1}{n-1} \sum_{i=1}^{n} (X_i - \overline{X})^2.$$

The sample mean is unbiased for μ, but the maximum likelihood estimator $\hat{\sigma}^2$ has a slight bias (see Example 3.7). Because of the small bias, the sample variance $S_X^2 = (n/(n-1))\hat{\sigma}^2$ is often preferred. However, the mean square error of S_X^2 is greater than that of $\hat{\sigma}^2$, and both are subordinate to $((n-1)/(n+1)) S_X^2$ in terms of the mean square error.[‡] Because the difference is small for large numbers of observations, it does not matter much which of these estimators is used.

We obtain another model if we assume μ known. The parameter is then $\theta = \sigma^2$, and the parameter space is $(0, \infty)$. We then find that the maximum likelihood estimator for σ^2 is equal to $n^{-1} \sum_{i=1}^{n} (X_i - \mu)^2$. Note that this is only an estimator if μ may be assumed known!

[‡] It takes some calculations to support this statement. Theorem 4.29 can be used to simplify them. See Exercise 4.27 in Chapter 4.

If the maximum of the (log-)likelihood function is not taken in the interior of the parameter space, then the maximum likelihood estimate $\hat{\theta}$ is usually not a stationary point of the derivative of the likelihood function, but rather a local maximum, and the likelihood equations do not hold. In yet other examples, the likelihood function is not everywhere differentiable (or even continuous), and the maximum likelihood estimate also does not satisfy the likelihood equations. Example 3.15 illustrates this situation. Moreover, it is possible that the likelihood function has several (local) maxima and minima. The likelihood equations can then have more than one solution. The maximum likelihood estimate is by definition the global maximum of the likelihood function.

Example 3.15 Uniform distribution

Let $x = (x_1, \ldots, x_n)$ be an observed sample from the uniform distribution on the interval $[0, \theta]$, where $\theta > 0$ is unknown. We want to estimate the parameter θ using the maximum likelihood estimator. Since the observations $x_1, \ldots, x_n$ lie in the interval $[0, \theta]$, we must have $\theta \geq x_i$ for $i = 1, \ldots, n$. It immediately follows that $\theta \geq x_{(n)}$, where $x_{(n)}$ is the largest observed order statistic.

The likelihood function of the observations $x_1, \ldots, x_n$ is equal to the joint density of $X_1, \ldots, X_n$ in $x_1, \ldots, x_n$, viewed as a function of θ. Because $X_1, \ldots, X_n$ are independent and identically distributed, the joint density is equal to the product of the marginal densities, which is equal to $1/\theta$ on the interval $[0, \theta]$ and 0 elsewhere. The likelihood function is therefore equal to

$$\theta \mapsto L(\theta; x_1, \ldots, x_n) = \prod_{i=1}^{n} \frac{1}{\theta} 1_{0 \leq x_i \leq \theta} = \left(\frac{1}{\theta}\right)^n 1_{x_{(1)} \geq 0} 1_{x_{(n)} \leq \theta}.$$

This function of θ is equal to 0 for $\theta < x_{(n)}$ because the indicator function $1_{x_{(n)} \leq \theta}$ is then equal to 0. In $\theta = x_{(n)}$, the function jumps to $1/\theta^n$. In $\theta = x_{(n)}$, the likelihood, and therefore also the log-likelihood, is not differentiable with respect to θ. A maximum can be found by plotting the likelihood function as a function of θ. For $\theta \geq x_{(n)}$, the likelihood function is equal to the decreasing function $\theta \to 1/\theta^n$. Figure 3.5 illustrates the course of the likelihood function (as a function of θ). In $x_{(n)}$, the likelihood function is upper semi-continuous and also maximal; the maximum likelihood estimate of θ is therefore equal to $x_{(n)}$ and the corresponding maximum likelihood estimator is $X_{(n)}$.

Example 3.16 Normal distribution with restriction

Suppose that the observations $X_1, \ldots, X_n$ are independent and normally distributed with expected value μ and variance 1, where we know $\mu \geq 0$. For a realization $x_1, \ldots, x_n$ of $X_1, \ldots, X_n$, on $\mathbb{R}$ the likelihood function takes on an absolute maximum in $\bar{x}$. Now, $\bar{x}$ can be negative, and $\mu \geq 0$, hence $\bar{x}$ is not the maximum likelihood estimate. If $\bar{x} \leq 0$, then on the parameter space $[0, \infty)$, the likelihood function takes on a local maximum in 0. The maximum likelihood estimate is $\bar{x}$ if it is nonnegative and 0 otherwise. The corresponding maximum likelihood estimator is $max(0, \bar{X})$.

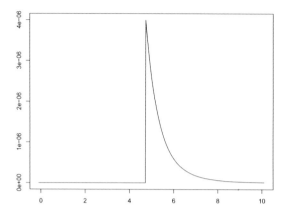

Figure 3.5. Realization of the likelihood function of a sample of size 8 from the uniform distribution on $[0, 5]$. The maximum likelihood estimate $x_{(n)}$ (the location of the spike) is 4.73.

The statistical model and the maximum likelihood estimator are determined by both the form of the density of the observation and the definition of the parameter space!

If $g: \Theta \to H$ is a $1 - 1$ (bijective) function with a set H as codomain, we can also parameterize the model using the parameter $\eta = g(\theta) \in H$ instead of $\theta \in \Theta$. It immediately follows from the definition that $g(\hat{\theta})$ is the maximum likelihood estimator for η if $\hat{\theta}$ is the maximum likelihood estimator for θ. Accordingly, for an arbitrary function g, we *define* the maximum likelihood estimator for $g(\theta)$ simply as $g(\hat{\theta})$. (This estimator maximizes the profile likelihood function $L_g(\tau; x) = \sup_{\theta \in \Theta: g(\theta) = \tau} p_\theta(x)$; see Definition 5.24.)

In Definition 3.11, the maximum likelihood estimator is based on the maximum likelihood estimate. In practice, the (log-)likelihood function is often written directly in terms of the random variable X instead of the realization x, and the estimator is deduced directly by maximizing this function with respect to θ. This shortened notation is used in the following examples of applications of the maximum likelihood method. Examples in which the method is applied to regression models can be found in Chapter 7.

Example 3.17 Exponential distribution, continued from Example 3.12

Let $X = (X_1, \ldots, X_n)$ be a sample from the exponential distribution with unknown parameter $\lambda > 0$. In Example 3.12, we showed that the maximum likelihood estimator for λ is equal to $\hat{\lambda} = 1/\overline{X}$. From this, we can easily deduce the maximum likelihood estimator for $\mathrm{E}_\theta X_i = 1/\lambda$, namely $\widehat{\mathrm{E}_\theta X_i} = 1/\hat{\lambda} = \overline{X}$.

Example 3.18 Gamma distribution

Let $X = (X_1, \ldots, X_n)$ be a sample from a gamma distribution with probability density

$$p_{\alpha, \lambda}(x) = \frac{x^{\alpha-1} \lambda^\alpha e^{-\lambda x}}{\Gamma(\alpha)}.$$

Here $\alpha > 0$ and $\lambda > 0$ are the unknown shape and inverse scale parameters, and Γ is the gamma function

$$\Gamma(\alpha) = \int_0^\infty s^{\alpha-1} e^{-s} ds.$$

The log-likelihood function for $X_1, \ldots, X_n$ is then equal to

$$(\alpha, \lambda) \mapsto \log \prod_{i=1}^n \frac{X_i^{\alpha-1} \lambda^\alpha e^{-\lambda X_i}}{\Gamma(\alpha)}$$

$$= (\alpha - 1) \sum_{i=1}^n \log X_i + n\alpha \log \lambda - \lambda \sum_{i=1}^n X_i - n \log \Gamma(\alpha).$$

As parameter space for $\theta = (\alpha, \lambda)$, we take $\Theta = [0, \infty) \times [0, \infty)$. To determine the maximum likelihood estimators for α and λ, we determine the partial derivatives of the log-likelihood function with respect to λ and α

$$\frac{\partial}{\partial \lambda} \log L(\alpha, \lambda; X_1, \ldots, X_n) = \frac{n\alpha}{\lambda} - \sum_{i=1}^n X_i,$$

$$\frac{\partial}{\partial \alpha} \log L(\alpha, \lambda; X_1, \ldots, X_n) = \sum_{i=1}^n \log X_i + n \log \lambda - n \frac{\int_0^\infty s^{\alpha-1} \log s \, e^{-s} ds}{\int_0^\infty s^{\alpha-1} e^{-s} ds}.$$

(In the derivative with respect to α, we have differentiated the gamma function $\alpha \mapsto \Gamma(\alpha)$ under the integral sign and used that $(\partial/\partial\alpha)s^\alpha = s^\alpha \log s$.) The partial derivatives are equal to 0 in the maximum likelihood estimators $(\hat{\alpha}, \hat{\lambda})$; this gives two likelihood equations. It immediately follows from the first equation that $\hat{\lambda} = \hat{\alpha}/\overline{X}$. We substitute this into the second likelihood equation. This gives

$$\sum_{i=1}^n \log X_i + n \log \hat{\alpha} - n \log \overline{X} - n \frac{\int_0^\infty s^{\hat{\alpha}-1} \log s \, e^{-s} ds}{\int_0^\infty s^{\hat{\alpha}-1} e^{-s} ds} = 0.$$

This equation does not have an explicit solution for $\hat{\alpha}$, but can be solved numerically, using an iterative method, when a realization of $X_1, \ldots, X_n$ has been observed. For most numeric algorithms, we need initial values as starting point for the search for a solution of the equation. The method of moments estimates can be used as initial values (see Section 3.4).

We substitute the resulting value $\hat{\alpha}$ in the equation $\hat{\lambda} = \hat{\alpha}/\overline{X}$ to determine $\hat{\lambda}$. To verify whether the log-likelihood function takes on a maximum in the solution, we must compute the eigenvalues of the Hessian matrix in $(\hat{\alpha}, \hat{\lambda})$. If both eigenvalues are negative in $(\hat{\alpha}, \hat{\lambda})$, then $(\hat{\alpha}, \hat{\lambda})$ is the maximum likelihood estimator for (α, λ).

Example 3.19 Application: counting bacteria

Bacteria in contaminated water are impossible to count either by the naked eye or using a microscope. To obtain an idea of the degree of contamination, we estimate the number of colony-forming units of bacteria in a centiliter of water. We proceed as follows. We assume that the number of colony-forming units of bacteria in a centiliter of contaminated water is Poisson-distributed with parameter μ. To obtain an indication of the number of colony-forming units of bacteria in the water, we want to estimate μ. We pour the contaminated water in a bucket with 100 liters of pure water, mix well, and divide the water over 100 Petri dishes with each a volume of 1 liter. We then check each dish to see whether a colony forms. If this is the case, then there was at least one colony-forming unit of bacteria in the centiliter; if it is not the case, then this centiliter was free of bacteria. Let X be the total number of colony-forming units of bacteria in the centiliter of contaminated water; then we can write X as $X = \sum_{i=1}^{100} X_i$, where X_i is the number of colony-forming units of bacteria in the ith Petri dish. The variables $X_1, \ldots, X_{100}$ are independent and Poisson-distributed with parameter $\mu/100$.

However, we cannot observe $X_1, \ldots, X_{100}$. Rather, we observe $Y_1, \ldots, Y_{100}$, where Y_i is defined by

$$Y_i = \begin{cases} 0 & \text{if no colony forms in the } i\text{th dish} \\ 1 & \text{otherwise.} \end{cases}$$

The observations Y_i are independent and have a Bernoulli distribution with

$$P(Y_i = 0) = P(X_i = 0) = e^{-\mu/100} \qquad \text{and} \qquad P(Y_i = 1) = 1 - e^{-\mu/100}.$$

Define $p := P(Y_i = 1) = 1 - e^{-\mu/100}$. The maximum likelihood estimator for the parameter p of the Bernoulli distribution can be deduced simply by drawing up the likelihood equations and solving them for p. Based on the sample $Y_1, \ldots, Y_{100}$, this estimator is equal to $\hat{p} = \sum_{i=1}^{100} Y_i/100$. Since $p = 1 - e^{-\mu/100}$, the parameter μ is equal to $-100 \log(1 - p)$, and the maximum likelihood estimator for μ is given by $\hat{\mu} = -100 \log(1 - \sum_{i=1}^{100} Y_i/100)$.

Example 3.20 Application: Poisson stocks

In Example 1.4, a statistical model is described for the total number of specimens of a certain item sold per week and per retailer. We observe $X = (X_{1,1}, X_{1,2}, \ldots, X_{I,J})$, where $X_{i,j}$ is the number of specimens sold by retailer i in week j. Suppose that $X_{1,1}, \ldots, X_{I,J}$ are independent and that $X_{i,j}$ has a Poisson distribution with unknown parameter μ_i. The parameter μ_i depends only on the retailer and not on the week. We estimate the parameters $\mu_1, \ldots, \mu_I$ using the maximum likelihood method.

The log-likelihood function for $X_{1,1}, \ldots, X_{I,J}$ is equal to

$$(\mu_1, \ldots, \mu_I) \mapsto \sum_{i=1}^{I} \sum_{j=1}^{J} \log\left(e^{-\mu_i} \frac{\mu_i^{X_{i,j}}}{X_{i,j}!}\right)$$

$$= -\sum_{i=1}^{I} J\mu_i + \sum_{i=1}^{I} \sum_{j=1}^{J} X_{i,j} \log \mu_i - \sum_{i=1}^{I} \sum_{j=1}^{J} \log(X_{i,j}!).$$

We take the natural parameter space $(0, \infty)^I$ for $(\mu_1, \ldots, \mu_I)$. Solving the likelihood equations gives $\hat{\mu}_k = J^{-1} \sum_{j=1}^{J} X_{k,j}$, provided $\sum_{j=1}^{J} X_{k,j} > 0$. It is easy to check that the Hessian matrix in an arbitrary point $(\mu_1, \ldots, \mu_I)$ is a diagonal matrix with only negative eigenvalues when $\sum_{j=1}^{J} X_{k,j} > 0$ for all k. If $\sum_{j=1}^{J} X_{k,j} = 0$ (which has a positive probability of happening), there in fact does not exist a maximum likelihood estimator for μ_k, because in that case, the likelihood function is strictly decreasing and therefore does not reach a maximum on $(0, \infty)$. If we define the Poisson distribution with parameter 0 as the probability distribution that is degenerate in the point 0 and extend the parameter space for μ_k to $[0, \infty)$ for every k, then $J^{-1} \sum_{j=1}^{J} X_{k,j}$ is the maximum likelihood estimator for μ_k.

If the number of items sold changes linearly over the weeks, we may assume $\mu_{i,j} = \mu_i(1 + \beta j)$. We assume that the change β is the same for all retailers. In that case, the log-likelihood function for $X_{1,1}, \ldots, X_{I,J}$ is equal to

$$(\mu_1, \ldots, \mu_I, \beta) \mapsto \sum_{i=1}^{I} \sum_{j=1}^{J} \left(-\mu_i(1 + \beta j) + X_{i,j} \log(\mu_i(1 + \beta j)) - \log(X_{i,j}!) \right).$$

The likelihood equations for μ_k and β are equal to

$$\sum_{j=1}^{J} \left(-(1 + \hat{\beta}j) + \frac{X_{k,j}}{\hat{\mu}_k} \right) = 0 \quad \text{for } k = 1, \ldots, I$$

$$\sum_{i=1}^{I} \sum_{j=1}^{J} \left(-\hat{\mu}_i j + \frac{j X_{i,j}}{1 + \hat{\beta}j} \right) = 0.$$

There are no explicit solutions for these equations, but the zeros of the derivatives can be found using an iterative algorithm. ▭

* Example 3.21 Autoregression

The maximum likelihood method is not restricted to independent observations. We illustrate this with a model that is often used to analyze a variable that changes over time, the autoregressive model:

$$X_i = \beta X_{i-1} + e_i.$$

Here β is an unknown parameter, and the variables $e_1, \ldots, e_n$ are unobservable random fluctuations, also called "noise" or "innovations" in this context. This model greatly resembles the linear regression model without intercept, except that the observation X_i is "explained" by regression on the observation X_{i-1}. If we view the index $i \in \{1, \ldots, n\}$ as indicating successive moments in time, then regression takes place from X_i to the past X_{i-1} of the sequence itself, thus explaining the term "autoregression." Here, we consider the autoregression model of order 1; the extension to regression on more than one variable in the past is obvious.

The order of the data points is now of great importance, and it is useful to depict the data as a function of time. Figure 3.6 gives three possible realizations $(x_0, x_1, \ldots, x_n)$ of the vector $(X_0, X_1, \ldots, X_n)$ as a plot of the index i along the horizontal axis against the value x_i along the vertical axis. All three realizations begin with $x_0 = 1$, but after that, they are generated according to the model $X_i = \beta X_{i-1} + e_i$ with independent innovations e_i but the same value of β. The statistical problem is to estimate the value of β based on the observed realization $(x_0, x_1, \ldots, x_n)$. We will solve this using the maximum likelihood method.

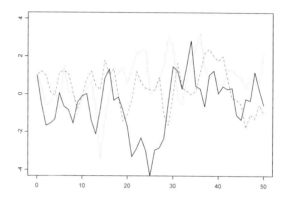

Figure 3.6. Three realizations of the vector $(X_0, X_1, \ldots, X_{50})$ distributed according to the autoregressive model with standard normal innovations, $x_0 = 1$ and $\beta = 0.7$. Each of the three graphs is a linear interpolation of the points $\{(i, x_i): i = 0, \ldots, 50\}$.

We complete the description of the model by assuming that X_0 is distributed according to the probability density p^{X_0} and that the innovations $e_1, \ldots, e_n$ are independent, normally $N(0, \sigma^2)$-distributed quantities that are independent of X_0. The likelihood function is the joint probability density of the observation vector $X = (X_0, \ldots, X_n)$. Because the observations $X_0, X_1, \ldots, X_n$ are stochastically dependent, the joint density is not the product of the marginal densities. However, we can use the general decomposition for a joint density:

$$p^{X_0, \ldots, X_n}(x_0, \ldots, x_n) = p^{X_0}(x_0) p^{X_1 | X_0}(x_1 | x_0) p^{X_2 | X_0, X_1}(x_2 | x_0, x_1) \times$$
$$\cdots \times p^{X_n | X_0, \ldots, X_{n-1}}(x_n | x_0, \ldots, x_{n-1}).$$

This formula gives a factorization of the joint density as the product of conditional densities and generalizes the product formula for the case of independent observations. The formula can be proved by repeatedly applying the formula $f^{X,Y}(x, y) = f^X(x) f^{Y|X}(y | x)$. In the autoregressive model, the conditional density of X_i given $X_0 = x_0, \ldots, X_{i-1} = x_{i-1}$ is equal to the density of $\beta x_{i-1} + e_i$, that is, the density of the normal distribution with expectation βx_{i-1} and variance $\operatorname{var} e_i = \sigma^2$. The likelihood function is therefore of the form

$$(\beta, \sigma) \mapsto L(\beta, \sigma; X_0, \ldots, X_n) = p^{X_0}(X_0) \prod_{i=1}^{n} \frac{1}{\sigma} \phi\left(\frac{X_i - \beta X_{i-1}}{\sigma}\right).$$

We have not yet specified the density of X_0. Because this density influences only one of the $n+1$ factors and n is usually large, the factor in question, $p^{X_0}(X_0)$, is left out of the likelihood function, and the analysis is carried out "conditionally on the value of X_0."

With this definition of the likelihood function, the (conditional) maximum likelihood estimator for the parameter (β, σ) can be determined using the same calculation as that used for the linear regression model (see Section 7.2.1). The maximum likelihood estimator $\hat{\beta}$ minimizes the sum of squares $\beta \mapsto \sum_{i=1}^{n}(X_i - \beta X_{i-1})^2$ and is equal to

$$\hat{\beta} = \frac{\sum_{i=1}^{n} X_i X_{i-1}}{\sum_{i=1}^{n} X_{i-1}^2}.$$

The maximum likelihood estimator for σ^2 is

$$\hat{\sigma}^2 = \frac{1}{n} \sum_{i=1}^{n} (X_i - \hat{\beta} X_{i-1})^2.$$

Depending on the modeling of the initial observation X_0, the maximum likelihood estimators based on the unconditional likelihood function take on slightly different forms.

* **Example 3.22** Application: compound Poisson process

A health insurance company refunds the incurred medical expenses to its clients and health care providers. At the beginning of the month, the company would like to have an estimate for how much money it needs to reserve for that month in order to be able to pay all approved claims. For this, a data set is drawn up containing all payments made in the past 120 months.

The number of approved claims varies from month to month and depends on the number of clients the health insurance company has in said month. We define N_i to be the number of approved claims in month i and assume that $N_1, \ldots, N_{120}$ are independent random quantities with

$$N_i \sim \text{Poisson}(\mu M_i), \qquad i = 1, \ldots, 120,$$

where $\mu > 0$ is an unknown parameter and M_i is the number of clients the company has at the beginning of month i. The numbers M_i are assumed known and not random.

We denote the size of the jth claim in month i by $C_{i,j}$. The payout in month i is then equal to $\sum_{j=1}^{N_i} C_{i,j}$. We assume that the sizes of the paid claims are independent random variables with

$$C_{i,j} \sim \exp(\theta), \qquad i = 1, \ldots, 120, \ j = 1, \ldots, N_i,$$

65

where $\theta > 0$ is an unknown parameter. We moreover assume that the sizes of the claims $C_{i,j}$ are independent of the number of claims N_i.

Under the assumptions on the model made above, it is possible to determine the expected payout for next month. If it were known that the number of claims for next month is n, then the expected payout would equal

$$\mathrm{E}_\theta \sum_{j=1}^{n} C_j = \frac{n}{\theta},$$

where $C_1, \ldots, C_n$ are the sizes of the claims approved next month. The total number of claims is, however, unknown, and has a $\mathrm{Poisson}(\mu M)$-distribution with M the number of clients next month. The expected payout then becomes

$$\mathrm{E}_{\mu,\theta}\left(\sum_{j=1}^{N} C_j\right) = \mathrm{E}_\mu\left(\mathrm{E}_\theta\left(\sum_{j=1}^{N} C_j | N\right)\right) = \mathrm{E}_\mu\left(\frac{N}{\theta}\right) = \frac{\mu M}{\theta}.$$

In this expression, we first compute the expectation of $\sum_{j=1}^{N} C_j$ for a given N, which gives N/θ, and then take the expectation of N/θ. When θ and μ are known, the expected payout next month is therefore equal to $\mu M/\theta$.

The parameters $\mu > 0$ and $\theta > 0$ are unknown and must be estimated using the entries in the data set. We use the maximum likelihood method. To deduce the likelihood function, we first determine the joint density of $(C_1, \ldots, C_N, N)$, the observations for one month. We denote this density by $f_{\theta,\mu}$,

$$f_{\theta,\mu}(c_1, \ldots, c_N, N = n) = f_{\theta,\mu}(c_1, \ldots, c_n | N = n) P_\mu(N = n)$$

$$= \left(\prod_{j=1}^{n} \theta e^{-\theta c_j}\right) e^{-\mu M} \frac{(\mu M)^n}{n!}.$$

We assume that the observations of different months and years are independent. The log-likelihood function for all observations in the data set of the past 10 years is then equal to the logarithm of the product of the joint probability densities of the different months:

$$(\mu, \theta) \mapsto \log\left(\prod_{i=1}^{120}\left(\prod_{j=1}^{N_i} \theta e^{-\theta C_{i,j}}\right) e^{-\mu M_i} \frac{(\mu M_i)^{N_i}}{N_i!}\right)$$

$$= \sum_{i=1}^{120} \log\left(\prod_{j=1}^{N_i} \theta e^{-\theta C_{i,j}}\right) + \sum_{i=1}^{120} \log\left(e^{-\mu M_i} \frac{(\mu M_i)^{N_i}}{N_i!}\right).$$

The first term does not depend on the parameter μ, and the second term does not contain the parameter θ. To determine the maximum likelihood estimators of θ and μ, it therefore suffices to maximize the first term with respect to θ and the second term with respect to μ. This gives

$$\hat{\theta} = \frac{\sum_{i=1}^{120} N_i}{\sum_{i=1}^{120} \sum_{j=1}^{N_i} C_{i,j}} \quad \text{and} \quad \hat{\mu} = \frac{\sum_{i=1}^{120} N_i}{\sum_{i=1}^{120} M_i}.$$

The maximum likelihood estimator for the payout is equal to

$$M\frac{\hat{\mu}}{\hat{\theta}} = M\frac{\sum_{i=1}^{120}\sum_{j=1}^{N_i} C_{i,j}}{\sum_{i=1}^{120} M_i}.$$

In this example, we assume that the parameters μ and θ are the same every month and every year. These assumptions are contestable. Indeed, on average the payout will increase because of inflation, and the number of claims will be greater during the winter than in the summer months. It is worth considering making the parameters dependent on the year and month. Instead of one parameter μ, we could take twelve parameters $\mu_1, \ldots, \mu_{12}$ for the different months. However, increasing the number of unknown parameters in the model decreases the precision of the estimates. ▭

* 3.3.1 Fisher's Scoring

Even though the previous examples of applying the maximum likelihood method might give another impression, it is often not possible to give an explicit formula in the data for the maximum likelihood estimator (see Example 3.18). In such a case, we need to apply a numerical approximation method. For a given observation x, the likelihood function $\theta \mapsto L(\theta; x)$ is a "normal" function of the parameter θ, and we are looking for the value of θ where this function is maximal. We can use, for example, the Newton–Raphson method or the variation of this method known in statistics as *Fisher's scoring*. This section contains a short description of these numerical methods.

In most cases, the desired value $\hat{\theta}$ is a stationary point of the derivative of the log-likelihood function with respect to θ. We therefore discuss finding a zero $\hat{\theta}$ of the function $\theta \mapsto \dot{\Lambda}(\theta; x)$, where $\dot{\Lambda}$ is the vector of partial derivatives of the log-likelihood function $\theta \mapsto \Lambda(\theta; x) = \log L(\theta; x)$. The idea behind the Newton–Raphson method is to start out with a reasonable "first estimate" $\tilde{\theta}_0$ for $\hat{\theta}$ and replace the function $\dot{\Lambda}$ by the linear approximation

$$\dot{\Lambda}(\theta; x) \approx \dot{\Lambda}(\tilde{\theta}_0; x) + \ddot{\Lambda}(\tilde{\theta}_0; x)(\theta - \tilde{\theta}_0).$$

Here, $\ddot{\Lambda}(\theta; x)$ is the matrix of the second derivatives of the log-likelihood function with respect to the parameter. Instead of looking for the value of θ where the equation $\dot{\Lambda}(\theta; x)$ equals 0, we now turn to solving the equation $\dot{\Lambda}(\tilde{\theta}_0; x) + \ddot{\Lambda}(\tilde{\theta}_0; x)(\theta - \tilde{\theta}_0) = 0$. The zero of this second equation is equal to

$$(3.1) \qquad \tilde{\theta}_1 = \tilde{\theta}_0 - \ddot{\Lambda}(\tilde{\theta}_0; x)^{-1}\dot{\Lambda}(\tilde{\theta}_0; x).$$

Since the linear approximation is not exact, the value $\tilde{\theta}_1$ will in general not be the desired zero $\hat{\theta}$. However, we do expect the value $\tilde{\theta}_1$ to be a better approximation for $\hat{\theta}$ than the initial value $\tilde{\theta}_0$. We then take $\tilde{\theta}_1$ as initial value and compute a third value, etc. This gives a sequence of approximations $\tilde{\theta}_0, \tilde{\theta}_1, \tilde{\theta}_2, \ldots$ that, under certain conditions, converges to a zero $\hat{\theta}$. The convergence is assured if the initial value $\tilde{\theta}_0$ lies sufficiently close to the target value $\hat{\theta}$ and the function $\dot{\Lambda}$ is sufficiently smooth, but in practice we, of course, do not have this guarantee. Different modifications of the algorithm can make the convergence more reliable. However, if the log-likelihood function has several local maxima and/or minima, then a word of caution is necessary, because the convergence can also take place toward another zero of $\dot\Lambda$ (corresponding to a local maximum of minimum), in addition to the possibility that the sequence $\tilde\theta_0, \tilde\theta_1, \tilde\theta_2, \dots$ diverge.

In Section 6.3, we will see that the second derivative $\ddot\Lambda(\hat\theta; x)$ of the log-likelihood function evaluated in the maximum likelihood estimator has a special significance. This second derivative is called the *observed information*, and is approximately equal to the Fisher information (see Lemma 5.10). Instead of the second derivative, another matrix is sometimes used in the Newton–Raphson algorithm (3.1). If the Fisher information is used, the algorithm is known as *Fisher's scoring*. This is especially interesting when the Fisher information can be computed analytically.

* 3.3.2 The EM Algorithm

Like Fisher's scoring algorithm, the *expectation-maximization algorithm*, or *EM algorithm*, is also a frequently used general algorithm to determine maximum likelihood estimators. The algorithm is meant to be used when the target data is only partially observed. In many practical applications, such a *missing data* model appears naturally, but the algorithm can also be applied by viewing the observations as part of an imaginary "complete observation" (see Example 3.24).

As usual, we denote the observation by X, but we assume that we observe "only" X instead of the "complete data" (X, Y), which could, theoretically, also be available. If $(x, y) \mapsto \bar p_\theta(x, y)$ is a probability density of the vector (X, Y), then we obtain the density of X through marginalization:

$$p_\theta(x) = \int \bar p_\theta(x, y)\, dy.$$

(In the case of discretely distributed observations, we take a sum instead of an integral.) The maximum likelihood estimator for θ based on the observation X maximizes the likelihood function $\theta \mapsto p_\theta(X)$. If the integral in the displayed equation can be evaluated explicitly, then computing the maximum likelihood estimator is a standard problem, which can be solved, for example, analytically or using an iterative algorithm. If the integral cannot be evaluated analytically, then computing the likelihood requires a numerical approximation of the integral in every value θ, and finding the maximum likelihood estimator may require many such approximations. The EM algorithm tries to circumvent these approximations.

If we had the "complete data" (X, Y) at our disposal, we would have determined the maximum likelihood estimator using (X, Y). This estimator, which will in general be better than the maximum likelihood estimator based on only X, is the point giving the maximum of the log-likelihood function $\theta \mapsto \log \bar p_\theta(X, Y)$, which is probably easy to evaluate. A natural procedure when Y is not available, is to replace this log-likelihood function with its conditional expectation

(3.2) $$\theta \mapsto \mathrm{E}_{\theta_0}\big(\log \bar p_\theta(X, Y)\,|\,X\big).$$

This is the conditional expectation of the log-likelihood for the complete data given the observation X. The idea is to replace the usual log-likelihood with the function (3.2) and determine the point giving the maximum of the latter.

Unfortunately, the expected value in (3.2) will usually depend on the true parameter θ_0, which is why we have included it as a subscript of the expectation operator E_{θ_0}. Since the true value of θ is unknown, the function in the displayed equation cannot be used as the basis for an estimation method. The EM algorithm solves this problem by using iteration. Given a suitably chosen first guess $\tilde{\theta}_0$ for the true value of θ, we determine an estimator $\tilde{\theta}_1$ by maximizing the criterion function in (3.2). We then replace $\tilde{\theta}_0$ in $E_{\tilde{\theta}_0}$ by $\tilde{\theta}_1$, maximize the new criterion, etc.

Initialize $\tilde{\theta}_0$.
E-step: given $\tilde{\theta}_i$, determine the function
$\theta \mapsto E_{\tilde{\theta}_i}\left(\log \overline{p}_\theta(X,Y)\mid X = x\right).$

M-step: define $\tilde{\theta}_{i+1}$ as the point where this function takes on its maximum.

The EM algorithm gives a sequence of values $\tilde{\theta}_0, \tilde{\theta}_1, \ldots$, and we hope that for increasing i, the value $\tilde{\theta}_i$ is an increasingly good approximation of the unknown maximum likelihood estimator.

This description gives the impression that the result of the EM algorithm is a new type of estimator. This is not true, because if the sequence $\tilde{\theta}_0, \tilde{\theta}_1, \ldots$ generated by the EM algorithm converges to a limit, then this limit is exactly the maximum likelihood estimator based on the observation X. Indeed, under certain regularity conditions, we have, for every i,

(3.3) $$p_{\tilde{\theta}_{i+1}}(X) \geq p_{\tilde{\theta}_i}(X)$$

(see Lemma 3.23). Thus, the iterates of the EM algorithm give a constantly increasing value for the likelihood function of the observation X. If the algorithm works "as desired," the values $p_{\tilde{\theta}_i}(X)$ will end up increasing up to the maximum of the likelihood, and $\tilde{\theta}_i$ will converge to the maximum likelihood estimator. Unfortunately, there is, in general, no guarantee for such a convergence, and it needs to be studied case by case. The sequence $\tilde{\theta}_i$ can, for example, converge to a local maximum. Moreover, carrying out the two steps of the algorithm is not necessarily easy.

Lemma 3.23

The sequence $\tilde{\theta}_0, \tilde{\theta}_1, \tilde{\theta}_2, \ldots$ generated by the EM algorithm gives an increasing sequence of likelihood values $p_{\tilde{\theta}_0}(X), p_{\tilde{\theta}_1}(X), p_{\tilde{\theta}_2}(X), \ldots$.

Proof. The density $\overline{p}_\theta$ of (X,Y) can be factored as

$$\overline{p}_\theta(x,y) = p_\theta^{Y\mid X}(y\mid x)p_\theta(x).$$

The logarithm changes this product into a sum, and so we have

$$\mathrm{E}_{\tilde{\theta}_i}\left(\log \overline{p}_{\theta}(X, Y)\mid X\right) = \mathrm{E}_{\tilde{\theta}_i}\left(\log p_{\theta}^{Y\mid X}(Y\mid X)\mid X\right) + \log p_{\theta}(X).$$

As the value $\tilde{\theta}_{i+1}$ maximizes this function with respect to θ, this expression is greater in $\theta = \tilde{\theta}_{i+1}$ than in $\theta = \tilde{\theta}_i$,

$$\mathrm{E}_{\tilde{\theta}_i}\left(\log \overline{p}_{\tilde{\theta}_{i+1}}(X, Y)\mid X\right) \geq \mathrm{E}_{\tilde{\theta}_i}\left(\log \overline{p}_{\tilde{\theta}_i}(X, Y)\mid X\right).$$

If we can show that $\mathrm{E}_{\tilde{\theta}_i}\left(\log p_{\theta}^{Y\mid X}(Y\mid X)\mid X\right)$ is smaller in $\theta = \tilde{\theta}_{i+1}$ than in $\theta = \tilde{\theta}_i$, then the converse must hold for $\log p_{\theta}(X)$ (and the difference must be compensated by this second term), from which follows that (3.3) holds. It therefore suffices to show that

$$\mathrm{E}_{\tilde{\theta}_i}\left(\log p_{\tilde{\theta}_{i+1}}^{Y\mid X}(Y\mid X)\mid X\right) \leq \mathrm{E}_{\tilde{\theta}_i}\left(\log p_{\tilde{\theta}_i}^{Y\mid X}(Y\mid X)\mid X\right).$$

This inequality is of the form $\int \log(q/p)\, dP \leq 0$ for p and q the conditional densities of Y given X for the parameters $\tilde{\theta}_i$ and $\tilde{\theta}_{i+1}$, respectively, and P the probability measure corresponding to the density p. Since $\log x \leq x - 1$ for every $x \geq 0$, every pair of probability densities p and q satisfies

$$\int \log(q/p)\, dP \leq \int (q/p - 1)\, dP = \int_{p(x)>0} q(x)\, dx - 1 \leq 0.$$

This implies the previous displayed equation, completing the proof. ∎

Example 3.24 Mixture distribution

Suppose that a number of objects or individuals can, in principle, be grouped in more or less uniform clusters. Unfortunately, we cannot observe the cluster labels, but instead of that, we measure a vector x_i for each object. We want to determine the clustering of the objects based on the observations $x_1, \ldots, x_n$.

We could assume that each observation x_i is the realization of a random vector X_i, with probability density f_j if the object belongs to the jth cluster. We can view the qualification of "more or less uniform" in the previous paragraph to mean that the probability densities $f_1, \ldots, f_k$ for the different clusters show little overlap. We will assume that the number k of clusters is known, even though we could also determine this from the data.

One way to determine the clusters is to maximize the likelihood

$$\prod_{j=1}^{k} \prod_{i \in I_j} f_j(X_i)$$

over all partitions $(I_1, \ldots, I_k)$ of $\{1, \ldots, n\}$ in k subsets and over all unknown parameters in the densities f_j. The partition then gives the clustering. For example, taking the normal density with expectation vector μ_j for f_j leads to *k-means clustering*: the best classification is given by the partition that minimizes

$$\min_{(\mu_1, \ldots, \mu_k) \in \mathbb{R}^k} \sum_{j=1}^{k} \sum_{i \in I_j} \|X_i - \mu_j\|^2.$$

Computationally, this is not a simple problem, but the clusters can be approximated using an iterative algorithm.

Another way is to assume that every object has been assigned to a cluster randomly (by "nature"). We can then speak of a random vector $(C_1, \ldots, C_n)$ that gives the clusters labels ($C_i = j$ if the ith object belongs to the cluster j), and view the density f_j as the conditional probability density of X_i given $C_i = j$. The class vector $(C_1, \ldots, C_n)$ is not observed. If we assume that $(C_1, X_1), \ldots, (C_n, X_n)$ are independent, identically distributed vectors with $P(C_i = j) = p_j$ for $j = 1, \ldots, k$ for all i, then we can determine the maximum likelihood estimator for the parameters $p = (p_1, \ldots, p_k)$ and the unknown parameters in $f = (f_1, \ldots, f_k)$ using the EM algorithm.

The complete data consist of $(C_1, X_1), \ldots, (C_n, X_n)$. The corresponding likelihood function can be written as

$$(p, f) \mapsto \prod_{i=1}^{n} \sum_{j=1}^{k} p_j f_j(X_i) 1_{\{C_i = j\}} = \prod_{i=1}^{n} \prod_{j=1}^{k} (p_j f_j(X_i))^{1_{\{C_i = j\}}}.$$

The E-step of the EM algorithm is therefore the computation of

$$E_{\tilde{p}, \tilde{f}} \left(\log \prod_{i=1}^{n} \prod_{j=1}^{k} (p_j f_j(X_i))^{1_{\{C_i = j\}}} \mid X_1, \ldots, X_n \right)$$

$$= \sum_{i=1}^{n} \sum_{j=1}^{k} E_{\tilde{p}, \tilde{f}} \left(\left(\log p_j + \log f_j(X_i) \right) 1_{\{C_i = j\}} \mid X_i \right).$$

Using Bayes's rule, we find the conditional probability density of C_i given X_i to be $P(C_i = j \mid X_i = x) = p_j f_j(x) / \sum_c p_c f_c(x)$. The last displayed equation is therefore equal to

$$\sum_{j=1}^{k} \sum_{i=1}^{n} \log p_j \frac{\tilde{p}_j \tilde{f}_j(X_i)}{\sum_c \tilde{p}_c \tilde{f}_c(X_i)} + \sum_{j=1}^{k} \sum_{i=1}^{n} \log f_j(X_i) \frac{\tilde{p}_j \tilde{f}_j(X_i)}{\sum_c \tilde{p}_c \tilde{f}_c(X_i)}.$$

In the M-step of the EM algorithm, we maximize this expression with respect to p and f. For the maximization with respect to p, only the first term matters. Arguments using calculus show that the maximum is reached for

$$p_j = \frac{1}{n} \sum_{i=1}^{n} \frac{\tilde{p}_j \tilde{f}_j(X_i)}{\sum_c \tilde{p}_c \tilde{f}_c(X_i)}.$$

(Compare this with the calculation in Exercise 3.15.) For the maximization over f, only the second term matters. Moreover, we maximize each of the j terms individually with respect to f_j if the parameters $f_1, \ldots, f_k$ vary independently from one another: in that case, f_j maximizes

$$f_j \mapsto \sum_{i=1}^{n} \log f_j(X_i) \frac{\tilde{p}_j \tilde{f}_j(X_i)}{\sum_c \tilde{p}_c \tilde{f}_c(X_i)}.$$

71

If, for example, we choose the normal density with expectation vector μ_j for f_j, so that $\log f_j(x)$ is equal to $-\frac{1}{2}\|x - \mu_j\|^2$ up to a constant, and maximize with respect to μ_j, we find

$$\mu_j = \frac{\sum_{i=1}^{n} \alpha_{ij} X_i}{\sum_{i=1}^{n} \alpha_{ij}}, \qquad \alpha_{ij} = \frac{\tilde{p}_j \tilde{f}_j(X_i)}{\sum_c \tilde{p}_c \tilde{f}_c(X_i)}.$$

This is a weighted average of the observations X_i, where the weights are equal to the conditional probabilities $\alpha_{ij} = P_{\tilde{p},\tilde{f}}(C_i = j|X_i)$ that the ith object belongs to the jth cluster, for $1 \leq i \leq n$, computed using the current approximation $(\tilde{p}, \tilde{f})$ of the parameters. We now repeatedly iterate these updating formulas until the result hardly changes.

From the maximum likelihood estimates of the parameters, we also deduce a maximum likelihood estimate of the probability $P_{p,f}(C_i = j|X_i)$ that the ith object belongs to the cluster j. We could assign the object to the cluster where this probability is the greatest.

3.4 Method of Moments Estimators

The *method of moments* is an alternative to the maximum likelihood method. Because the method of moments often does not use all the available information from the statistical model, method of moments estimators are often less efficient than maximum likelihood estimators. On the other hand, the method is sometimes easier to implement. Moreover, the method only requires the theoretical form of the moments and not the complete probability distribution of the observations. Since these moments are often easier to model realistically than the full probability distribution, this can be a great advantage. Using a wrong model to construct estimators can thus be avoided.

Definition 3.25 Moment and sample moment

The jth moment of a random variable X with a distribution that depends on the unknown parameter θ, is $E_\theta(X^j)$, provided that this expectation exists. The jth sample moment of a sample of independent and identically distributed variables $X_1, \ldots, X_n$ is $\overline{X^j} = n^{-1}\sum_{i=1}^{n} X_i^j$.

The jth moment can be estimated using the jth sample moment of a sample with the same distribution. It follows from the law of large numbers (Theorem A.26) that this is a good estimator for $E_\theta(X^j)$.

Definition 3.26 Method of moments estimator

Let $X_1, \ldots, X_n$ be a sample from a distribution with unknown parameter θ. The method of moments estimator for θ is the value $\hat{\theta}$ where the jth moment corresponds to the jth sample moment:
$$E_{\hat{\theta}}(X^j) = \overline{X^j}.$$
The method of moments estimator for $g(\theta)$ with $g: \Theta \to H$ a function with codomain H is $g(\hat{\theta})$.

In practice, we prefer the method of moments estimator given by taking j as small as possible. For a 1-dimensional parameter θ, it suffices to take $j = 1$, provided that the expected value of the marginal distribution depends on θ. When the first moment does not depend on θ, we choose $j = 2$, etc. If θ has dimension greater than 1, we need more than one equation to obtain a unique solution for $\hat{\theta}$. In that case, the method of moments estimator $\hat{\theta}$ is solved from the equations for $j = 1, \ldots, k$ with k the smallest integer for which the system of equations has a unique solution.

Example 3.27 Exponential distribution

Let $X_1, \ldots, X_n$ be a sample from an exponential distribution with unknown parameter λ. Then $E_\lambda X_i = 1/\lambda$. The method of moments estimator for λ can now be found by solving the equation $\overline{X} = 1/\hat{\lambda}$ for $\hat{\lambda}$. This gives $\hat{\lambda} = 1/\overline{X}$ as a method of moments estimator for λ. This estimator is also the maximum likelihood estimator for λ (see Example 3.12).

Example 3.28 Uniform distribution

Let $X_1, \ldots, X_n$ be a sample from the $U[0, \theta]$-distribution with unknown parameter θ. Then $E_\theta X_i = \theta/2$ and the method of moments estimator for θ is equal to $\hat{\theta} = 2\overline{X}$. The maximum likelihood estimator for θ is equal to $X_{(n)}$ (see Example 3.15). We saw in Example 3.6 that the mean square error of $X_{(n)}$ is less than that of $2\overline{X}$. In this case, we therefore prefer to use the maximum likelihood estimator.

Example 3.29 Normal distribution

Let $X_1, \ldots, X_n$ be a sample from the $N(0, \sigma^2)$-distribution with unknown parameter $\sigma^2 > 0$. Then $E_{\sigma^2} X_i = 0$, and therefore the first moment cannot be used to determine the method of moments estimator for σ^2. The second moment of X_i is equal to $E_{\sigma^2} X_i^2 = \sigma^2$. The method of moments estimator for σ^2 is then equal to $\hat{\sigma}^2 = \overline{X^2}$. If the expectation of X_i were unknown or nonzero, then we would have found a different method of moments estimator for σ^2 (see Example 3.31).

Example 3.30 Gamma distribution

Let $X_1, \ldots, X_n$ be random variables with a gamma distribution with unknown shape and inverse scale parameters α and λ, respectively. Then $E_{\alpha,\lambda} X_i = \alpha/\lambda$ and $\mathrm{var}_{\alpha,\lambda} X_i = \alpha/\lambda^2$, and therefore the second moment is equal to $E_{\alpha,\lambda} X_i^2 = \mathrm{var}_{\alpha,\lambda} X_i + (E_{\alpha,\lambda} X_i)^2 = \alpha(1 + \alpha)/\lambda^2$. The method of moments estimators for α and λ can be found by solving the equations

$$E_{\hat{\alpha},\hat{\lambda}} X_i = \hat{\alpha}/\hat{\lambda} = \overline{X}$$
$$E_{\hat{\alpha},\hat{\lambda}} X_i^2 = \hat{\alpha}(1 + \hat{\alpha})/\hat{\lambda}^2 = \overline{X^2}$$

for $\hat{\alpha}$ and $\hat{\lambda}$. This gives

$$\hat{\alpha} = \frac{(\overline{X})^2}{\overline{X^2} - (\overline{X})^2} \quad \text{and} \quad \hat{\lambda} = \frac{\overline{X}}{\overline{X^2} - (\overline{X})^2}.$$

Since no explicit expressions for the maximum likelihood estimators are known, the mean square error cannot be determined. In order to choose between the two estimators anyway based on their performance (bias and variance), we can carry out a simulation as described in Section 3.2.

Example 3.31 Expectation and variance

Let $X_1, \ldots, X_n$ be a sample with expectation μ and variance σ^2. Solving for $\hat{\mu}$ and $\hat{\sigma}^2$ in the equations

$$E_{\hat{\mu},\hat{\sigma}^2} X_i = \hat{\mu} = \overline{X},$$
$$E_{\hat{\mu},\hat{\sigma}^2} X_i^2 = \hat{\mu}^2 + \hat{\sigma}^2 = \overline{X^2}$$

gives the method of moments estimators for $\hat{\mu}$ and $\hat{\sigma}^2$:

$$\hat{\mu} = \overline{X}, \qquad \hat{\sigma}^2 = \overline{X^2} - (\overline{X})^2 = \frac{1}{n} \sum_{i=1}^{n} (X_i - \overline{X})^2.$$

If the underlying distribution is the $N(\mu, \sigma^2)$-distribution, then these method of moments estimators are equal to the maximum likelihood estimators for μ and σ^2 (see Example 3.14).

* 3.4.1 Generalized Method Of Moments Estimators

The method of moments can be generalized in several ways. For example, instead of using the sample moments $n^{-1}\sum_{i=1}^{n} X_i^j$, we can use averages of the type $n^{-1}\sum_{i=1}^{n} g(X_i)$ for suitably chosen functions g. Furthermore, the observation X need not be a sample, and we can also use more general functions of X instead of averages. The essence is solving a system of equations of the type $g(X) = e(\theta)$ for suitably chosen functions and $e(\theta) = E_\theta g(X)$.

If the parameter is k-dimensional, then it seems natural to use k equations for the definition of the method of moments estimator. The question is then: which functions? In fact, the method of moments first reduces the observations to the values of k functions of those observations, and the method of moments estimator is based on this reduced data. If the original data cannot be reconstructed from the k values, this reduction leads to a loss of information. The choice of which functions to use is therefore important for the efficiency of the resulting estimators.

A possible way to avoid this loss of information is to use more moments than there are unknown parameters. Because this leads to more equations than unknowns, in this case, it will in general not be possible to find a parameter value for which the sample moments are exactly equal to the theoretical moments. Instead of this, we could minimize a measure of distance between these two types of moments, for example an expression of the form

$$\sum_{j=1}^{l}\left(\frac{1}{n}\sum_{i=1}^{n} g_j(X_i) - E_\theta g_j(X_1)\right)^2.$$

The functions $g_1, \ldots, g_l$ are known, fixed functions. The estimator $\hat{\theta}$ is the value of θ that minimizes this expression. This method is known (especially in econometrics) as the *generalized method of moments*.

3.5 Bayes Estimators

Bayes's method is the oldest method for constructing estimators; it was suggested by Thomas Bayes at the end of the 18th century. This method is guided by a philosophy on the way to express uncertainty. The starting point of this philosophy (in its strictest form) is that the statistical model does not contain a unique parameter value that corresponds to the "true" state of reality. However, every parameter value has a probability, which can, if necessary, be determined in a subjective, personal way. This subjective element of the method has lead to much criticism. Bayesian methods in a more objective sense, however, have been widely accepted, and have known a great popularity since the 1990s, because initial problems with the computations can now be solved using computer simulation (see Section 3.5.1).

A Bayesian approach begins with the specification of a so-called *prior probability distribution* on the parameter space Θ, in addition to the specification of a statistical model (or likelihood function). The prior distribution is chosen either using ad hoc arguments or as an expression of the a priori, possibly subjective, estimate of the probability of the different parameter values. For example, given a binomial variable X with success parameter $\theta \in [0, 1]$, we could choose the uniform distribution as prior distribution for θ.

This prior distribution is then adjusted to the available data by applying Bayes's rule from probability theory. This adjusted distribution is called the *posterior probability distribution*. We will first describe Bayes's method as a method for constructing estimators, and will describe this adjustment of the probability distribution in more detail in Section 3.5.1.

For simplicity, we take the prior distribution to be continuous with density π, an arbitrary probability density on Θ. The *Bayes risk* of an estimator T for a real-valued parameter $g(\theta)$ is defined as the weighted average of the $\mathrm{MSE}(\theta; T)$, with weight π,

$$R(\pi; T) = \int \mathrm{E}_\theta (T - g(\theta))^2 \, \pi(\theta) \, d\theta.$$

This is a measure for the quality of the estimator T, which awards a higher weight to those values θ that are deemed, a priori, more probable. The Bayes estimator is defined as the best estimator for this quality criterion. The aim is still to find an estimator for which the $\mathrm{MSE}(\theta; T)$ are small for all θ; we make the criterion more concrete by giving weights to the different values of θ.

Definition 3.32 Bayes estimator

The Bayes estimator with respect to the prior density π is the estimator T that minimizes $R(\pi; T)$ over all estimators T.

In the following theorem, the Bayes estimator is specified as a quotient of two integrals. Let $x \mapsto p_\theta(x)$ be the probability density of the random vector X.

Theorem 3.33

The Bayes estimate for $g(\theta)$ with respect to the prior density π is given by

$$T(x) = \frac{\int g(\theta) p_\theta(x) \, \pi(\theta) \, d\theta}{\int p_\vartheta(x) \, \pi(\vartheta) \, d\vartheta}.$$

The Bayes estimate therefore depends on both the likelihood function $\theta \mapsto p_\theta(x)$ and the prior density π. Whereas the maximum likelihood estimator is defined as the point where the likelihood function takes on its maximum, the Bayes estimator is some kind of weighted average of this function.

Example 3.34 Exponential distribution

Let $X = (X_1, \ldots, X_n)$ be a sample from the exponential distribution with unknown parameter θ. As prior distribution for θ, we also take the exponential distribution, but this time with known parameter λ. The Bayes estimate $T_\lambda(x)$ for θ based on $x = (x_1, \ldots, x_n)$ and with respect to the given prior distribution is

$$\frac{\int_0^\infty \theta \left(\prod_{i=1}^n \theta e^{-\theta x_i} \right) \lambda e^{-\lambda \theta} d\theta}{\int_0^\infty \left(\prod_{i=1}^n \vartheta e^{-\vartheta x_i} \right) \lambda e^{-\lambda \vartheta} d\vartheta} = \frac{\int_0^\infty \theta^{n+1} \lambda e^{-\theta(\lambda + \sum_{i=1}^n x_i)} d\theta}{\int_0^\infty \vartheta^n \lambda e^{-\vartheta(\lambda + \sum_{i=1}^n x_i)} d\vartheta}.$$

Computing the integrals in the numerator and denominator of this fraction explicitly is not the best way to determine $T_\lambda(x)$. We will see that this becomes easier if we first determine the posterior density; see Example 3.37. In that example, we deduce that $T_\lambda(x) = (n+1)/(\lambda + \sum_{i=1}^n x_i)$ is the Bayes estimate. The Bayes estimator for θ is therefore equal to $T_\lambda(X) = (n+1)/(\lambda + \sum_{i=1}^n X_i)$. For large values of n, the Bayes estimator $T_\lambda(X)$ and the maximum likelihood estimator $\hat{\theta} = 1/\overline{X}$ are approximately equal.

    The proof of Theorem 3.33 is an exercise in the manipulation of conditional distributions. The following "Bayesian" notation and notions are useful for this, and of great importance in their own right. They describe the Bayesian method in a more comprehensive framework, where the so-called posterior distribution forms the end point of the analysis.

    Normally, we view the parameter θ as being deterministic, and there is a single "true" parameter value that determines the density $x \mapsto p_\theta(x)$ of the observation X. In this section, we deviate from this and view p_θ as the conditional density $p_{X|\overline{\Theta}=\theta}$ of a variable X given that a (hypothetical) random variable $\overline{\Theta}$ takes on the value θ. We give this quantity $\overline{\Theta}$ the (marginal) probability density π. The joint density of $(X, \overline{\Theta})$ is then equal to

$$p_{X,\overline{\Theta}}(x, \theta) = p_{X|\overline{\Theta}=\theta}(x) p_{\overline{\Theta}}(\theta) = p_\theta(x) \pi(\theta).$$

The marginal density of X in this Bayesian setting is obtained by integrating the joint density with respect to θ and is therefore equal to

$$p_X(x) = \int p_{X,\overline{\Theta}}(x, \theta) \, d\theta = \int p_\theta(x) \pi(\theta) \, d\theta.$$

Hence, the conditional density of $\overline{\Theta}$ given $X = x$ is equal to

$$p_{\overline{\Theta}|X=x}(\theta) = \frac{p_{X,\overline{\Theta}}(x, \theta)}{p_X(x)} = \frac{p_\theta(x) \pi(\theta)}{\int p_\vartheta(x) \pi(\vartheta) \, d\vartheta}.$$

(This formula is exactly Bayes's rule from probability theory; see Section A.6.)

Definition 3.35 Posterior density

The *posterior density* of $\overline{\Theta}$ is

$$p_{\overline{\Theta}|X=x}(\theta) = \frac{p_\theta(x)\pi(\theta)}{\int p_\vartheta(x)\pi(\vartheta)\,d\vartheta}.$$

The term in the denominator of the posterior density is just a normalization constant such that

$$\int p_{\overline{\Theta}|X=x}(\theta)\,d\theta = 1.$$

Before the observation was known, we awarded the prior density π to $\overline{\Theta}$. Once we know the observation, the posterior density gives the adjusted probability distribution. This way, the observation leads us to adjust our assumptions concerning the parameter.

These computations show that the expression $T(x)$ in Theorem 3.33 is exactly the expectation of $g(\overline{\Theta})$ for the posterior probability distribution, the conditional expectation of $g(\overline{\Theta})$ given $X = x$. We can therefore reformulate the theorem as follows.

Theorem 3.36

Using the Bayesian notation, the Bayes estimate for $g(\theta)$ with respect to the prior density π is given by

$$T(x) = \mathrm{E}\big(g(\overline{\Theta})|\,X = x\big) = \int g(\theta)p_{\overline{\Theta}|X=x}(\theta)\,d\theta.$$

Proof. First, we write the Bayes risk in the Bayesian notation. The term $\mathrm{E}_\theta\big(T - g(\theta)\big)^2$ in the usual notation is the conditional expectation

$$\mathrm{E}\big[(T(X) - g(\overline{\Theta}))^2|\,\overline{\Theta} = \theta\big]$$

in the Bayesian notation. From this, we deduce that

$$R(\pi;T) = \int \mathrm{E}\Big((T(X) - g(\theta))^2|\,\overline{\Theta} = \theta\Big)\pi(\theta)\,d\theta$$

$$= \mathrm{E}\big(T(X) - g(\overline{\Theta})\big)^2$$

$$= \int \mathrm{E}\Big((T(x) - g(\overline{\Theta}))^2|\,X = x\Big)p_X(x)\,dx.$$

We have used the decomposition rule for expectations $\mathrm{E}Z = \int \mathrm{E}(Z|\,Y = y)\,f_Y(y)\,dy$ with $Z = \big(T(X) - g(\overline{\Theta})\big)^2$ twice: in the second equality with $Y = \overline{\Theta}$ and in the third equality with $Y = X$.

To minimize $R(\pi; T)$ with respect to T, we can minimize the integrand with respect to every x, because the integrand is everywhere nonnegative. Therefore, for every x, we are looking for the number $t = T(x)$ such that

$$\mathrm{E}\left(\left(t - g(\overline{\Theta}) \right)^2 \big| X = x \right) p_X(x)$$

is minimal. Because for given x, the term $p_X(x)$ is a nonnegative constant, minimizing the integrand with respect to t is equivalent to minimizing

$$\mathrm{E}\left(\left(t - g(\overline{\Theta}) \right)^2 \big| X = x \right)$$

with respect to t. Consequently, for every x, we can find the number $t = T(x)$ that minimizes the last expression. Minimizing $\mathrm{E}(t - Y)^2$ with respect to t gives the value $t = \mathrm{E}Y$, the minimum of the parabola $t \mapsto \mathrm{E}(t - Y)^2 = t^2 - 2t\,\mathrm{E}\,Y + \mathrm{E}Y^2$. Here, we must apply this principle with a random variable Y that has the conditional distribution of $g(\overline{\Theta})$ given $X = x$. We find $t = \mathrm{E}\big(g(\overline{\Theta})\big| X = x\big)$; that is, the Bayes estimate is given by $T(x) = \mathrm{E}\big(g(\overline{\Theta})\big| X = x\big)$. ∎

Example 3.37 Exponential distribution, continued from Example 3.34

Let $X = (X_1, \ldots, X_n)$ be a sample from the exponential distribution with unknown parameter θ. Assume that the prior density for θ is the exponential distribution with known parameter λ. Example 3.34 gives an expression for the Bayes estimate for θ. By first determining the posterior distribution, we can more easily determine the Bayes estimate explicitly.

The posterior distribution is given by

$$\theta \mapsto p_{\overline{\Theta}|X=x}(\theta) = \frac{\left(\prod_{i=1}^n \theta e^{-\theta x_i}\right)\lambda e^{-\lambda \theta}}{\int_0^\infty \left(\prod_{i=1}^n \vartheta e^{-\vartheta x_i}\right)\lambda e^{-\lambda \vartheta}\,d\vartheta}$$

$$= \frac{\theta^n \lambda e^{-\theta(\lambda + \sum_{i=1}^n x_i)}}{\int_0^\infty \vartheta^n \lambda e^{-\vartheta(\lambda + \sum_{i=1}^n x_i)}\,d\vartheta} = \frac{\theta^n e^{-\theta(\lambda + \sum_{i=1}^n x_i)}}{C(x, \lambda)},$$

where $C(x, \lambda)$ is a normalization constant depending on $x = (x_1, \ldots, x_n)$ and λ such that $p_{\overline{\Theta}|X=x}$ is a density. We see that this posterior distribution is the gamma distribution with shape parameter $n + 1$ and inverse scale parameter equal to $\lambda + \sum_{i=1}^n x_i$. In general, the expected value corresponding to the gamma distribution with shape parameter α and inverse scale parameter λ is equal to α/λ (see Example A.13). The Bayes estimate for θ is the expected value of the posterior distribution, and is therefore equal to $T_\lambda(x) = (n+1)/(\lambda + \sum_{i=1}^n x_i)$. The corresponding Bayes estimator is $T_\lambda(X) = (n + 1)/(\lambda + \sum_{i=1}^n X_i)$.

We determine the Bayes estimator for θ^2 similarly. By Theorem 3.36, it is equal to the second moment of the posterior distribution, in this case the gamma distribution with shape parameter $n + 1$ and inverse scale parameter $\lambda + \sum_{i=1}^n x_i$. The second moment of a gamma(α, λ)-distributed random variable is equal to $\alpha/\lambda^2 + (\alpha/\lambda)^2 = (\alpha + 1)\alpha/\lambda^2$. The Bayes estimator for θ^2 is therefore equal to $(n + 2)(n + 1)/(\lambda + \sum_{i=1}^n X_i)^2$.

Example 3.38 Binomial distribution

Let X be a random variable with a binomial distribution with parameters n and θ, where n is known and $0 \leq \theta \leq 1$ is unknown. A useful class of prior densities on $[0,1]$ is the class of *beta densities*, parameterized by α and β (see Example A.14):

$$\pi(\theta) = \frac{\theta^{\alpha-1}(1-\theta)^{\beta-1}}{B(\alpha,\beta)} 1_{[0,1]}(\theta).$$

When we take the beta distribution with parameters α and β as prior distribution for $\overline{\Theta}$, the posterior density is given by

$$p_{\overline{\Theta}|X=x}(\theta) = \frac{\binom{n}{x}\theta^x(1-\theta)^{n-x}\pi(\theta)}{\int_0^1 \binom{n}{x}\vartheta^x(1-\vartheta)^{n-x}\pi(\vartheta)\,d\vartheta} = \frac{\theta^{x+\alpha-1}(1-\theta)^{n-x+\beta-1}}{C(x,\alpha,\beta)},$$

with $C(x,\alpha,\beta)$ a normalization constant such that $p_{\overline{\Theta}|X=x}$ is a density. In other words, the posterior distribution of $\overline{\Theta}$ is the beta distribution with parameters $x+\alpha, n-x+\beta$ and with $C(x,\alpha,\beta) = B(x+\alpha, n-x+\beta)$ for the beta function B. Figure 3.7 shows two times three realizations of the posterior density. In all cases, the true parameter value is equal to $\theta = \frac{1}{2}$ and the prior density (dashed in the figure) is the beta density with parameters $\alpha = 25$ and $\beta = 5$. In the top figure $n = 20$, while in the bottom figure $n = 100$. The prior density gives a relatively large probability to values of $\overline{\Theta}$ near 1, and is therefore not suitable for estimating the true parameter value $\theta = \frac{1}{2}$. The figure shows that this incorrect prior density is corrected well if sufficient data is available, but influences the posterior density if this is not the case.

The Bayes estimate for θ is now given by the expected value corresponding to the beta distribution with parameters $x + \alpha$ and $n - x + \beta$. In general, the expected value corresponding to the beta distribution with parameters α and β is equal to $\alpha/(\alpha+\beta)$, so that the Bayes estimator for θ is equal to

$$T_{\alpha,\beta}(X) = \frac{X+\alpha}{n+\alpha+\beta}.$$

We find a different estimator for each combination of parameters (α,β) with $\alpha > 0$ and $\beta > 0$. The natural estimator X/n is not in the class of Bayes estimators; rather, it is the limit case $(\alpha,\beta) \to (0,0)$.

Which estimator should we use? If we feel strongly about a prior distribution, we can use the corresponding Bayes estimator. A problem is that another researcher may have other "feelings," leading to another prior distribution and therefore another estimator. No Bayes estimator is "wrong." After all, any Bayes estimator is best if we decide to use the corresponding Bayes risk as quality criterion. Still, it would be wise to compare the estimators further, for example by computing the mean square errors.

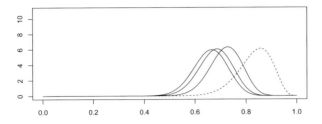

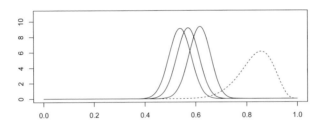

Figure 3.7. Three realizations of the posterior density in the cases $n = 20$ (top) and $n = 100$ (bottom). In both cases, the prior density (dashed) is equal to the beta density with $\alpha = 25$ and $\beta = 5$. The realizations (solid) are based on samples from the binomial distribution with parameters n and $\frac{1}{2}$.

These are equal to

$$
\begin{aligned}
\mathrm{MSE}(\theta; T_{\alpha,\beta}) &= \mathrm{E}_\theta \left(\frac{X + \alpha}{n + \alpha + \beta} - \theta \right)^2 \\
&= \frac{\mathrm{var}_\theta\, X}{(n + \alpha + \beta)^2} + \left(\frac{\mathrm{E}_\theta X + \alpha}{n + \alpha + \beta} - \theta \right)^2 \\
&= \frac{\theta^2 \left((\alpha + \beta)^2 - n \right) + \theta(n - 2\alpha(\alpha + \beta)) + \alpha^2}{(n + \alpha + \beta)^2}.
\end{aligned}
$$

Figure 3.8 shows the mean square error of several estimators as a function of θ. Every estimator is better than another at some point, and there is no absolutely best estimator. Interesting special cases are $\alpha = \beta = \frac{1}{2}\sqrt{n}$ (constant mean square error) and $\alpha = \beta = 0$ (estimator X/n). The choice $\alpha = \beta = 1$ corresponds to the uniform prior distribution, which a priori gives all $\theta \in [0, 1]$ the same probability. The latter seems reasonable, but this estimator is nevertheless seldom used. Fortunately, the differences are small when n is large, and even disappear as $n \to \infty$. Note that in the bottom graph (corresponding to $n = 100$) in Figure 3.7 the three realizations of the posterior distribution lie closer to the true value $1/2$, but are also more concentrated. The posterior densities seem surprisingly normal. We will come back to this in Section 5.7, where we will see that Bayes and maximum likelihood estimators often differ little when the number of observations is large.

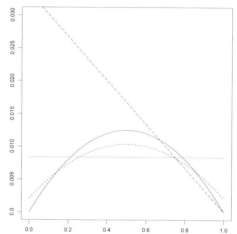

Figure 3.8. Mean square error of the Bayes estimators $T_{\alpha,\beta}$ with $n = 20$ and $\alpha = \beta = \frac{1}{2}\sqrt{n}$ (constant), $\alpha = \beta = 0$ (curved, solid), $\alpha = \sqrt{n}$, $\beta = 0$ (linear, dashed), and $\alpha = \beta = 1$ (small dashes) as functions of θ.

Example 3.39 Geometric distribution

Let $X = (X_1, \ldots, X_n)$ be a sample from the geometric distribution with parameter θ,

$$P_\theta(X_1 = x) = (1 - \theta)^{x-1}\theta, \qquad x = 1, 2, \ldots,$$

where $0 \leq \theta \leq 1$ is unknown. As prior distribution for θ, we choose the beta distribution with parameters $\alpha = \beta = 2$ with probability density

$$\pi(\theta) = 6(1 - \theta)\theta, \qquad \theta \in (0, 1).$$

Then, the posterior distribution is given by

$$p_{\overline{\Theta}|X=x}(\theta) = \frac{\prod_{i=1}^{n} P_\theta(X_i = x_i)\pi(\theta)}{\int_0^1 \prod_{i=1}^{n} P_\vartheta(X_i = x_i)\pi(\vartheta)\,d\vartheta} = \frac{\theta^{n+1}(1 - \theta)^{n(\overline{x}-1)+1}}{C(x_1, \ldots, x_n)}.$$

This posterior distribution of θ is the beta distribution with parameters $n + 2$ and $n(\overline{x} - 1) + 2$. As in the previous example, we determine the Bayes estimator for θ using the expectation of the beta distribution $T(X) = (n + 2)/(n\overline{X} + 4)$.

* 3.5.1 MCMC Methods

The principle behind Bayes's method is simple: from a model and a prior distribution, we compute the posterior distribution using Bayes's rule. However, the computation in the last step is not always simple. Traditionally, the prior distribution is chosen in such a way that it simplifies the computation for the given model. The combination of the binomial distribution with the beta prior distribution is an example. A more recent approach is to replace the analytic computation by numerical approximations, or *Markov Chain Monte Carlo* (or *MCMC*) methods. In principle, such methods allow us to combine an arbitrary prior distribution with a given random model. In this section, we give a short introduction to these methods.

Given an observation X with realization x with probability density p_θ and a prior density π, the posterior density is proportional to the function

$$\theta \mapsto p_\theta(x)\pi(\theta).$$

In most cases, this expression is easy to compute, because this function is directly related to the specification of the random model and the prior distribution. In general, however, it is not easy to compute the posterior density or the Bayes estimate: for this, we need to evaluate the integral of $p_\theta(x)\pi(\theta)$ or $\theta p_\theta(x)\pi(\theta)$, respectively, with respect to θ for given x. The fact that this can be difficult has decreased the popularity of Bayes estimators. It is not very attractive to have to choose a certain prior density for the sake of simpler computations.

If the dimension of the parameter θ is low, for example if θ is real, implementing the computations numerically is reasonable straightforward, for example by approximating the integrals by sums. For higher-dimensional parameters, for example of dimension greater than or equal to 4, the problems are more important. Simulation methods have been used to reduce these problems since the 1990s. MCMC methods are a general process used to simulate a Markov chain $Y_1, Y_2, \ldots$ whose marginal distributions are approximately equal to the posterior distribution. Before we describe the MCMC algorithms, we discuss a number of essential notions from the theory of Markov chains.

A Markov chain is a sequence $Y_1, Y_2, \ldots$ of random variables such that the conditional distribution of Y_{n+1} given the previous variables $Y_1, \ldots, Y_n$ depends only on Y_n. An equivalent formulation is that given the "present" variable Y_n, the "future" variable Y_{n+1} is independent of the "past" $Y_1, \ldots, Y_{n-1}$. We can then see the variable Y_n as the state at "time" n, and to simulate the next state Y_{n+1}, it suffices to know the current state Y_n; knowledge of the prior states is irrelevant. We will consider only Markov chains that are "time-homogeneous." This means that the conditional distribution of Y_{n+1} given Y_n does not depend on n, so that the transition from one state to the next always follows the same mechanism. The behavior of the chain is then completely determined by the *transition kernel* Q given by

$$Q(y, B) = P(Y_{n+1} \in B \mid Y_n = y).$$

For a fixed y, the map $B \mapsto Q(y, B)$ gives the probability distribution at the next time given the current state y. Often, Q is described using the corresponding *transition density* q. This is the conditional density of Y_{n+1} given Y_n; it satisfies $Q(y, B) = \int_B q(y, z) \, dz$, where in the discrete case, the integral must be replaced by a sum.

A probability distribution Υ is called a *stationary distribution* for the transition kernel Q if, for every event B,

$$\int Q(y, B) \, d\Upsilon(y) = \Upsilon(B).$$

This equation says precisely that the stationary distribution is preserved under the transition from Y_n to Y_{n+1}. If Y_1 has a stationary distribution, then Y_2 also has a stationary distribution, etc. If Q has transition density q and Υ has density υ (which is then called a *stationary density*), then

$$\int q(y, z) \, \upsilon(y) \, dy = \upsilon(z)$$

is an equivalent equation. This gives a simple way to characterize stationary distributions. (The stationary distribution and density of a Markov chain are customarily called Π and π, respectively. However, in the context of Bayesian estimates, this notation may lead to confusion, which is why we use the symbols given above.) When a density υ satisfies the *detailed balance* relation

$$\upsilon(y)q(y, z) = \upsilon(z)q(z, y),$$

υ is a stationary density. This can be seen by integrating both sides of the relation with respect to y and using the fact that $\int q(z, y) \, dy = 1$ for every z. The detailed balance relation requires that a transition from y to z be as probable as a transition from z to y when in both cases, the first point is chosen from the density υ. A Markov chain with this property is called *reversible*.

The introduction to Markov chains we just gave suffices to understand the principle of MCMC algorithms. In MCMC algorithms, Markov chains are generated with a transition kernel whose stationary density is equal to the desired posterior density. In the application to MCMC, the stationary density $y \mapsto \upsilon(y)$ in the general discussion of Markov chains is replaced by a posterior density that is proportional to $\theta \mapsto p_\theta(x)\pi(\theta)$ for observed data x. Fortunately, in simulation schemes, the proportionality constant is unimportant, so that the fact that the integrals are difficult to evaluate is not relevant. Consequently, MCMC algorithms can, in principle, be used for any prior distribution.

Because it is usually not easy to generate the first value Y_1 of the chain from the stationary density (in the MCMC context, this is the posterior density), an MCMC chain is usually not stationary. The chain does converge to a stationary one as $n \to \infty$. In practice, the chain is simulated over a great number (N) of steps, and the first simulated data $Y_1, \ldots, Y_b$ are thrown out; this is called the "burn-in." The remaining variables $Y_{b+1}, Y_{b+2}, \ldots, Y_N$ can be viewed as a realization of a Markov chain with the posterior distribution as stationary distribution. Using, for example, a histogram of $Y_{b+1}, \ldots, Y_N$, we obtain a good idea of the posterior distribution, and the average of $Y_{b+1}, \ldots, Y_N$ is a good approximation of the Bayes estimator, the posterior expectation. The motivation for using this "empirical approach" is the same as in Section 2.2.1, except that the variables $Y_1, Y_2, \ldots$ now form a Markov chain and are therefore not independent. However, many Markov chains also follow a law of large numbers, which guarantees that now, too, averages behave asymptotically as expectations. The convergence rate does turn out to depend strongly on the transition kernel, so that in practice, it can still be quite difficult to set up an MCMC algorithm that leads to a good approximation within a reasonable (CPU) time.

There now exist many types of MCMC algorithms. The two most important ones, which are often used together, are the *Metropolis–Hastings* algorithm and the *Gibbs sampler*.

Example 3.40 Metropolis–Hastings algorithm

The Metropolis–Hastings algorithm generates a Markov chain using a so-called *proposal* transition density q (with associated transition kernel Q). This transition density is chosen in such a way that it is easy to simulate using the probability density $z \mapsto q(y, z)$, for every given y. At the end of this example, we will come back to the choice of the proposal density. Next, define

$$\alpha(y, z) = \frac{v(z)q(z, y)}{v(y)q(y, z)} \wedge 1$$

with v the posterior density we want to approximate and $a \wedge b = \min(a, b)$. Note that to determine $\alpha(y, z)$, it suffices to know the form of v and q; the proportionality constant disappears. In the Metropolis–Hastings algorithm, for every transition from Y_n to Y_{n+1} in the Markov chain, a state Z_{n+1} is generated following the proposal transition kernel Q, which acts as a candidate value (whence the name proposal). This state is accepted (that is, $Y_{n+1} = Z_{n+1}$) with probability $\alpha(Y_n, Z_{n+1})$ and rejected with probability $1 - \alpha(Y_n, Z_{n+1})$, in which case the current state is kept (that is, $Y_{n+1} = Y_n$). The simulation algorithm is then as follows:

Take a fixed initial value Y_0 and then continue recursively as follows:
given Y_n, generate Z_{n+1} from $Q(Y_n, \cdot)$
generate U_{n+1} from the uniform distribution on $[0, 1]$
if $U_{n+1} < \alpha(Y_n, Z_{n+1})$, let $Y_{n+1} := Z_{n+1}$
else let $Y_{n+1} := Y_n$.

85

The transition kernel P of the resulting Markov chain $Y_1, Y_2, \ldots$ consists of two pieces, corresponding to the "if-else" split. The kernel is given by

$$
\begin{aligned}
P(y, B) &= P(Y_{n+1} \in B \mid Y_n = y) \\
&= P(Z_{n+1} \in B, U_{n+1} < \alpha(Y_n, Z_{n+1}) \mid Y_n = y) \\
&\quad + P(Y_n \in B, U_{n+1} \geq \alpha(Y_n, Z_{n+1}) \mid Y_n = y) \\
&= \int_B \alpha(y, z) q(y, z) \, dz \\
&\quad + \big(1 - P(U_{n+1} < \alpha(Y_n, Z_{n+1}) \mid Y_n = y)\big) 1_{y \in B} \\
&= \int_B \alpha(y, z) q(y, z) \, dz + \big(1 - \mathrm{E}(\alpha(y, Z_{n+1}) \mid Y_n = y)\big) 1_{y \in B} \\
&= \int_B \alpha(y, z) q(y, z) \, dz + \Big(1 - \int \alpha(y, z) q(y, z) \, dz\Big) 1_{y \in B},
\end{aligned}
$$

where the last integral is taken over the entire state space. The moves of the chain corresponding to the first term in the last expression are governed by the transition density $r(y, z) = \alpha(y, z) q(y, z)$. The function α is chosen in such a way that the codomain contains the interval $[0, 1]$ and that the detailed balance relation

(3.4) $$\qquad\qquad\qquad v(y) r(y, z) = v(z) r(z, y)$$

is satisfied. This part of the Markov chain is therefore reversible. In the second part of the chain, given $Y_n = y$, we stay in y with probability

$$
1 - \int \alpha(y, z) q(y, z) \, dz.
$$

This movement from y to y is trivially symmetric. It easily follows from these statements that v is a stationary density for the Markov chain $Y_1, Y_2, \ldots$.

A popular choice for the proposal density q is the *random walk kernel* $q(y, z) = f(z - y)$ for a given density f. If we choose f symmetric about 0, then $\alpha(y, z)$ reduces to $v(z)/v(y)$. Choosing a good kernel is not easy. The general principle is to choose a transition kernel Q that represents "movements" toward variables Z_{n+1} in the full domain of v in the first step of the algorithm, and at the same time, does not lead too often to the step "else", because this would negatively influence the efficiency of the algorithm. In MCMC jargon, we say that we are looking for a proposal transition kernel Q that "is sufficiently mixing," "searches the space sufficiently well," and "does not linger too much."

To illustrate this, we apply the algorithm given above to the situation of Example 3.39, where the posterior density can easily be derived analytically. For this, we generate a sample of size $n = 25$ from the geometric distribution with parameter $\theta = 0.2$ and find $\bar{x} = 5.88$. For the prior density, we take $\pi(\theta) = 6\theta(1 - \theta)$, the beta$(2, 2)$-density.

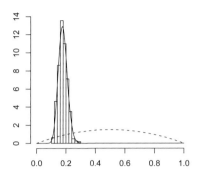

Figure 3.9. Histogram of the values $Y_{201}, \ldots, Y_{1000}$ in the Markov chain generated by the Metropolis–Hastings algorithm based on a geometric sample of size $n = 25$ with $\theta = 0.2$ and $\bar{x} = 5.88$. The posterior beta$(27, 124)$-density (solid line) and the prior beta$(2, 2)$-density (dashed line) are also shown.

Figure 3.9 shows the histogram of $Y_{201}, \ldots, Y_{1000}$ for the chain generated using the Metropolis–Hastings algorithm with a normal random walk kernel. The dashed line in the figure depicts the prior density. The computations in Example 3.39 imply that the posterior density for this case is equal to the density of the beta$(27, 124)$-distribution, which is also drawn in Figure 3.9 (solid line). We see that the histogram of the values of Y gives a good representation of the posterior density, as expected. Moreover, the average (0.18) of the values of Y is equal to the Bayes estimate: $27/(27+124)=0.18$.

Example 3.41 Gibbs sampler

The Gibbs sampler reduces the problem of approximating a high-dimensional posterior density to repeatedly approximating lower-dimensional distributions. The algorithm is often used in combination with the Metropolis–Hastings sampler if no suitable proposal transition density q is available for the Metropolis–Hastings algorithm.

Let v be a density for the m-dimensional variable Y from which we want to generate a sample. Suppose that we have at our disposal a procedure to generate variables from each of the conditional densities

$$v_i\big(y_i \,\big|\, y_1, \ldots, y_{i-1}, y_{i+1}, \ldots y_m\big) = \frac{v(y)}{\int v(y)\,dy_i}, \quad i = 1, \ldots, m,$$

where $y = (y_1, \ldots, y_m)$. The following algorithm yields a chain $Y_1, \ldots, Y_n$ with stationary density v:

Choose an initial value $Y_0 = (Y_{0,1}, \ldots, Y_{0,m})$, then continue recursively as follows:

> Given $Y_n = (Y_{n,1}, \ldots, Y_{n,m})$,

```
generate Yₙ₊₁,₁ using υ₁(·| Yₙ,₂, . . . , Yₙ,ₘ)
generate Yₙ₊₁,₂ using υ₂(·| Yₙ₊₁,₁, Yₙ,₃ . . . , Yₙ,ₘ)
```

$$\vdots$$

```
generate Yₙ₊₁,ₘ using υₘ(·| Yₙ₊₁,₁, . . . , Yₙ₊₁,ₘ₋₁).
```

One by one, the coordinates are replaced by a new value, each time conditionally on the latest available value of the other coordinates. We can check that the density υ is stationary for each of the steps of the algorithm individually (see Exercise 3.42). The resulting chain $Y_1, \ldots, Y_n$ has stationary density υ. The Gibbs sampler can be used, for example, when there is a high-dimensional proposal transition density in the Metropolis–Hastings algorithm.

Example 3.42 Missing data

Suppose that instead of the "complete data" (X, Y), we can only observe the data X. If $(x, y) \mapsto p_\theta(x, y)$ is a probability density of (X, Y) that depends on the parameter θ, then $x \mapsto \int p_\theta(x, y) \, dy$ is a probability density of the observation X. Given a prior density $\theta \mapsto \pi(\theta)$, the posterior density based on the observed value x is therefore proportional to

$$\theta \mapsto \pi(\theta) \int p_\theta(x, y) \, dy.$$

We can apply the MCMC algorithms described earlier to this posterior density. However, if the marginal density of X (the integral in the display above) cannot be computed analytically, then implementing the MCMC algorithms is difficult.

An alternative to computing the marginal distribution is to also approximate the unobserved values Y. In the Bayesian notation, the posterior distribution is the conditional distribution of an imaginary variable $\overline{\Theta}$ given the observation X. This is the marginal distribution of the conditional distribution of the pair $(\overline{\Theta}, Y)$ given X. If we could generate a sequence of variables $(\overline{\Theta}_1, Y_1), \ldots, (\overline{\Theta}_n, Y_n)$ using the last conditional distribution, then the first coordinates $\overline{\Theta}_1, \ldots, \overline{\Theta}_n$ of this sequence would be samples from the desired posterior distribution. Marginalizing an empiric distribution is the same as "forgetting" variables, and that is very easy to do computationally.

Thus, we can apply an MCMC algorithm to simulate variables $(\overline{\Theta}_i, Y_i)$ from the probability density that is proportional to the map $(\theta, y) \mapsto p_\theta(x, y)\pi(\theta)$, with x equal to the observed value of X. Next, we throw out the Y-values and view the remaining $\overline{\Theta}$-values as a sample from the posterior distribution of the parameter.

* 3.6 M-Estimators

Let $M(\theta; X)$ be an arbitrary function of the parameter and the observation. An *M-estimator* for a parameter θ is the value of θ that maximizes (or minimizes) the criterion function $\theta \mapsto M(\theta; X)$. Another term is *maximum (or minimum) contrast estimator*.

If we take M equal to the likelihood function, we find the maximum likelihood estimator for θ. There are many other possibilities. The most common criterion functions for independent observations $X = (X_1, \ldots, X_n)$ have a sum structure:

$$M(\theta; X) = \sum_{i=1}^{n} m_\theta(X_i)$$

for suitably chosen functions m_θ.

Maximizing a function is often the same as solving the system of equations obtained by setting the derivative equal to 0. The term "M-estimator" is therefore also used for estimators that solve an equation $\Psi(\theta; X) = 0$. Such equations are called *estimating equations*. Because not every vector-valued function is a gradient of a function, estimating equations are more general than contrast functions. The most common criterion functions for independent observations $X = (X_1, \ldots, X_n)$ have a sum structure:

$$\Psi(\theta; X) = \sum_{i=1}^{n} \psi_\theta(X_i)$$

for suitably chosen vector-valued functions ψ_θ. The equation $\Psi(\theta; X) = 0$ is understood as a system of equations. The number of equations is equal to the dimension of the range of ψ_θ, and would typically be chosen equal to the number of parameters to be estimated.

Example 3.43 Median

The average $\overline{X}$ of random variables $X_1, \ldots, X_n$ minimizes the function $\theta \mapsto \sum_{i=1}^{n}(X_i - \theta)^2$. The average is an estimate for the "center" of the probability distribution of the observations. An alternative estimator with roughly the same interpretation is obtained by minimizing the function $\theta \mapsto \sum_{i=1}^{n}|X_i - \theta|$. We can show that this leads to the *sample median*

$$\text{med}\{X_1, \ldots, X_n\} = \begin{cases} X_{((n+1)/2)} & \text{if } n \text{ is odd,} \\ \frac{1}{2}(X_{(n/2)} + X_{(n+2)/2)}) & \text{if } n \text{ is even.} \end{cases}$$

This is the "middle observation."

Replacing the square by the absolute value has the effect of reducing the influence of very large or very small observations. Indeed, the sample median does not change if the big and small observations are made even bigger or smaller. This property is referred as the *robustness* of the median or, more precisely, robustness against outliers. By making different choices of contrast function, we may define other robust estimators. For instance, the Huber estimator is given by $m_\theta(x) = (x-\theta)^2 1_{|x-\theta|<c} + c|x - \theta|1_{|x-\theta|\geq c}$ and is a compromise between mean and median. The parameter c is typically estimated, to reflect the scale of the data.

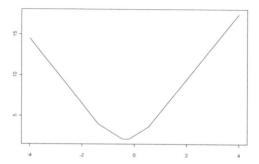

Figure 3.10. The function $\theta \mapsto \Sigma_{i=1}^n |x_i - \theta|$ for a sample $x_1, \ldots, x_n$ of size 4 from the standard normal distribution.

Example 3.44 Least-squares estimator

In Example 1.5, we briefly described the simple linear regression model (see also Section 7.2). For dependent variables $Y_1, \ldots, Y_n$ and predictor variables $x_1, \ldots, x_n$, we have $Y_i = \alpha + \beta x_i + e_i$. The measurement errors $e_1, \ldots, e_n$ are often assumed to be independent and have a normal distribution with expectation 0 and variance σ^2. The unknown parameters α and β can be estimated using the least-squares estimators (LS-estimators); these are the values that minimize

$$\sum_{i=1}^n (Y_i - \alpha - \beta x_i)^2$$

with respect to α and β. If the measurement errors are normally distributed, the least-squares estimators correspond to the maximum likelihood estimators for α and β (see Section 7.2). The LS-estimators can also be used without the normality assumption. They are then not maximum likelihood estimators, but general M-estimators.

More generally, we can use the least-squares method in a *nonlinear regression model* $Y_i = g_\theta(x_i) + e_i$, where g_θ is a nonlinear function of θ, the terms $e_1, \ldots, e_n$ are unobservable measurement errors, and $x \mapsto g_\theta(x)$ is a function that is known up to the parameter θ. The LS-estimator for θ minimizes the criterion

$$\theta \mapsto \sum_{i=1}^n (Y_i - g_\theta(x_i))^2.$$

If the measurement errors are normally distributed, this again leads to the maximum likelihood estimator for θ. For a nonlinear function g_θ, we often need a numerical algorithm to compute the least-squares estimate.

An example of nonlinear regression is fitting a time curve when we have observations $y_1, \ldots, y_n$, including measurement errors, of the curve at certain times $x_1, \ldots, x_n$. If a parameterized curve is of the form $t \mapsto g_\theta(t)$, for example $g_\theta(t) = \theta_0 + \theta_1 t + \theta_2 e^{-\theta_3 t}$ with 4-dimensional parameter $\theta = (\theta_0, \theta_1, \theta_2, \theta_3)$, then we can estimate the parameter θ using the observations (x_i, y_i) for $i = 1, \ldots, n$. ▭

Example 3.45 Generalized estimating equations

Suppose that we measure each of n experimental units or individuals repeatedly, obtaining the observations $Y_i = (Y_{i,1}, \ldots, Y_{i,T_i})^T$, for $i = 1, 2 \ldots, n$, which we wish to model by a linear regression model of the type as described in Example 1.5 (see also Section 7.2). We use a common set of parameters $\beta = (\beta_1, \ldots, \beta_p)^T$ for all observations and hence obtain the model, for $i = 1, \ldots, n$ and $t = 1, \ldots, T_i$,

$$Y_{i,t} = x_{i,t}^T \beta + e_{i,t},$$

where the vectors $x_{i,t} \in \mathbb{R}^p$ are known explanatory variables. Since the observations $Y_{i,t}$ for the same value of i refer to the same experimental unit, it is often not reasonable to model the errors $e_{i,t}$ as independent random variables, as in the ordinary regression model. On the other hand, if the units themselves are a sample of possible units, then it is reasonable to model the n error vectors $e_i := (e_{i,1}, \ldots, e_{i,T_i})^T$ as independent. A standard model is then to assume that every e_i follows a multivariate normal distribution $N_{T_i}(0, \sigma^2 \Lambda_i)$, where $\sigma^2 > 0$ and Λ_i is a positive-definite $(T_i \times T_i)$-matrix. The logarithm of the likelihood for observing the independent vectors $Y_1, \ldots, Y_n$ is then, up to an additive constant,

$$-\frac{1}{2} \sum_{i=1}^n \log \det(\sigma^2 \Lambda_i) - \frac{1}{2\sigma^2} \sum_{i=1}^n (Y_i - X_i\beta)^T \Lambda_i^{-1} (Y_i - X_i\beta).$$

Here X_i is the $(T_i \times p)$-matrix with rows the vectors $x_{i,t}^T$, for $t = 1, \ldots, T_i$. The expression in the display may be maximized with respect to the unknown parameters to obtain the maximum likelihood estimators.

The main interest is usually in the vector β of regression parameters. Within the maximum likelihood setup, the estimation of this vector is confounded by the presence of the additional parameters σ^2 and Λ_i. The matrices Λ_i may contribute many unknowns, even in the simplest case that we choose to restrain them to be equal for different i (and $T_i = T$ is large). The method of estimating equations is helpful to overcome this problem.

We start by noting that the maximum likelihood estimator for β solves the stationary equation obtained by setting the partial derivative of the log likelihood with respect to β equal to 0. This takes on the form

(3.5)
$$\sum_{i=1}^n X_i^T \Lambda_i^{-1} (Y_i - X_i\beta) = 0.$$

This still contains the matrices Λ_i, but only as weight factors. The reason that solving the equation gives a good estimator for β is that the expectations of the terms of the sum vanish, as $E(Y_i - X_i\beta) = 0$, by the fact that the errors $e_{i,t}$ in the model have expectation 0. One says that the preceding display gives an *unbiased estimating equation* for β. This fact does not depend on the matrix Λ_i, but remains true if this matrix is replaced by a different one. Further analysis will reveal that the weight matrices Λ_i are optimal in the sense of leading to the smallest possible (asymptotic) mean square error for the estimator for β. However, if we do not know these matrices, then we could only use them in the equation after estimating them from the data, and this might introduce considerable additional variance, particularly when the dimensions of the matrices are large relative to n and p. The method of *generalized estimating equations*, or *GEE*, is to replace the matrices by either fixed matrices or matrices of a particular form, given by a low-dimensional parametric model. For instance, popular choices are the *autoregressive* and *exchangeable* matrices, which in the case that $T_i = 4$ take on the forms

$$\begin{pmatrix} 1 & \rho & \rho^2 & \rho^3 \\ \rho & 1 & \rho & \rho^2 \\ \rho^2 & \rho & 1 & \rho \\ \rho^3 & \rho^2 & \rho & 1 \end{pmatrix}, \qquad \begin{pmatrix} 1 & \rho & \rho & \rho \\ \rho & 1 & \rho & \rho \\ \rho & \rho & 1 & \rho \\ \rho & \rho & \rho & 1 \end{pmatrix}.$$

The parameter $\rho \in (-1, 1)$ in these matrices determines the dependence between the multiple observations on a given experimental unit. The value 0 corresponds to independence and a value close to the ends of the interval $(-1, 1)$ gives strong negative or positive dependence. It works best to choose ρ such that the corresponding matrix is close to the true matrix Λ_i. In practice, we may estimate an appropriate value from the data. We next substitute the estimated matrix for the matrix Λ_i in the equation (3.5), and finally solve for β.

3.7 Summary

Let $X = (X_1, \ldots, X_n)$ be an observation with distribution P_θ that depends on the unknown parameter θ. An *estimator* $T = T(X)$ for $g(\theta)$ is a random variable that depends only on the observation X (and therefore not on the unknown parameter θ!). (If $g(\theta) = \theta$, then T is an estimator for θ.) The corresponding *estimate* is denoted by $T(x)$, where x is the vector of the observed values.

Measures for the quality of estimators:
- The *bias* of an estimator T for $g(\theta)$ is $E_\theta T - g(\theta)$.
- The *mean square error* (MSE) is a measure for the accuracy of an estimator. The MSE is defined as the expected square difference between T and $g(\theta)$:

$$\mathrm{MSE}(\theta; T) = E_\theta \|T - g(\theta)\|^2.$$

We prefer an estimator with a small MSE for all values of θ. If $g(\theta) \in \mathbb{R}$, then $\mathrm{MSE}(\theta; T) = \mathrm{var}_\theta T + (E_\theta T - g(\theta))^2$; that is, the MSE is the sum of the variance and the square bias.

Different types of estimators:
- The *maximum likelihood estimate* for θ is the value $\hat\theta$ that maximizes the likelihood function. The *likelihood function* is the (joint) probability density p_θ of X viewed as a function of θ, for a given observation x: $\theta \mapsto L(\theta, x) = p_\theta(x)$. If $X = (X_1, \ldots, X_n)$ is a sample from a distribution with marginal probability density f_θ, then the likelihood function is equal to $L(\theta; x = (x_1, \ldots, x_n)) = \prod_{i=1}^n f_\theta(x_i)$. The maximum likelihood estimate is often found as a solution of the *likelihood equations*, but can also be a value of θ where the likelihood function is discontinuous. The maximum likelihood estimate for $g(\theta)$ is defined as $g(\hat\theta)$. The *maximum likelihood estimator* is the corresponding random variable.
- A *method of moments estimator* for θ based on a sample $X = (X_1, \ldots, X_n)$ is a random variable $\hat\theta$ for which the first k theoretical moments are equal to the first k sample moments: $E_{\hat\theta} X_i^j = \overline{X^j}$ for $j = 1, \ldots, k$, with k the least possible. The method of moments estimator for $g(\theta)$ is defined as $g(\hat\theta)$.
- The *Bayes estimator* for $g(\theta)$ with respect to a prior density π is the estimator that minimizes the Bayes risk $\int_\Theta E_\theta(T - g(\theta))^2 \pi(\theta)\, d\theta$ over all estimators T. This Bayes risk is the probability-weighted average of $\mathrm{MSE}(\theta; T)$ for the probability density π. For a given observation x, the *posterior density* of the parameter random variable $\overline\Theta$ is

$$p_{\overline\Theta | X = x}(\theta) = \frac{p_\theta(x)\pi(\theta)}{\int p_\vartheta(x)\pi(\vartheta)\, d\vartheta}.$$

The Bayes estimate for $g(\theta)$ is equal to the expected value of $g(\overline\Theta)$ with respect to the posterior distribution: $E(g(\overline\Theta)| X = x)$. In general, the Bayes estimate for $g(\theta)$ is not equal to the transformation $g(\hat\theta)$ of the Bayes estimate $\hat\theta$ for θ.

Exercises

1. Give a theoretical explanation for the forms of the (exponential and normal) histograms in Figure 3.1.

2. Let $X_1, \ldots, X_n$ be independent and $U[0, \theta]$-distributed, with $\theta > 0$ unknown. Determine the mean square errors of the estimators $cX_{(n)}$ for θ, for every value of $c > 0$. Which value for c gives the best estimator?

3. Let X be binomially distributed with parameters n and p, with n known and $p \in [0, 1]$ unknown. Let $T_c = cX/n$ be an estimator for p, where $c > 0$ is yet to be determined.
    (i) For which value of c is T_c unbiased?
    (ii) Determine the mean square error of T_c.
    (iii) For which value of c is this estimator optimal? Is this optimal estimator usable in practice? Explain.
    (iv) Determine the limit of the optimal value for c as $n \to \infty$. Which estimator T_c do you obtain?

4. Let $X_1, \ldots, X_n$ be a sample from the Poisson(θ)-distribution. We want to estimate θ^2.
    (i) Is $(\overline{X})^2$ an unbiased estimator for θ^2?
    (ii) Determine an unbiased estimator for θ^2.

5. Let $X_1, \ldots, X_m$ and $Y_1, \ldots, Y_n$ be independent samples from the Bernoulli distribution with unknown parameter $p \in [0, 1]$.
    (i) Prove that $(\overline{X} + \overline{Y})/2$ and $(\sum_{i=1}^{m} X_i + \sum_{j=1}^{n} Y_j)/(m + n)$ are unbiased estimators for p.
    (ii) Which of these two estimators is preferable (if $m \neq n$)?

6. In a study on discrimination in Amsterdam, the subjects are asked whether they have experienced discrimination (based on race, skin color, gender, or religion). A stratified sample is taken: 50 men and 50 women are chosen randomly from the adult population of Amsterdam. Let X be the number of men and Y the number of women in the sample that have experienced discrimination. Define:

$$p_M = \text{proportion of male Amsterdammers having experienced discrimination}$$
$$p_V = \text{proportion of female Amsterdammers having experienced discrimination}$$
$$p = \text{proportion of Amsterdammers having experienced discrimination}$$

Assume $p_V = 2p_M$ and that there are as many men as women living in Amsterdam.
    (i) Compute the mean square error of the estimator $(X + Y)/100$ for p.
Now, define Z as the number of persons having experienced discrimination in a normal (nonstratified = simple) sample of 100 adult Amsterdammers.
    (ii) Compute the mean square error of the estimator $Z/100$ for p.
    (iii) Compare the two mean square errors. What do you conclude?

7. We want to study how many Dutch households have a tablet. Let Π be the total population of all Dutch households. Let k be the number of towns in the Netherlands, and let $1000m_i$ be the number of households in the ith town, for $i = 1, 2, \ldots, k$. For convenience, we assume $m_i \in \mathbb{N}$. So in Π, there are $M = \sum_i m_i$ thousands of households. We then take a sample as follows. First, randomly choose 100 thousands from all these thousands, without replacement. Let Y_i be the number of chosen thousands in the ith town. Next, randomly choose $10Y_i$ households in the ith town, without replacement. Let p_i be the proportion of households with a tablet in the ith town, and p the proportion of the total population. Approximate p with $X/1000$, where X is the total number of chosen households with a tablet. Is $X/1000$ an unbiased estimator for p?

8. Compute the maximum likelihood estimator for θ based on a sample $X_1, \ldots, X_n$ from the Poisson(θ)-distribution.

9. Let $X_1, \ldots, X_n$ be a sample from a *Weibull distribution*, whose probability density is given by
$$p_\theta(x) = \theta a x^{a-1} e^{-\theta x^a} \quad \text{for } x > 0$$
and 0 otherwise. Here a is a known number, and $\theta > 0$ is an unknown parameter.
   (i) Determine the maximum likelihood estimator for θ.
   (ii) Determine the maximum likelihood estimator for $1/\theta$.

10. Let $X_1, \ldots, X_n$ be a sample from a distribution with probability density
$$p_\theta(x) = \theta x^{\theta-1} \quad \text{for } x \in (0, 1)$$
and 0 otherwise. Here $\theta > 0$ is an unknown parameter.
   (i) Compute $\mu = g(\theta) = E_\theta X_1$.
   (ii) Determine the maximum likelihood estimator for μ.

11. An urn contains white and black balls in the ratio $p : 1 - p$. We draw balls one by one with replacement, continuing until we draw a white ball. Let Y_i be the number of draws necessary. We repeat this process n times, giving numbers $Y_1, \ldots, Y_n$. Determine the maximum likelihood estimator for p.

12. Let $X_1, \ldots, X_n$ be a sample from a distribution with probability density
$$p_\theta(x) = \theta x^{-2} \quad \text{for } x \geq \theta$$
and 0 for $x < \theta$, with $\theta > 0$ unknown.
   (i) Determine the maximum likelihood estimator for θ.
   (ii) Is this estimator unbiased?
   (iii) Determine the mean square error of this estimator.

13. Let $X_1, \ldots, X_n$ be a sample from a distribution with probability density
$$p_\theta(x) = \theta(1 + x)^{-(1+\theta)} \quad \text{for } x \geq 0$$
and 0 elsewhere, with $\theta > 0$ unknown. Determine the maximum likelihood estimator for θ.

14. Let $X_1, \ldots, X_m$ and $Y_1, \ldots, Y_n$ be two independent samples from the normal distributions with parameters (μ_1, σ^2) and (μ_2, σ^2), respectively. Determine the maximum likelihood estimator for $\theta = (\mu_1, \mu_2, \sigma^2)$.

15. Suppose that the vector $X = (X_1, \ldots, X_m)$ has a multinomial distribution with parameters n and $(p_1, \ldots, p_m)$, where $p_1 + \ldots + p_m = 1$. We assume that n is known and that the probabilities $p_1, \ldots, p_m$ are unknown. Show that the maximum likelihood estimator for $(p_1, \ldots, p_m)$ is equal to $(X_1/n, \ldots, X_m/n)$.

16. Let $X_1, \ldots, X_n$ be a sample from the shifted exponential distribution with intensity parameter 1 and unknown shift parameter $\theta \in (-\infty, \infty)$. The corresponding density is given by $p_\theta(x) = e^{\theta-x}$ for $x \geq \theta$ and $p_\theta(x) = 0$ for $x < \theta$. Determine the maximum likelihood estimator for θ.

17. We want to estimate the number N of fish in a pond. We proceed as follows. We catch r fish and mark them. We then set them free. After some time, we catch n fish (without putting them back). Let X_i be equal to 0 if the ith fish we catch is marked and 1 if it is not ($i = 1, .., n$).

   (i) Determine the probability distribution of $\sum X_i$ expressed in r, n, and N.

   (ii) Determine the maximum likelihood estimator for N based on $\sum_{i=1}^{n} X_i$.

18. Let $X_1, \ldots, X_n$ be a sample from a distribution with an unknown distribution function F. We denote the empirical distribution function of the sample by $\hat{F}$.

   (i) Which distribution does $n\hat{F}(x)$ have?

   (ii) Is $\hat{F}(x)$ an unbiased estimator for $F(x)$?

   (iii) Determine the variance of $\hat{F}(x)$.

   (iv) Show that $\mathrm{cov}(\hat{F}(u), \hat{F}(v)) = n^{-1}(F(m) - F(u)F(v))$ with $m = \min\{u, v\}$. It follows that $\hat{F}(u)$ and $\hat{F}(v)$ have a positive correlation.

19. (**k-means clustering.**) Let $X_1, \ldots, X_n$ be independent random variables such that for an unknown partition $\{1, \ldots, n\} = \cup_{j=1}^{k} I_j$, the variables $(X_i; i \in I_j)$ are normally distributed with expectation μ_j and variance 1. Show that the maximum likelihood estimator for the partition and parameter vector $(\mu_1, \ldots, \mu_k)$ minimizes the sum of squares $\sum_{j=1}^{k} \sum_{i \in I_j} (X_i - \mu_j)^2$. Give an interpretation of this procedure in words.

20. Let $X_1, \ldots, X_n$ be a sample from the exponential distribution with parameter λ, where $\lambda > 0$ is an unknown parameter.

   (i) Determine the maximum likelihood estimator for $1/\lambda^2$.

   (ii) Determine a method of moments estimator for $1/\lambda^2$.

   (iii) Determine an unbiased estimator for $1/\lambda^2$.

21. Let $X_1, \ldots, X_n$ be a sample from the binomial distribution with parameters n and p, where $p \in [0, 1]$ is unknown. Determine the maximum likelihood estimator and the method of moments estimator for p.

22. Let $X_1, \ldots, X_n$ be a sample from the Bernoulli distribution with unknown parameter $p \in [0, 1]$.

   (i) Determine the method of moments estimator T for p.

   (ii) Show that the estimator T^2 is biased for p^2, and then determine an unbiased estimator for p^2.

23. Let $X_1, \ldots, X_n$ be a sample from the geometric distribution with unknown parameter $p \in (0, 1]$. Determine the method of moments estimator for p.

24. Let $X_1, \ldots, X_n$ be a sample from a probability distribution with density

$$p_\theta(x) = \theta(1 + x)^{-(1+\theta)} \quad \text{for } x > 0$$

and 0 elsewhere, with $\theta > 1$ unknown. Determine the method of moments estimator for θ.

25. Let $X_1, \ldots, X_n$ be a sample from a probability distribution with density

$$p_\theta(x) = \frac{2x}{\theta^2} 1_{\{0 \le x \le \theta\}},$$

where $\theta > 0$ is an unknown parameter.

   (i) Determine the method of moments estimator T for θ.

   (ii) Show that T is unbiased for θ.

   (iii) Give the method of moments estimator for θ^2.

(iv) Show that the method of moments estimator for θ^2 is biased for θ^2, and then determine an unbiased estimator for θ^2.

26. Let $X_1, \ldots, X_n$ be a sample from a probability distribution given by $P_\theta(X = x) = 1/\theta$ for $x \in \{1, 2, \ldots, \theta\}$, where $\theta \in \mathbb{N}$ is unknown.
   (i) Determine the method of moments estimator for θ.
   (ii) Determine the maximum likelihood estimator for θ.

27. Let $X_1, \ldots, X_n$ be sample from the $U[\sigma, \tau]$-distribution with $\sigma < \tau$ unknown.
   (i) Determine the maximum likelihood estimator for the vector (σ, τ).
   (ii) Determine the method of moments estimator for the vector (σ, τ).

28. Let $X_1, \ldots, X_n$ be a sample from the uniform distribution on $[-\theta, \theta]$, with $\theta > 0$ unknown.
   (i) Determine the maximum likelihood estimator for θ.
   (ii) Determine the method of moments estimator for θ.

29. Let X be a random variable with finite second moment. Show that the function $b \mapsto E(X-b)^2$ is minimal at $b = EX$.

30. Let X be a continuously distributed random variable with finite first moment. Show that the function $b \mapsto E|X - b|$ is minimal at a point b such that $P(X < b) = P(X > b) = 1/2$; we call b the *population median*.

31. Let $X_1, \ldots, X_n$ be a sample from the *Laplace distribution* (or *double exponential distribution*), which has probability density

$$p_\theta(x) = \frac{1}{2} e^{-|x-\theta|} \quad \text{for } \theta \in \mathbb{R}.$$

   (i) Determine the population median (see previous exercise).
   (ii) Determine the maximum likelihood estimator for θ.
   (iii) Determine the method of moments estimator for θ.

32. The method of moments estimator and maximum likelihood estimator for the parameter of a Laplace distribution are very different. Use a simulation to determine which estimator is preferable. The R-program in Table 3.1 can be used for this.
   Explanation: in the first line, we declare two vectors (arrays) of length 1000, which we fill with 1000 realizations of the two estimators. In the last two lines, we compute the mean square deviation of these two vectors from the true value of the parameter (equal to 0 in this case). These are not the true mean square errors, but they are good approximations thereof. In the first line of the for-loop, a sample of size n ($n = 100$) is taken from the standard Laplace distribution. Next, both estimates are computed based on this sample. This is repeated 1000 times.

33. Let $X_1, \ldots, X_n$ be a sample from a probability distribution with density

$$p_\theta(x) = \theta x^{\theta-1} \quad \text{for } 0 \leq x \leq 1$$

and 0 elsewhere, with $\theta > 0$ unknown.
   (i) Determine the method of moments estimator for θ.
   (ii) Determine the maximum likelihood-estimator for θ.
   (iii) Determine the Bayes estimator for θ with respect to the prior density π given by $\pi(\theta) = e^{-\theta}$ for $\theta > 0$ and 0 elsewhere.

```
moments = mls = numeric(1000)
n = 100
for (i in 1:1000) {
     x = rexp(n)*(2*rbinom(n,1,0.5)-1)
     moments[i] = mean(x)
     mls[i] = median(x) }
msemoments = mean(moments^2)
msemls = mean(mls^2)
```

Table 3.1. R-code for comparing the moment and maximum likelihood estimators.

34. Let $X_1, \ldots, X_n$ be a sample from the distribution with probability density p_θ given by

$$p_\theta(x) = \frac{1}{2}\theta e^{-\theta|x|} \quad \text{for } x \in (-\infty, \infty),$$

where $\theta > 0$ is an unknown parameter. To determine a Bayes estimator for the parameter, we use a gamma distribution with fixed parameters $r, \lambda > 0$ as prior distribution.
  (i) Suppose that we have very little prior knowledge of θ. Explain how you would choose the parameters λ and r in this case.
  (ii) Determine the Bayes estimator for θ based on this sample, for general λ and r.

35. Determine the posterior distribution and the Bayes estimator for θ based on an observation X with negative binomial distribution with parameters r (known) and θ, with respect to a beta prior distribution.

36. Compute the Bayes estimator for θ based on a sample $X_1, \ldots, X_n$ from the U$[0, \theta]$-distribution with respect to a U$[0, M]$ prior distribution.

37. Compute the Bayes estimator for θ based on an observation X from the Poisson distribution with parameter θ with respect to a gamma distribution with parameters α and λ,
  (i) for $\alpha = 1$,
  (ii) for general $\alpha > 0$.

38. Compute the posterior distribution and Bayes estimator for θ based on a sample $X_1, \ldots, X_n$ from the distribution with probability density

$$p_\theta(x) = 2\theta x e^{-\theta x^2} \quad \text{for } x > 0$$

and 0 elsewhere, with respect to the gamma distribution with parameters α and λ.

39. Compute the posterior distribution and Bayes estimator for θ based on a sample $X_1, \ldots, X_n$ from the $N(\theta, 1)$-distribution with respect to an $N(0, \tau^2)$ prior distribution. Which estimator do we find as $\tau \to \infty$? How can we characterize the prior distribution for $\tau \approx \infty$?

40. Let $X_1, \ldots, X_n$ be a sample from a Bernoulli distribution with unknown parameter $p \in [0, 1]$. We want to give a Bayesian estimate for the variance $\text{var}_p(X_i) = p(1 - p)$ with respect to a beta(α, β) prior distribution with parameter p.
  (i) Determine the posterior density for p.
  (ii) Determine the Bayes estimators for p and $\text{var}_p(X_i)$.

41. Suppose that instead of the mean square error, we use the *mean absolute deviation* (MAD) to define a Bayes estimator: in Section 3.5, we replace $R(\pi; T)$ with $\int E_\theta |T - \theta| \pi(\theta) \, d\theta$ and define a Bayes estimator to be an estimator T that minimizes this expression. Show that in this case, the median of the posterior distribution is a Bayes estimator.

42. Let $Y = f(X)$ be a function of a random vector X with distribution Υ, and let $Q(y, B) = P(X \in B | Y = y)$ be the conditional distribution of X given $Y = y$. If we generate X from Υ, compute $Y = f(X)$, and then generate Z from the probability density $Q(Y, \cdot)$, then Z has distribution Υ.
    (i) Prove this.
    (ii) Apply this with $f(x) = (x_1, \dots, x_{i-1}, x_{i+1}, \dots, x_m)$ to prove that the Gibbs sampler has stationary density v.

Parents with blue eyes have children with blue eyes. On the other hand, parents with obesity do not necessarily have children with obesity. Some characteristics, like eye color, are determined fully genetically and are fixed at birth. Other characteristics, like having obesity, are only partially genetically determined and are also influenced by environmental factors like diet and lifestyle. Studies involving identical and fraternal twins can provide insight into degree to which characteristics in people are determined by genetic or environmental factors, or an interaction between the two.

Identical (monozygotic) twins occur when during the first cell division of a fertilized egg, two separate groups of cells form that each grow into an embryo. Identical twins are identical genetically and therefore always have the same gender. Fraternal (dizygotic) twins occur when the mother has a double ovulation and both eggs are fertilized. On average, fraternal twins have 50% of their genetic material in common; genetically, they are simply siblings. Twins usually grow up in the same family, go to the same school, and have the same lifestyle; they are exposed to more or less the same environmental factors. If for some characteristic, the correlation is higher in pairs of identical twins than it is in pairs of fraternal twins, then this difference can be attributed to the degree to which the genetic material corresponds; indeed, the environmental factors are virtually identical. The characteristic is therefore partially genetic. If, on the other hand, the correlations are more or less the same (and nonzero), then the characteristic is determined mostly by environmental factors.

The Netherlands Twin Register (see www.tweelingenregister.org/en) contains data on twins and their family members for the sake of scientific research in the fields of health, lifestyle, and personality. The register contains, among other things, the heights of the twins. Based on this data, we want to obtain an indication of the degree in which individual differences in adult height are determined genetically.

On average, men are taller than women. In doing research into the hereditary component in height, we must therefore take the gender into account. To simplify the notation, we restrict ourselves to male identical and fraternal twins; extending this to female or mixed-gender twins is simple as far as the method is concerned, but greatly complicates the notation. We denote the heights of a pair of adolescent male twins by (X_1, X_2) and suppose that the heights X_1 and X_2 can be written as sums

$$X_1 = \mu + G_1 + C + E_1$$
$$X_2 = \mu + G_2 + C + E_2$$

of an average height μ and three random components that represent the deviation from the average height of the male population by genetic influences (G_1 and G_2), by environmental factors that the twins have in common (C), and by factors specific to the individuals, both genetic and environmental (E_1 and E_2). We often assume that the variables for the genetic, environmental, and individually specific factors are independent from one another: $(G_1, G_2), C$, and (E_1, E_2) are independent. This *means that we assume that there is no interaction between the environmental and genetic factors (it is doubtful that this is true for the height).*

We assume that G_1 and G_2 are equally distributed with expectation 0 and unknown variance σ_g^2. These variables describe the genetic factors influencing the variation of the height. In twins, the genetic material is partially or fully equal; G_1 and G_2 are therefore correlated. Identical twins are genetically identical; for them, $G_1 = G_2$ (with probability 1), and the correlation between G_1 and G_2 is equal to $\text{cor}(G_1, G_2) = 1$. Fraternal twins only share part of their genes, so that G_1 and G_2 are not equal to each other, but are correlated. On average, fraternal twins have 50% of their genetic material in common. Under the assumption of the additive model given above (and a few other assumptions), we can show that in fraternal twins, the correlation between G_1 and G_2 is equal to $\text{cor}(G_1, G_2) = 1/2$. The individually specific factors E_1 and E_2 are assumed to be independent and equally distributed, with expectation 0 and unknown variance σ_e^2. The expectation and variance of C are 0 and σ_c^2, respectively. Under the assumptions made above, X_1 and X_2 are equally distributed with expectation $\mathbb{E}X_i = \mu$ and variance $\text{var}\, X_i$ equal to

$$\sigma^2 := \text{var}(\mu + G_i + C + E_i)$$
$$= \text{var}\, G_i + \text{var}\, C + \text{var}\, E_i$$
$$= \sigma_g^2 + \sigma_c^2 + \sigma_e^2, \qquad i = 1, 2,$$

where the second equality holds because of the independence of the various components.

The term $h^2 := \text{var}\, G_i / \text{var}\, X_i = \sigma_g^2 / \sigma^2$ is also called the "heritability." The heritability describes to what degree the variation in, in this case, the height of individuals, is due to genetic differences. Heritability is at least 0 and at most 1, because $0 \leq \sigma_g^2 \leq \sigma^2$. If the heritability of the height is equal to 1, then $\sigma_g^2 = \sigma^2$ and σ_c^2 and σ_e^2 must both equal 0. Because the expected values of C, E_1, and E_2 are also equal to 0, we see that C, E_1, and E_2 are equal to 0 with probability 1. The variation in the height of individuals is then completely due to genetics. If the heritability is equal to 0, then $\sigma_g^2 = 0$ and G_1 and G_2 are equal to 0 with probability 1; the variation in the body length of individuals is then not due to genetics at all.

The aim is to estimate h^2 based on a sample of heights of identical and fraternal twins. To do this, we first write h^2 in terms of the correlations between the heights within pairs of identical and fraternal twins, and estimate these parameters using the sample correlations. The correlations between the heights within both pairs of identical and fraternal twins are equal to

$$\frac{\text{cov}(X_1, X_2)}{\sqrt{\text{var}\, X_1 \, \text{var}\, X_2}} = \frac{\text{cov}(\mu + G_1 + C + E_1, \mu + G_2 + C + E_2)}{\sqrt{\text{var}\, X_1 \, \text{var}\, X_2}}$$
$$= \frac{\text{cov}(G_1, G_2)}{\sigma^2} + \frac{\text{cov}(C, C)}{\sigma^2} = \frac{\text{cov}(G_1, G_2)}{\sigma^2} + \frac{\sigma_c^2}{\sigma^2},$$

where the second equality follows from the independence assumptions made earlier. The covariance of the genetic components G_1 and G_2 within pairs of identical twins

101

is equal to $\mathrm{cov}(G_1, G_2) = \mathrm{var}\, G_1 = \sigma_g^2$ *because* $G_1 = G_2$ *with probability 1. Within pairs of fraternal twins, this covariance is equal to*

$$\mathrm{cov}(G_1, G_2) = \mathrm{cor}(G_1, G_2)\sqrt{\mathrm{var}\, G_1 \,\mathrm{var}\, G_2} = \frac{1}{2}\,\mathrm{var}\, G_1 = \frac{1}{2}\sigma_g^2.$$

It follows from these calculations that the correlations ρ_1 *and* ρ_2 *between identical and fraternal twins, respectively, are equal to*

$$\rho_1 = \frac{\sigma_g^2}{\sigma^2} + \frac{\sigma_c^2}{\sigma^2}, \qquad\qquad \rho_2 = \frac{\sigma_g^2}{2\sigma^2} + \frac{\sigma_c^2}{\sigma^2}.$$

It immediately follows that $\rho_1 \geq \rho_2$, *with equality if* $\sigma_g^2 = 0$. *In other words, the correlation between the heights within pairs of identical twins is greater than or equal to the correlation between the heights within pairs of fraternal twins. Equality occurs only if there is no genetic influence on the variation in the heights, and the difference is maximal if the variation in the heights is fully due to genetics, that is, if* $\sigma_c^2 = 0$.

It follows from the expressions for the correlations ρ_1 *and* ρ_2 *that the heritability is equal to*

$$h^2 = \frac{\sigma_g^2}{\sigma^2} = 2(\rho_1 - \rho_2).$$

To estimate h^2, *we can estimate* ρ_1 *and* ρ_2 *using their sample correlations,*

$$r_{X_1, X_2} = \frac{\sum_{i=1}^{n}(X_{1,i} - \overline{X}_1)(X_{2,i} - \overline{X}_2)}{(n-1)\sqrt{S_{X_1}^2}\sqrt{S_{X_2}^2}}$$

based on only identical and fraternal twins, respectively. In this formula, $X_{1,i}$ *and* $X_{2,i}$ *denote the first and second individual of the* i*th pair of identical or fraternal twins, respectively,* $\overline{X}_1$ *and* $\overline{X}_2$ *are the respective sample means of the first and second individuals within the pairs of identical or fraternal twins, and* $S_{X_1}^2$ *and* $S_{X_2}^2$ *are the corresponding sample variances. Since the marginal distribution for the height of all individuals in the data set is equal, it makes sense to replace* $\overline{X}_1$ *and* $\overline{X}_2$ *by the average height of all individuals, both identical and fraternal twins, and both the first and the second individual in a pair. The same can be considered for the sample variations in the denominator of* r_{X_1, X_2}. *This method for estimating the heritability has many similarities with the method of moments; namely, the unknown parameters are found by setting a theoretical quantity, in this case the correlation, equal to the sample value of this same quantity.*

Figures 2.11 and 2.12 show the heights of identical (Figure 2.11) and fraternal (Figure 2.12) twins set out against each other. It is clear that the correlation between the heights within pairs of identical twins is greater than that within pairs of fraternal twins. The sample correlations for identical twins are equal to 0.87 and 0.96 for male and female twins, respectively, and those for fraternal twins are equal to 0.55 and 0.50 for male and female twins, respectively. The heritability is estimated to be 0.64 for men and 0.92 for women.[b]

---

[b] The data can be found on the book's webpage at http://www.aup.nl under twindata.

Another method for estimating the heritability is the maximum likelihood method. Assume that the heights (X_1, X_2) of male adult twins have a 2-dimensional normal distribution (for information on the multi-dimensional normal distribution, see Appendix B) with expectation vector $\nu = (\mu, \mu)^T$ and covariance matrices Σ_1 and Σ_2 for identical and fraternal twins, respectively, where

$$\Sigma_1 = \begin{pmatrix} \sigma^2 & \sigma_g^2 + \sigma_c^2 \\ \sigma_g^2 + \sigma_c^2 & \sigma^2 \end{pmatrix} \qquad \Sigma_2 = \begin{pmatrix} \sigma^2 & \frac{1}{2}\sigma_g^2 + \sigma_c^2 \\ \frac{1}{2}\sigma_g^2 + \sigma_c^2 & \sigma^2 \end{pmatrix}$$

with $\sigma^2 = \sigma_g^2 + \sigma_c^2 + \sigma_e^2$. The diagonal elements of the covariance matrices are equal to the variances of X_1 and X_2; the other two terms are equal to the covariance of X_1 and X_2. The probability density of the height of a pair of twins is equal to

$$x \mapsto \frac{1}{2\pi\sqrt{\det \Sigma}} e^{-\frac{1}{2}(x-\nu)^T \Sigma^{-1}(x-\nu)},$$

where $x = (x_1, x_2)^T$ and Σ equals Σ_1 or Σ_2 depending on the type of twin, while $\nu = (\mu, \mu)^T$ is the vector described earlier. Moreover, $\det \Sigma$ denotes the determinant of Σ. We assume that the heights of different pairs of twins are independent, so that the likelihood is equal to a product of 2-dimensional densities and the log-likelihood is equal to

$$l_{\mu,\sigma_g^2,\sigma_c^2,\sigma_e^2}(X_1, \ldots, X_{n_1}, Y_1, \ldots, Y_{n_2}) =$$
$$- (n_1 + n_2) \log 2\pi - \frac{n_1}{2} \log(\det \Sigma_1) - \frac{n_2}{2} \log(\det \Sigma_2)$$
$$- \frac{1}{2} \sum_{i=1}^{n_1} (X_i - \nu)^T \Sigma_1^{-1} (X_i - \nu) - \frac{1}{2} \sum_{i=1}^{n_2} (Y_i - \nu)^T \Sigma_2^{-1} (Y_i - \nu),$$

with $X_1, \ldots, X_{n_1}$ the heights of the pairs of identical twins and $Y_1, \ldots, Y_{n_2}$ those of the pairs of fraternal twins. So we have $X_i = (X_{i,1}, X_{i,2})^T$ and $Y_i = (Y_{i,1}, Y_{i,2})^T$ with $X_{i,1}$ and $X_{i,2}$ the heights of the first and second individuals in the ith pair of identical twins, and likewise for Y_i. Maximizing the log-likelihood for $(\mu, \sigma_g^2, \sigma_c^2, \sigma_e^2)$ over the parameter space $[0, \infty)^4$ gives the maximum likelihood estimates. The heritability σ_g^2/σ^2 is estimated by substituting the estimates of σ_g^2 and σ^2 in the definition of h^2: $\hat{h}^2 = \hat{\sigma}_g^2/\hat{\sigma}^2 = 0.61$.

We can carry out the same computations for female pairs of identical and fraternal twins. This gives an estimated heritability of 0.93. When a joint likelihood is set up for the data on men and women, the assumption is often made that the expected height of women is different from that of men, but that the covariance matrices, and therefore the heritability, are equal. Maximizing the likelihood for the height for men and women gives an estimate of 0.79 for the heritability.

Heritability is a measure for the variation of a characteristic within a population, in our case height. That the heritability is large does not mean that the height is determined completely genetically; it does mean that the variation of the height within the population giving our data, is predominantly determined by differences in genetic material. Environmental factors certainly influence height (see Example 1.5), but they are probably so uniform over the population from which the data was drawn, that only genetic differences are observable in the variation of the height.

4 Hypothesis Testing

4.1 Introduction

In scientific research, in the industry, and in daily life, we often want to check whether certain questions have an affirmative answer or not. Does a particular type of therapy help? Does the age or gender of the patient play a role? Is one type of car safer than another? Does a batch contain an excessive number of defective items? Does one type of lamp have a longer life span than another? Does the DNA profile of the suspect correspond to the DNA profile found at the crime scene? Are the log returns of stock market values on different days independent? Etcetera.

Answers to such questions are based on the results of experiments or studies. In many cases, however, the results of those experiments do not lead to an unequivocal answer. If a new form of therapy is tested on 100 patients and gives good results in 64 of them, while this only holds for 50% of the patients with the old therapy, is the new therapy truly better than the old one, or were we just "lucky"? If 75 of the 100 patients improve, then we can no longer talk of luck, or can we? Is a sample correlation coefficient of 0.17 "significantly" different from 0?

The theory of testing is aimed at formalizing this type of decision-making process where we must choose between two conflicting hypotheses.

4.2 Null Hypothesis and Alternative Hypothesis

The decision between conflicting hypotheses is based on a suitable statistical model for the observation X. The hypotheses are coded in parameter values that index the probability distributions in the statistical model. Here, we will restrict ourselves to two hypotheses. The parameter θ belongs either to a set Θ_0 corresponding to the one hypothesis or to the complement $\Theta_1 = \Theta \setminus \Theta_0$, where $\Theta = \Theta_0 \cup \Theta_1$ is a disjoint partition of the full parameter space Θ. We call the hypothesis $H_0: \theta \in \Theta_0$ the *null hypothesis* and the hypothesis $H_1: \theta \in \Theta_1$ the *alternative hypothesis*.

In the standard approach to testing (followed by most users of statistics), the null and alternative hypotheses are not treated symmetrically. We, in particular, want to know whether the alternative hypothesis is correct. If the data do not give sufficient indication to support this, this does not necessarily imply that the alternative hypothesis is incorrect (and the null hypothesis correct); it is also possible that there is not sufficient proof for either of the hypotheses. The statistical analysis can thus lead to two conclusions:

- Reject H_0 (and accept H_1 as being correct).
- Do not reject H_0 (but do not accept H_0 as being correct).

The first is a strong conclusion, the second is not truly a conclusion. The second should be seen as the statement that more information is needed to reach a conclusion.

By basing our statements concerning the hypotheses on our observations, we can make two types of mistakes, corresponding to mistakenly coming to one of the two possible conclusions:

- A *type I error* consists of rejecting H_0 when it is correct.
- A *type II error* consists of not rejecting H_0 when it is incorrect.

A type I error corresponds to falsely choosing the strong conclusion. This is very undesirable. A type II error corresponds to falsely choosing the weak conclusion. This is also undesirable, but since the weak conclusion is not truly a conclusion, it is not as bad. Because of the asymmetric handling of the hypotheses H_0 and H_1 when choosing a test, we should not attach too much value to not rejecting H_0. It is therefore of great importance to choose the null hypothesis and the alternative hypothesis wisely. *In principle, we choose the statement we want to show as the alternative hypothesis.* We then argue for H_0: we only reject H_0 if there is strong evidence against it.

Example 4.1 Binomial test

Suppose that we wish to compare a new therapy against depression to an existing therapy. This existing therapy is successful in only half of the cases. Let p be the probability of success for the new therapy when applied to an arbitrary patient. Since we are only interested in the new therapy when it is better than the old one, we compare the unknown probability of success p of the new therapy to 0.5, the (known) probability of success of the existing therapy. We want to "prove" that the new therapy is better than the old one. We therefore take the statement "$p > 0.5$" as the alternative hypothesis. The null and alternative hypotheses are then $H_0: p \leq 0.5$ and $H_1: p > 0.5$.

When we can reject H_0, we assume that the new therapy is better than the existing one.

Example 4.2 Multinomial distribution

When rolling dice that are not fair, the probabilities $p_1, \ldots, p_6$ for throwing the different face values are not all exactly equal to $1/6$. The face value thrown at each roll X is in general multinomially distributed with parameters $(1, \theta)$, with $\theta = (p_1, \ldots, p_6)$. In the statistical model, we can take the parameter space for θ equal to $\Theta = \{(p_1, p_2, p_3, p_4, p_5, p_6) \in [0, 1]^6 : \sum_{i=1}^{6} p_i = 1\}$.

Suppose that we do not trust our opponent's dice in a game of backgammon. We suspect that he has tampered with the probabilities of the different outcomes. The null hypothesis to formally test whether a die is crooked, is then $H_0 : p_i = 1/6$ for $i = 1, \ldots, 6$, and the alternative hypothesis is $H_1 : p_i \neq 1/6$ for *at least one* $i \in \{1, \ldots, 6\}$. The null hypothesis space Θ_0 is then a subset of Θ consisting of one point: $\Theta_0 = \{(1/6, 1/6, 1/6, 1/6, 1/6, 1/6)\}$. When we are only interested in the outcome consisting of the value 6, we can test the null hypothesis $H_0 : p_6 = 1/6$ against $H_1 : p_6 \neq 1/6$. In that case, the null hypothesis space is equal to $\Theta_0 = \{(p_1, p_2, p_3, p_4, p_5, 1/6) \in [0, 1]^6 : \sum_{i=1}^{5} p_i = 5/6\}$.

Example 4.3 Two samples

Figure 4.1 shows boxplots for the level of expression of a gene in different types of tumors. The samples consist of 26 and 15 tumors, respectively. The question is whether the expression of the gene is greater in one type of tumor than in the other.

The boxplot does not directly answer this question. Although the box of the second sample is higher than that of the first, there is a clear overlap and the spread of the second sample clearly lies within the spread of the first sample. The latter may be significant, but may also be due to the different sizes of the samples.

A formal test can help answer the question. A reasonable statistical model is that the two samples $X_1, \ldots, X_{26}$ and $Y_1, \ldots, Y_{15}$ are independent samples from normal distributions with respective parameters (μ, σ^2) and (ν, τ^2). We want to test the null hypothesis $H_0 : \mu = \nu$ against the alternative hypothesis $H_1 : \mu \neq \nu$. We can take the parameter equal to $\theta = (\mu, \nu, \sigma^2, \tau^2)$, with parameter space $\Theta = \mathbb{R}^2 \times (0, \infty)^2$. The null hypothesis space is the subset $\Theta_0 = \{(\mu, \mu) : \mu \in \mathbb{R}\} \times (0, \infty)^2$.

4.3 Sample Size and Critical Region

Based on the observation X, we must decide whether there is sufficient evidence against the null hypothesis H_0, so that we want to reject H_0 and view the statement of the alternative hypothesis as the correct one. The values of X for which the evidence is strong enough form the critical region K. For these values of X, we have sufficient confidence in the alternative hypothesis to reject H_0.

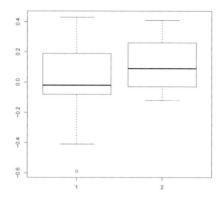

Figure 4.1. Boxplots of the measure of expression of a gene measured in two groups of 26 (on the left) and 15 (on the right) tumors.

Definition 4.4 Statistical test

Given a null hypothesis H_0, a statistical test consists of a set K of possible values for the observation X, the critical region. Suppose that we have an observation x. If $x \in K$, we reject H_0; if $x \notin K$, we do not reject H_0.

When $X = (X_1, \ldots, X_n)$ is a vector of observations, in particular, it is often difficult to decide based on X whether the statement of the alternative hypothesis can be true. We therefore often summarize the data in a *test statistic*. A test statistic is a real-valued quantity $T = T(X)$ based on the data that gives information on the correctness of the null and alternative hypotheses; so the test statistic does not depend on the unknown parameter.

Example 4.5 Binomial test, continued from Example 4.1

Example 4.1 describes a test situation. We want to test whether the probability of success p of a new therapy is higher than 0.5, the probability of success of the existing therapy. In all, 100 patients received the new therapy. Let X be the number of patients for whom the new therapy is successful, and assume that X has the $\mathrm{bin}(100, p)$-distribution.

It makes sense to take $T(X) = X$ as test statistic and to choose the critical domain of the form

$$K = \{c, c + 1, \ldots, 100\},$$

where we still need to determine the value c. Namely, a large value of X gives an indication that H_0 may be incorrect. The value c should therefore be chosen such that we do not have sufficient confidence in the correctness of the statement of the null hypothesis if $x \geq c$.

108

The critical region K is often of the form $\{x\colon T(x) \in K_T\}$, or $\{T \in K_T\}$ for short, for a test statistic T and a set K_T in the codomain of T. In practice, the set K_T is often also called the critical region. How to determine the critical region K or K_T is discussed in the next section.

Example 4.6 Gauss test

Let $X_1, \ldots, X_n$ be a sample from the normal distribution with unknown expectation μ and known variance σ^2. We want to test the null hypothesis $H_0\colon \mu \leq \mu_0$ against the alternative hypothesis $H_1\colon \mu > \mu_0$, for μ_0 a fixed number, for example $\mu_0 = 0$. This problem comes up, for example, with the quality control of products in a factory. Since the manufacturer finds it too expensive to control all products, the quality is measured for a sample. Earlier research has shown that the measure of quality is normally distributed. The manufacturer wants to verify that the average quality of the total production is greater than μ_0. (Assuming σ^2 known is unrealistic, but simplifies the example. In practice, σ^2 is assumed unknown, and the t-test from Example 4.30 is almost always used.) The average $\overline{X}$ is the maximum likelihood estimator for μ and can therefore be used to give an idea of the correctness of the null and alternative hypotheses. If the observed average $\overline{x}$ is greater than μ_0, this indicates that the alternative hypothesis may be true, and the greater $\overline{x}$, the stronger this indication. We can therefore use the average $\overline{X}$ as test statistic, and we reject H_0 for large values of this test statistic. The critical region is then of the form

$$ K = \{(x_1, \ldots, x_n)\colon \overline{x} \geq c\} $$

for some c. But, how large must we take c to have sufficient confidence in the correctness of the alternative hypothesis if $\overline{x} \geq c$ and have a sufficiently small probability of making a type I error?

Suppose that a statistical test has critical region $K = \{x\colon T(x) \in K_T\}$, where T is a test statistic and K_T is a subset of the codomain of T. The set K_T depends on the choice of the test statistic T. In general, another test statistic T' leads to a different set $K_{T'}$. However, the critical region K can be the same in both cases; the same critical region K can correspond to two different test statistics (see Exercise 4.10).

4.3.1 Size and Power Function

When, in testing $H_0\colon \theta \in \Theta_0$ against $H_1\colon \theta \in \Theta_1$, the true value of θ belongs to Θ_0, the null hypothesis is true. If in that case $x \in K$, then we falsely reject H_0 and make a type I error. For a good test, the probability $P_\theta(X \in K)$ for $\theta \in \Theta_0$ must therefore be small. On the other hand, when the null hypothesis is false ($\theta \in \Theta_1$), we want $P_\theta(X \in K)$ to be large. The quality of a test can therefore be measured using the function $\theta \mapsto P_\theta(X \in K)$.

Definition 4.7 Power function

The *power function* of a test with critical region K is

$$\theta \mapsto \pi(\theta; K) = P_\theta(X \in K).$$

We are looking for a critical region for which the power function takes on "small values" (close to 0) when $\theta \in \Theta_0$, and "large values" (close to 1) when $\theta \in \Theta_1$. Figure 4.2 shows the power functions of two tests (as functions of θ along the horizontal axis), an "ideal test" with probability of both types of errors equal to 0 and a real test.

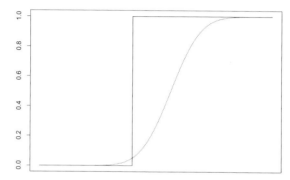

Figure 4.2. Power function of an ideal test (solid) and of a real test (dotted). The parameter spaces under the null and alternative hypotheses (Θ_0 and Θ_1) are the sections of the horizontal axis where the power function of the ideal test is equal to 0 and 1, respectively.

Definition 4.8 Size

The *size* of a test with critical region K with power function $\pi(\cdot; K)$ is the number

$$\alpha = \sup_{\theta \in \Theta_0} \pi(\theta; K).$$

A test has *significance level* or *level* α_0 if $\alpha \le \alpha_0$.

The asymmetry between the two hypotheses is now made formal by an assumption that ensures that the probability of a type I error is at most α_0.

Convention 4.9

In every practical situation, we first choose a fixed number α_0, the *level*. We then only use tests of level α_0. In other words, we only allow tests whose power function $\pi(\cdot; K)$ under the null hypothesis is at most α_0:

$$\sup_{\theta \in \Theta_0} \pi(\theta; K) \le \alpha_0.$$

It seems appealing to choose the level α_0 extremely small, so that making a type I error is rare. We can only achieve this by making K very small. In that case, however, the power function for $\theta \in \Theta_1$ also becomes small. The probability of a type II error,

$$P_\theta(X \notin K) = 1 - \pi(\theta; K), \qquad \theta \in \Theta_1,$$

therefore becomes very large, which is also inadvisable. The requirements for making both the type I and type II errors small work against each other. We do not treat the two types of errors symmetrically; for example, we do not try to minimize the sum of the maximum probabilities of errors of types I and II.

In practice, α_0 is often chosen equal to the magical number 0.05. With this choice, it should not surprise us that if we carry out many tests, 1 out of 20 times, we will falsely reject the null hypothesis (making a type I error). We should, in fact, choose α_0 depending on the possible consequences of a type I error. If these are disproportionately serious, $\alpha_0 = 0.05$ may be much too large.

As far are type I errors are concerned, we see Convention 4.9 as giving sufficient guarantee that the probability of these is small. Many tests (possibly with different test statistics) will satisfy this condition. Of these tests, we prefer the test with the smallest probability of making a type II error. How small this probability is depends on the situation, among other things on the number of observations and the chosen level α_0. If the probability of making a type II error is too large, the test is, of course, not very meaningful, because we then almost never reject H_0 and instead choose the second, weak (non-)conclusion.

Convention 4.10

Given the level α_0, we prefer a test of level α_0 with the greatest possible power function $\pi(\theta; K)$ for $\theta \in \Theta_1$.

Under this assumption, for a given level α_0, we prefer a test with critical region K_1 to a test with critical region K_2 if both have level α_0 and the first has a greater power function than the second for all $\theta \in \Theta_1$:

$$\sup_{\theta \in \Theta_0} \pi(\theta; K_i) \le \alpha_0, \quad i = 1, 2 \quad \text{and} \quad \pi(\theta; K_1) \ge \pi(\theta; K_2), \quad \forall \theta \in \Theta_1,$$

with strict inequality for at least one $\theta \in \Theta_1$. We call the test with critical region K_1 *more powerful* than the test with critical region K_2 in some $\theta \in \Theta_1$ if $\pi(\theta; K_1) > \pi(\theta; K_2)$. We call the test with critical region K_1 *uniformly more powerful* if the inequality holds for all $\theta \in \Theta_1$. In principle, we are now looking for the *uniformly most powerful test* of level α_0; this is a test whose power function (at a given level) is maximal for all $\theta \in \Theta_1$. We are comparing two functions, and it is possible that one test is more powerful for certain $\theta \in \Theta_1$, and the other test is more powerful for other $\theta \in \Theta_1$. It is then not immediately clear which test we should choose. We do not discuss this question in this book. In exceptional cases, a uniformly most powerful test exists for all tests of level α_0. There is then an absolutely best test. We will see examples of this in Chapter 6.

Example 4.11 Binomial test, continued from Example 4.5

In Example 4.5, we showed that for a test statistic $T(X) = X$, it makes sense to take the critical region of the form

$$K = \{c_{\alpha_0}, c_{\alpha_0} + 1, \ldots, 100\}.$$

The value c_{α_0} must be chosen such that the size of a test is at most α_0. The size of the test is given by

$$\alpha = \sup_{p \leq 0.5} P_p(X \geq c_{\alpha_0}) = P_{0.5}(X \geq c_{\alpha_0}).$$

The supremum is taken in $p = 0.5$, because as a function of p, the probability $P_p(X \geq c_{\alpha_0})$ is monotonically increasing. We can prove the latter analytically with some difficulty, but it is also clear intuitively. The function $p \mapsto P_p(X \geq c_{\alpha_0})$ has been drawn in Figure 4.3 for $c_{\alpha_0} = 59$.

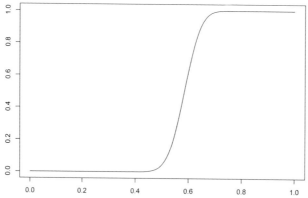

Figure 4.3. The function $p \mapsto P_p(X \geq 59)$ for X with a $\mathrm{bin}(100, p)$-distribution.

Suppose that we choose $\alpha_0 = 0.05$. If we take $c_{0.05} = 59$, then the size $\alpha = P_{0.5}(X \geq 59) = 0.044$ is less than $\alpha_0 = 0.05$, while for $c_{0.05} = 58$, the size satisfies $P_{0.5}(X \geq 58) = 0.067 > 0.05$. For $c_{0.05} \leq 58$, the test therefore does not have level 0.05 and is therefore not admissible at this value of the level. We must therefore choose $c_{0.05} \geq 59$. As an example, Figure 4.4 shows the function $x \mapsto P_{0.5}(X \geq x)$. According to Convention 4.10, we must choose the critical region such that the power function is the greatest possible. This corresponds to choosing the critical region as large as possible such that under H_1, the probability of (correctly) rejecting the null hypothesis, $P_p(X \in K)$, is the greatest possible. We therefore choose $K = \{59, 60, \ldots, 100\}$. Under all tests of the given form, this is the test of level 0.05 with the greatest power function. The function $p \mapsto P_p(X \geq 59)$ in Figure 4.3 is exactly the power function of this test.

If we find 64 successes using the new therapy, then H_0 is therefore rejected at level 0.05, and the conclusion is that the new therapy has a higher probability of success than the existing one. With 58 successes, we could not have drawn this conclusion: H_0 would then not have been rejected.

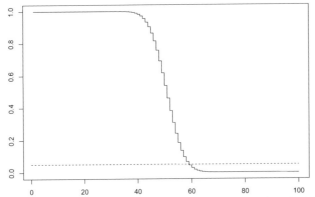

Figure 4.4. The function $x \mapsto P_{0.5}(X \geq x)$ for X binomially distributed with parameters 100 (and 0.5). This function is left continuous in points where x takes on a value in $\mathbb{N}$. The dashed horizontal line is at height 0.05.

In the case of a one-dimensional parameter θ, we speak of a *one-sided null hypothesis* when the null hypothesis is of the form $H_0: \theta \leq \theta_0$ or $H_0: \theta \geq \theta_0$, where θ_0 is a fixed number. The alternative hypothesis is then of the form $H_1: \theta > \theta_0$ or $H_1: \theta < \theta_0$, respectively. We call the first hypothesis *right one-sided* and the second hypothesis *left one-sided*. When the null and alternative hypotheses are of the form $H_0: \theta = \theta_0$ and $H_1: \theta \neq \theta_0$, respectively, we speak of a *two-sided null hypothesis*.

For a test statistic T, the critical region often takes on one of the following forms:

$$K_T = \{T \geq c_{\alpha_0}\},$$
$$K_T = \{T \leq c_{\alpha_0}\},$$
$$K_T = \{T \leq c_{\alpha_0}\} \cup \{T \geq d_{\alpha_0}\},$$

for numbers c_{α_0} and d_{α_0} with $c_{\alpha_0} < d_{\alpha_0}$ in the last critical region. Which form the critical region takes on depends on the chosen hypotheses and the choice of test statistic. The first two forms of K_T are called one-sided, the last two-sided. The numbers c_{α_0} and d_{α_0} are called the *critical values*. If the value of the test statistic surpasses the critical value, then the null hypothesis is rejected. Note that "to surpass" can mean both "to be greater than" and "to be less than," depending on the context and the test statistic. The Gauss test in Example 4.12 is an example of a test where a one-sided null hypothesis leads to a one-sided critical region K_T and a two-sided null hypothesis leads to a two-sided critical region K_T. This is, however, not true in general; the form of the critical region depends on the hypotheses and the choice of test statistic. In Section 4.7 (likelihood ratio tests) we see, for example, a two-sided null hypothesis with a one-sided critical region K_T.

Example 4.12 Gauss test, continued from Example 4.6

Let $X_1, \ldots, X_n$ be a sample from the $N(\mu, \sigma^2)$-distribution, where σ^2 is a known constant. Consider the problem of testing $H_0: \mu \leq \mu_0$ against $H_1: \mu > \mu_0$, where μ_0 is a fixed number (for example $\mu_0 = 0$).

We saw in Example 4.6 that the average $\overline{X}$ might be a suitable test statistic. However, it turns out to be better to normalize this quantity to

$$T = \sqrt{n}\frac{\overline{X} - \mu_0}{\sigma},$$

so that under the assumption $\mu = \mu_0$, the quantity T has the $N(0,1)$-distribution. Both μ_0 and σ^2 are known, so that T is indeed a test statistic. The test statistic $T' = \overline{X}$ leads to another region $K_{T'}$, but to the same critical region K (see Exercise 4.10).

Large values of $\overline{X}$ (greater than μ_0) and therefore of T are more probable under H_1 than under H_0. After all, $\overline{X}$ is normally distributed with expectation μ and variance σ^2/n, and this distribution shifts to the right when μ increases. We therefore choose a critical region, based on the test statistic T, of the form $K = \{(x_1, \ldots, x_n): T \geq c_{\alpha_0}\}$. In the next two paragraphs, we argue that the correct choice for c_{α_0} is the $(1 - \alpha_0)$-quantile $\xi_{1-\alpha_0}$ of the standard normal distribution. (We denote by ξ_α the number for which $\Phi(\xi_\alpha) = \alpha$, where Φ is the standard normal distribution function.)

According to Convention 4.9, we are looking for a test of size at most α_0, that is,

$$(4.1) \qquad \sup_{\mu \leq \mu_0} P_\mu((X_1, \ldots, X_n) \in K) = \sup_{\mu \leq \mu_0} P_\mu(T \geq c_{\alpha_0}) \leq \alpha_0.$$

Since $\sqrt{n}(\overline{X} - \mu)/\sigma$, when μ is the true value of the parameter, has the standard normal distribution, we see that the probability $P_\mu(T \geq c_{\alpha_0})$ is equal to

$$P_\mu\left(\sqrt{n}\frac{\overline{X} - \mu_0}{\sigma} \geq c_{\alpha_0}\right) = P_\mu\left(\sqrt{n}\frac{\overline{X} - \mu}{\sigma} \geq c_{\alpha_0} + \sqrt{n}\frac{\mu_0 - \mu}{\sigma}\right)$$

$$= 1 - \Phi\left(c_{\alpha_0} + \sqrt{n}\frac{\mu_0 - \mu}{\sigma}\right).$$

This probability is an increasing function of μ (which is also evident intuitively from the fact that the normal distribution with expectation μ shifts to the right when μ increases), so that the supremum $\sup_{\mu \leq \mu_0} P_\mu(T \geq c_{\alpha_0})$ is taken on at the greatest possible value of μ, that is, $\mu = \mu_0$. Condition (4.1) that the size is at most α_0 reduces to

$$P_{\mu_0}(T \geq c_{\alpha_0}) \leq \alpha_0.$$

Since T has the standard normal distribution under the assumption that $\mu = \mu_0$, it follows that $c_{\alpha_0} \geq \xi_{1-\alpha_0}$.

Every critical region $K_T = [c_{\alpha_0}, \infty)$ with $c_{\alpha_0} \geq \xi_{1-\alpha_0}$ gives a size of at most α_0. Among these tests, we are now looking for the most powerful one; see Convention 4.10. This is, of course, the test with the largest critical region, that is, with the smallest possible critical value c_{α_0}. In combination with the inequality of the previous paragraph, we take $c_{\alpha_0} = \xi_{1-\alpha_0}$. Note that the size is now exactly equal to the level α_0.

In summary, the test rejects the null hypothesis $H_0: \mu \leq \mu_0$ for values of $(X_1, \ldots, X_n)$ such that $T = \sqrt{n}(\overline{X} - \mu_0)/\sigma \geq \xi_{1-\alpha_0}$. This is the usual test for this problem, the *Gauss test* (named after the mathematician who was one of the first to work with the normal distribution). The corresponding critical region is equal to

$$K = \{(x_1, \ldots, x_n): T \in K_T\} = \left\{(x_1, \ldots, x_n): \sqrt{n}\frac{\overline{x} - \mu_0}{\sigma} \geq \xi_{1-\alpha_0}\right\}.$$

The set K_T is therefore equal to $[\xi_{1-\alpha_0}, \infty)$. Note that the resulting critical value $c_{\alpha_0} = \xi_{1-\alpha_0}$ does not depend on the values of μ_0 and σ^2. The same critical region $K_T = [\xi_{1-\alpha_0}, \infty)$ is found for all values of μ_0 and σ^2. This is the advantage of taking the normalized T over $\overline{X}$ as test statistic. It is therefore common to use the normalized test statistic for the Gauss test. The set $K_T = [\xi_{1-\alpha_0}, \infty)$ is often called the critical region of a right-sided Gauss test.

We can test the null hypothesis $H_0 : \mu \geq \mu_0$ against the alternative hypothesis $H_1 : \mu < \mu_0$ analogously. We use the same test statistic T. The null hypothesis H_0 is rejected at level α_0 if $T = \sqrt{n}(\overline{X} - \mu_0)/\sigma \leq \xi_{\alpha_0} = -\xi_{1-\alpha_0}$.

The critical region for testing the null hypothesis $H_0 : \mu = \mu_0$ against the two-sided alternative $H_1 : \mu \neq \mu_0$ at level α_0 can be found by combining the critical regions of the two one-sided tests of size $\alpha_0/2$ each. This leads to rejecting the null hypothesis if $\sqrt{n}(\overline{X} - \mu_0)/\sigma \leq \xi_{\alpha_0/2}$ or $\sqrt{n}(\overline{X} - \mu_0)/\sigma \geq \xi_{1-\alpha_0/2}$ or, equivalently, if $\sqrt{n}|\overline{X} - \mu_0|/\sigma \geq \xi_{1-\alpha_0/2}$.

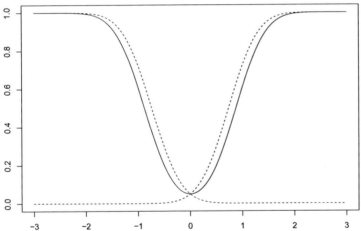

Figure 4.5. Power functions as functions of μ for the two one-sided Gauss tests (dashed) and the two-sided Gauss test (solid) for $\mu_0 = 0$ at $\alpha_0 = 0.05$ and $n = 5$.

Naturally, the value of the power function of the two-sided test is smaller than that of the left one-sided test for $\mu < \mu_0$ and smaller than that of the right one-sided test for $\mu > \mu_0$; see Figure 4.5. If we are only interested in one of these types of alternatives, then a suitable one-sided test is preferable to a two-sided test. This can be the case, for example, if we consider using a new production method or buying a new machine. In such a case, we are not so much interested in whether this innovation leads to a setback as whether we can expect an improvement. The choice between one-sided and two-sided tests therefore depends on the practical problem. If we want to take the idea behind "size" seriously, then we should not be influenced by the outcome of the experiments! In particular, it would be wrong to choose a right one-sided test after it has been established that $\overline{X} > \mu_0$.

Above, we introduced the Gauss test using an ad hoc argument. In addition to being intuitively reasonable, these tests are also the best possible. Indeed, we can prove that the one-sided Gauss tests are uniformly most powerful; that is, for these tests the power function is maximal in all possible values under the alternative hypothesis (see Section 6.4). The two-sided Gauss test is uniformly most powerful among the *unbiased tests*. Unbiased test are tests with $\pi(\theta_0) \le \alpha_0 \le \pi(\theta_1)$ for all $\theta_0 \in \Theta_0$ and $\theta_1 \in \Theta_1$, where α_0 is the level.

Example 4.13 Binomial test, continued from Example 4.11

Example 4.11 concerns a special case of the following binomial test. Suppose that for a fixed number $p_0 \in (0, 1)$, we want to test the null hypothesis $H_0: p \le p_0$ against $H_1: p > p_0$ based on a $\mathrm{bin}(n, p)$-distributed observation X. We choose X itself as test statistic and reject H_0 for large values of X. The critical region is therefore of the form $\{x \in \{0, 1, \ldots, n\}: x \ge c_{\alpha_0}\} = \{c_{\alpha_0}, \ldots, n\}$. We choose the critical value $c_{\alpha_0} \in \{0, \ldots, n\}$ such that the size of the test is less than or equal to α_0 and, under this secondary condition, the power function is maximal (compare with Example 4.11). The size of this test is equal to

$$\alpha = \sup_{p \le p_0} \mathrm{P}_p(X \ge c_{\alpha_0}) = \mathrm{P}_{p_0}(X \ge c_{\alpha_0}),$$

as the probability $\mathrm{P}_p(X \ge x)$ is increasing in p for fixed x. To make the power function as large as possible under the alternative hypothesis, we take the critical region as large as possible, that is, take the critical value as small as possible:

$$c_{\alpha_0} = \min\left\{t \in \{0, \ldots, n\}: \mathrm{P}_{p_0}(X \ge t) \le \alpha_0\right\}.$$

Naturally, we then have $\alpha \le \alpha_0$. Because of the jumps in the binomial distribution function, this inequality will be strict for most values of α_0.

For sufficiently large values of n, we can approximate the probability $\mathrm{P}_{p_0}(X \ge t)$ using the normal distribution, in which case the jumps in the distribution function of X are negligible. For the size of the binomial test, this gives

$$\alpha_0 \ge \mathrm{P}_{p_0}\left(X \ge c_{\alpha_0}\right) = \mathrm{P}_{p_0}\left(X \ge c_{\alpha_0} - \tfrac{1}{2}\right)$$

$$= \mathrm{P}_{p_0}\left(\frac{X - np_0}{\sqrt{np_0(1 - p_0)}} \ge \frac{c_{\alpha_0} - np_0 - \tfrac{1}{2}}{\sqrt{np_0(1 - p_0)}}\right) \approx 1 - \Phi\left(\frac{c_{\alpha_0} - np_0 - \tfrac{1}{2}}{\sqrt{np_0(1 - p_0)}}\right),$$

where the sign $\approx$ is necessary because of the approximation of the binomial distribution function by the normal distribution function and the term $1/2$ in the numerator is the continuity correction (see Appendix A). For given α_0, the value of c_{α_0} is the smallest integer for which

$$(4.2) \qquad \qquad \xi_{1-\alpha_0} \le \frac{c_{\alpha_0} - np_0 - \tfrac{1}{2}}{\sqrt{np_0(1 - p_0)}}.$$

It is evident how this one-sided test can be adapted to the case of a different one-sided problem, $H_1: p < p_0$, or to the two-sided problem $H_1: p \ne p_0$.

Example 4.14 Shifted exponential distribution

Let $X_1, \ldots, X_n$ be a sample from the shifted exponential distribution with intensity parameter 1 and unknown shift parameter $\theta \in (-\infty, \infty)$. The corresponding marginal density is given by $p_\theta(x) = e^{\theta-x}$ for $x \geq \theta$ and $p_\theta(x) = 0$ for $x < \theta$. Suppose that we wish to test the null hypothesis $H_0: \theta \leq 0$ against the alternative hypothesis $H_1: \theta > 0$ at level α_0. The maximum likelihood estimator for θ is given by the first order statistic $X_{(1)} = \min\{X_1, \ldots, X_n\}$ (see Exercise 3.16). It makes sense to use $X_{(1)}$ as test statistic T and to reject the null hypothesis for large values of T; indeed, when T is positive, this is an indication that the alternative hypothesis may be true. The critical region is therefore of the form $K = \{(x_1, \ldots, x_n): x_{(1)} \geq c_{\alpha_0}\}$. The next step is to determine the critical value c_{α_0} for which the size of the test is at most α_0 and the power function is maximal. The size of the test is given by

$$\sup_{\theta \leq 0} P_\theta((X_1, \ldots, X_n) \in K) = \sup_{\theta \leq 0} P_\theta(X_{(1)} \geq c_{\alpha_0}).$$

For $\theta < c_{\alpha_0}$, the probability $P_\theta(X_{(1)} \geq c_{\alpha_0}) = \left(P_\theta(X_1 \geq c_{\alpha_0})\right)^n = e^{n(\theta - c_{\alpha_0})}$ is increasing in θ. The supremum in the expression for the size of the test is taken in $\theta = 0$. The critical value c_{α_0} must now satisfy the inequality $e^{-nc_{\alpha_0}} \leq \alpha_0$, that is, $c_{\alpha_0} \geq -n^{-1} \log \alpha_0$. To maximize the power function, we must maximize the critical region. It follows that $c_{\alpha_0} = -n^{-1} \log \alpha_0$. The critical region is therefore equal to

$$K = \left\{(x_1, \ldots, x_n): x_{(1)} \geq -\frac{1}{n} \log \alpha_0\right\},$$

and the size of the test is exactly equal to α_0. The test rejects the null hypothesis when $X_{(1)} \geq -n^{-1} \log \alpha_0$. Note that $-n^{-1} \log \alpha_0 > 0$ for $\alpha_0 \in (0,1)$. Using the theory from Chapter 6, we can prove that the test above is uniformly most powerful. This means that for testing the null hypothesis $H_0: \theta \leq 0$ against the alternative hypothesis $H_1: \theta > 0$, for every value of $\theta > 0$, the test above is the most powerful among all tests of level α_0.

We could, of course, have chosen another test statistic, for example the method of moments estimator for θ, which is $\overline{X} - 1$, which leads to a different critical region. This test turns out to have a smaller power function for $\theta > 0$, which is why we do not prefer it.

4.3.2 Sample Size

The power function of a test, in general, depends strongly on the amount of available data. Obviously, with more data, we have a greater power function. Generally, with "infinitely much data," we can obtain the ideal power function from Figure 4.2. The null and alternative hypotheses can then be distinguished without any errors. In practical situations, we cannot avoid type I and type II errors, but we can positively influence the slope of the power function as in Figure 4.2 by basing the test procedure on more data.

117

In practice, this leads to the question of the so-called *minimal sample size*. This is the minimal size of the sample such that the corresponding test in a certain alternative $\theta \in \Theta_1$ has a greater power function than a given lower bound. It is clear from this description that the minimal sample size is only well defined if both the particular alternative θ and the desired probability of a type II error are fixed, in addition, of course, to the desired test size α. In most cases, this means that an honest statistician will not be able to give a simple answer to the question of what a minimal sample size is.

We illustrate this using several examples in which the computations are more or less explicit.

Example 4.15 Gauss test, continued from Example 4.12

The Gauss test rejects the null hypothesis $H_0\colon \mu \leq \mu_0$ for values of the test statistic $T = \sqrt{n}(\overline{X} - \mu_0)/\sigma$ greater than or equal to $\xi_{1-\alpha_0}$; the critical region for T is $K_T = [\xi_{1-\alpha_0}, \infty)$. The power function of the Gauss test is the function

$$\mu \mapsto \pi(\mu; K) = P_\mu\left(\sqrt{n}\frac{\overline{X} - \mu_0}{\sigma} \geq \xi_{1-\alpha_0}\right)$$

$$= P_\mu\left(\sqrt{n}\frac{\overline{X} - \mu}{\sigma} \geq \xi_{1-\alpha_0} - \sqrt{n}\frac{\mu - \mu_0}{\sigma}\right) = 1 - \Phi\left(\xi_{1-\alpha_0} - \sqrt{n}\frac{\mu - \mu_0}{\sigma}\right).$$

Using the fact that $x \mapsto \Phi(x)$ is a monotonically increasing function, and $\xi_{1-\alpha}$ is therefore decreasing in α, we deduce the following properties:

- the greater n, the greater the power function in $\mu > \mu_0$ (more information is available)
- the greater μ, the greater the power function in μ (μ lies further from the null hypothesis)
- the greater σ, the smaller the power function in $\mu > \mu_0$ (the greater spread in the observations makes it harder to say something about their expectation)
- the greater α_0, the greater the power function in $\mu > \mu_0$, but also the greater the probability of a type I error

Now, suppose that for a given level α_0 and a given alternative $\mu > \mu_0$, we want a power function of at least $1 - \beta$, that is, we want the probability of a type II error in μ to be less than β. It follows from the formula for the power function that this is the case provided

$$\Phi\left(\xi_{1-\alpha_0} - \sqrt{n}\frac{\mu - \mu_0}{\sigma}\right) \leq \beta,$$

that is, provided

$$\sqrt{n}\frac{\mu - \mu_0}{\sigma} \geq \xi_{1-\alpha_0} - \xi_\beta$$

with $\beta = \Phi(\xi_\beta)$. The minimal value of $\sqrt{n}$ satisfying this condition is equal to $(\xi_{1-\alpha_0} - \xi_\beta)\sigma/(\mu - \mu_0)$. Note that all natural choices for α_0 and β satisfy $1 - \alpha_0 > \beta$, so that $\xi_{1-\alpha_0} - \xi_\beta$ is positive.

Example 4.16 Binomial test, continued from Example 4.13

The standard test for the null hypothesis $H_0: p \leq p_0$ based on a variable X with the binomial distribution with parameters n and p rejects H_0 for values of X in the critical region $K = \{c_{\alpha_0}, \ldots, n\}$, where c_{α_0} can be approximated from the equation (4.2),

$$\frac{c_{\alpha_0} - np_0 - \frac{1}{2}}{\sqrt{np_0(1 - p_0)}} \approx \xi_{1-\alpha_0}.$$

The power function of the test is equal to the function

$$p \mapsto P_p(X \geq c_{\alpha_0}) \approx 1 - \Phi\left(\frac{c_{\alpha_0} - np - \frac{1}{2}}{\sqrt{np(1 - p)}}\right).$$

This function is sketched in Figure 4.6 for $n = 10$ and $n = 25$, with $\alpha_0 = 0.05$ and $p_0 = \frac{1}{2}$. It is clear that for $p > 0.5$, the power function for $n = 25$ is much greater than that for $n = 10$: when we have more observations, we can better determine whether H_1 is true and reject H_0 with a greater probability when H_1 is true. (Note that the size of the test for $n = 25$ is also slightly bigger. In both cases, we have chosen a value c_{α_0} satisfying our two conditions.)

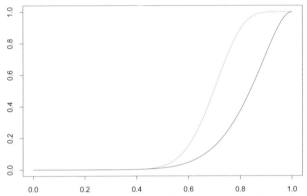

Figure 4.6. Power function of the test for $H_0: p \leq \frac{1}{2}$ at level $\alpha_0 = 0.05$ based on an observation from the binomial distribution for $n = 10$ (solid curve) and $n = 25$ (dotted curve).

The normal approximation is well suited to determine the minimal sample size for obtaining the prescribed power function. Suppose, for example, that we wish to test $H_0: p \leq \frac{1}{2}$ against $H_1: p > \frac{1}{2}$ at level $\alpha_0 = 0.05$, in such a way that the power function in $p = 0.6$ is at least 0.8. This leads to the system of equations

$$\frac{c_{0.05} - n0.5 - \frac{1}{2}}{\sqrt{n0.5(1 - 0.5)}} \approx \xi_{0.95} = 1.64,$$

$$\frac{c_{0.05} - n0.6 - \frac{1}{2}}{\sqrt{n0.6(1 - 0.6)}} \leq \xi_{0.2} = -0.84.$$

The equality gives $c_{0.05} \approx n/2 + 1.64\sqrt{n}/2 + 1/2$, and substituting this value for $c_{0.05}$ in the inequality gives $\sqrt{n} \geq 12.32$ and therefore $n \geq 152$.

Example 4.17 Shifted exponential distribution, continued from Example 4.14

Let $X_1, \ldots, X_n$ be a sample from a shifted exponential distribution with intensity parameter 1 and unknown shift parameter $\theta \in \mathbb{R}$. In Example 4.14, we deduced that the null hypothesis $H_0: \theta \leq 0$ is rejected when $X_{(1)} \geq -n^{-1} \log \alpha_0$, with α_0 the size of the test. We can determine the minimal sample size for a power function of at least 0.8 in $\theta = 0.1$. By expressing the distribution function of $X_{(1)}$ in the marginal distribution function (of X_1), we can determine the power function for every θ,

$$\pi(\theta, K) = P_\theta \left(X_{(1)} \geq -\frac{1}{n} \log \alpha_0 \right) = \left(P_\theta \left(X_1 \geq -\frac{1}{n} \log \alpha_0 \right) \right)^n = \alpha_0 e^{n\theta}.$$

The requirement that $\pi(0.1, K) \geq 0.8$ for $\alpha_0 = 0.05$ leads to the inequality $0.05 e^{n0.1} \geq 0.80$. It immediately follows that $n \geq 27.7$. ◻

Example 4.18 Application: contaminated pool water

The guidelines for the number of Legionella bacteria in pool water in the Netherlands is: *at most 100 colony-forming units of Legionella bacteria per liter*.[‡] Because the number of colony-forming units of bacteria cannot be determined exactly, in this example we use as norm that the probability of more than 100 colony-forming units of bacteria per liter should be at most 5%. In Example 3.19, we presented a procedure that can be used to estimate the number of colony-forming units of bacteria in (contaminated) water. Let X be the number of colony-forming units of Legionella bacteria in a 1 liter sample of pool water. We assume that X has a Poisson distribution with unknown parameter μ. For $p_\mu = P_\mu(X > 100)$, the norm can be tested formally using the hypotheses

$$H_0: p_\mu \leq 0.05 \qquad \text{and} \qquad H_1: p_\mu > 0.05.$$

The probability $p_\mu = P_\mu(X > 100)$ is monotonically increasing in μ. We have $P_\mu(X > 100) \leq 0.05$ for $\mu \leq 85.05$, while $P_\mu(X > 100) > 0.05$ for $\mu > 85.05$. Consequently, testing these hypotheses is equivalent to testing the hypotheses

$$H_0': \mu \leq 85.05 \qquad \text{against} \qquad H_1': \mu > 85.05.$$

We partition the sample mixed with 100 liters of pure water over 100 Petri dishes of 1 liter each. As in Example 3.19, we define X_i as the number of colony-forming units of bacteria in the ith liter and Y_i as the variable that indicates whether a colony forms in the ith Petri dish. We assume that the variables $X_1, \ldots, X_{100}$ are independent and Poisson-distributed with parameter $\mu/100$. The variables $Y_1, \ldots, Y_{100}$ are assumed to be identical and have the Bernoulli distribution with parameter $q_\mu = P_\mu(Y_i = 1) = 1 - e^{-\mu/100}$. Since q_μ is monotonically increasing in μ and $1 - e^{-85.05/100} = 0.5728$, the hypotheses H_0' and H_1' are equivalent to

$$H_0'': q_\mu \leq 0.5728 \qquad \text{and} \qquad H_1'': q_\mu > 0.5728.$$

[‡] Source: Decree "Besluit Hygiëne en Veiligheid Badinrichtingen en Zwemgelegenheden," January 27, 2011. In practice, things do not go exactly as described in this example; instead, *several* samples are taken at different places in the pool.

This null hypothesis H_0'' can be tested using the test statistic $T = \sum_{i=1}^{100} Y_i$, which is binomially distributed with parameters 100 and q_μ. The null hypothesis can therefore be tested with the one-sided binomial test, as described in Example 4.13. The normal approximation is justified because $100 \times 0.5728 \times (1 - 0.5728) = 24.47 > 5$ (see Appendix A). We can solve the critical value from equation (4.2). It follows that at level $\alpha_0 = 0.05$, the null hypothesis H_0'' is rejected when $\sum_{i=1}^{100} Y_i \geq 66$. When we find a colony in at least 66 Petri dishes, we assume $q_\mu > 0.5728$, that is, $\mu > 85.05$ (we then also reject H_0' and accept H_1'), and therefore in that case reject our initial null hypothesis H_0 and assume $p_\mu = P_\mu(X > 100) > 0.05$. We conclude that the pool water does not meet the norm when a colony forms in at least 66 Petri dishes.

4.4 Testing with p-Values

In the previous section, we described tests using test statistics and critical regions. We can also describe tests using so-called p-values. The relation between the critical region and p-values is as follows.

Suppose that the critical region is of the form $K = \{x : T(x) \geq d_{\alpha_0}\}$, where the constant d_{α_0} is the smallest number for which a test with a critical region of this form has level α_0. In other words,

$$(4.3) \qquad\qquad d_{\alpha_0} = \min\left\{ t \colon \sup_{\theta \in \Theta_0} P_\theta(T \geq t) \leq \alpha_0 \right\}.$$

Often, minimizing d_{α_0} corresponds to maximizing the power function in Θ_1. The formula is therefore a consequence of Convention 4.10. The equality (4.3) implies that for every $t \in \mathbb{R}$,

$$\sup_{\theta \in \Theta_0} P_\theta(T \geq t) \leq \alpha_0 \qquad \Longleftrightarrow \qquad t \geq d_{\alpha_0}.$$

Definition 4.19 p-Value

For a test that rejects the null hypothesis for large values of a test statistic T and observed value t for T, the p-value is equal to

$$\sup_{\theta \in \Theta_0} P_\theta(T \geq t).$$

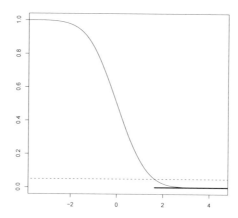

Figure 4.7. p-Value $t \mapsto \sup_{\mu \leq \mu_0} \mathrm{P}_\mu(T \geq t) = \mathrm{P}_{\mu_0}(T \geq t)$ (solid curve) for the Gauss test for $H_0 \colon \mu \leq \mu_0$ with $\mu_0 = 0$. A dashed line was drawn at the level $\alpha_0 = 0.05$. The thick solid line shows the corresponding critical region.

We can therefore also carry out the test as follows: when the p-value is less than or equal to α_0, we reject H_0; otherwise, we do not reject H_0. This rule gives exactly the test with critical region $K = \{x \colon T(x) \geq d_{\alpha_0}\}$, because the p-value is less than or equal to α_0 if and only if $t \geq d_{\alpha_0}$. The above is illustrated in Figure 4.7 using the Gauss test. It is clear in the figure that for values t in the critical region, we have $\sup_{\mu \leq \mu_0} \mathrm{P}_{\mu_0}(T \geq t) \leq \alpha_0$, and vice versa.

In words, the p-value is the maximum, under the null hypothesis, over all possible values of the probability that an identical experiment gives a more extreme value for the test statistic than the value t found in the experiment that has been carried out. Having to take the supremum, under the null hypothesis, over all possibilities makes determining the p-value somewhat complicated. In many cases, taking the supremum is unnecessary because one of the parameters $\theta_0 \in \Theta$ (often a boundary point of Θ_0) always has the maximum probability. In that case, the p-value equals $\mathrm{P}_{\theta_0}(T \geq t)$.

The p-value as we just defined it is specific for critical regions of the form $\{x \colon T(x) \geq d_{\alpha_0}\}$. It is evident how to extend it to critical regions of the form $\{x \colon T(x) \leq c_{\alpha_0}\}$, where the assumption is now that

$$c_{\alpha_0} = \max\left\{t \colon \sup_{\theta \in \Theta_0} \mathrm{P}_\theta(T \leq t) \leq \alpha_0\right\}.$$

Definition 4.20 p-Value

For a test that rejects the null hypothesis for small values of a test statistic T and observed value t for T, the p-value is equal to

$$\sup_{\theta \in \Theta_0} \mathrm{P}_\theta(T \leq t).$$

Two-sided critical regions of the form $\{x\colon T(x) \leq c\} \cup \{x\colon T(x) \geq d\}$ often consist of combinations of one-sided regions in the sense that $c = c_{\alpha_0/2}$ and $d = d_{\alpha_0/2}$ for c_{α_0} and d_{α_0} as defined earlier. The level α_0 is then divided into two equal parts of $\alpha_0/2$ each in the left and right tail. In this case, the p-value is defined as follows.

Definition 4.21 p-Value

For a test that rejects the null hypothesis for small and large values of a test statistic T and observed value t for T, the p-value is equal to

$$2 \min\left(\sup_{\theta \in \Theta_0} P_\theta(T \leq t),\ \sup_{\theta \in \Theta_0} P_\theta(T \geq t) \right).$$

Again, when the p-value is less than or equal to α_0, we reject H_0; otherwise we do not reject H_0. This corresponds to checking whether one of the two "one-sided p-values" is less than or equal to $\alpha_0/2$, since $2 \min(a, b) \leq \alpha_0$ if and only if $a \leq \alpha_0/2$ or $b \leq \alpha_0/2$.

Testing using p-values is often preferable to testing using a critical region because the resulting statements gives more information. Namely, when stating the p-value, it is also possible to still test (in a very simple way) the hypothesis against any desired level α_0, whereas when stating the critical region and the value of the test statistic for a fixed α_0, this is not possible. Moreover, for example, a very small p-value immediately tells us that H_0 must clearly be rejected.

p-Values can also be defined for tests with a critical region of a general form. To emphasize this, we give the general definition, of which Definitions 4.19, 4.20, and 4.21 are examples.

Definition 4.22 p-Value

For a collection of tests that contains a test of level α for every $\alpha \in (0, 1)$, the observed significance level or p-value is the smallest value of α for which the corresponding test rejects H_0.

Example 4.23 Binomial test, continued from Example 4.11

In Example 4.11, we concluded that in the case of 64 successes, the null hypothesis is rejected at $\alpha_0 = 0.05$, while it is not rejected in the case of 58 successes. The p-values for 64 and 58 successes are, respectively,

$$\sup_{p \leq 0.5} P_p(X \geq 64) = P_{0.5}(X \geq 64) = 0.0033$$

$$\sup_{p \leq 0.5} P_p(X \geq 58) = P_{0.5}(X \geq 58) = 0.0666.$$

The first probability is very small and is indeed less than 0.05, and the second is greater than 0.05. We moreover see that with 64 successes, the null hypothesis is rejected at all levels $\alpha_0 \geq 0.0033$. So the p-value gives more information than only the fact that the null hypothesis is rejected at $\alpha_0 = 0.05$, which was the conclusion of Example 4.11.

Example 4.24 Binomial test, continued from Example 4.23

The p-value of the binomial test for the null hypothesis $H_0\colon p \leq p_0$, for an observed value x, is equal to

$$\sup_{p \leq p_0} \mathrm{P}_p(X \geq x) = \mathrm{P}_{p_0}(X \geq x).$$

We reject $H_0\colon p \leq p_0$ when this probability is less than or equal to α_0.

For known p_0, α_0, n, and x, we can either look up the p-value in a table or compute it using a statistical computer package. For large n, we can also apply the normal approximation,

$$\mathrm{P}_{p_0}(X \geq x) \approx 1 - \Phi\left(\frac{x - np_0 - \frac{1}{2}}{\sqrt{np_0(1 - p_0)}}\right).$$

For the null hypothesis $H_0\colon p \geq p_0$, the p-value $\mathrm{P}_{p_0}(X \leq x)$ can also be computed using the normal approximation, this time with the continuity correction in the other direction.

Example 4.25 Gauss test, continued from Example 4.12

The Gauss test rejects the null hypothesis $H_0\colon \mu \leq \mu_0$ for large values of $T = \sqrt{n}(\overline{X} - \mu_0)/\sigma$. The critical value $\xi_{1-\alpha_0}$ of the test satisfies (4.3). For an observed value $\overline{x}$, the p-value of the test is therefore equal to

$$\sup_{\mu \leq \mu_0} \mathrm{P}_\mu\left(T \geq \sqrt{n}\frac{\overline{x} - \mu_0}{\sigma}\right) = \mathrm{P}_{\mu_0}\left(T \geq \sqrt{n}\frac{\overline{x} - \mu_0}{\sigma}\right) = 1 - \Phi\left(\sqrt{n}\frac{\overline{x} - \mu_0}{\sigma}\right).$$

When this probability is less than or equal to α_0, the null hypothesis H_0 is rejected at level α_0.

The p-value for testing the other one-sided null hypothesis $H_0\colon \mu \geq \mu_0$ against the alternative hypothesis $H_1\colon \mu < \mu_0$ is given by the probability $\mathrm{P}_{\mu_0}(T \leq \sqrt{n}(\overline{x} - \mu_0)/\sigma)$. We reject the null hypothesis when this probability is less than or equal to α_0.

The two-sided Gauss test is nothing more than the combination of the two one-sided tests, each with level $\alpha_0/2$. We can therefore carry out this test by computing the p-value of both the left-sided and the right-sided tests. The p-value of the two-sided test is then equal to twice the minimum of the two one-sided p-values. If one of the two p-values is less than or equal to $\alpha_0/2$, then the p-value of the two-sided test is less than or equal to α_0 and we reject the null hypothesis $H_0\colon \mu = \mu_0$.

Example 4.26 Application: Poisson-distributed stocks

Suppose that a distribution center stocks a certain perishable product weekly to supply different retailers (see Example 1.4). Since the product has a limited shelf life, they do not want to stock too much; products that are not sold are thrown out at the end of the week. On the other hand, when too little is stocked and consumer demand is not met, this gives dissatisfaction and a loss of clientele. The center has therefore decided to stock a fixed number of items such that the probability of having a shortage is at most 10%. However, lately, the stock has regularly been insufficient to meet the demands of the retailers. Apparently, the weekly demand has increased. We want to check this using a statistical test.

We assume that the total weekly demand Z has a Poisson distribution with parameter θ. If C is the number of items stocked weekly, we can determine the maximal parameter value θ_0 for which the 10% norm is satisfied: $\theta_0 = \max\{\theta \colon P_\theta(Z > C) \leq 0.10\}$. To test whether the current weekly demand has surpassed the number the purchasing policy is based on, we want to test the null hypothesis $H_0 \colon \theta \leq \theta_0$ against the alternative hypothesis $H_1 \colon \theta > \theta_0$. To do this, we register the weekly demand for n weeks. This gives observations $Z_1, \ldots, Z_n$. Assume that $Z_1, \ldots, Z_n$ are independent and have a Poisson distribution with parameter θ. To test the null hypothesis, we take as test statistic $T = \sum_{i=1}^n Z_i$, which has a Poisson distribution with parameter $n\theta$.

We carry out the test by determining the *p*-value. For the observed value $T = t$, the *p*-value is $\sup_{\theta \leq \theta_0} P_\theta(T \geq t) = P_{\theta_0}(T \geq t)$. If this *p*-value is less than or equal to the chosen level α_0, then the null hypothesis is rejected, and we can conclude that the current demand is too high to meet the 10% norm with the current purchasing policy. We can determine the *p*-value exactly using a statistical computer package, but we can also approximate it. When $n\theta$ is large, the test statistic T is approximately normally distributed with expectation and variance both equal to $n\theta$; see Section A.7. The *p*-value can then be approximated as follows:

$$P_{\theta_0}(T \geq t) = P_{\theta_0}\left(\frac{T - n\theta_0}{\sqrt{n\theta_0}} \geq \frac{t - n\theta_0}{\sqrt{n\theta_0}}\right) \approx 1 - \Phi\left(\frac{t - n\theta_0}{\sqrt{n\theta_0}}\right).$$

This test problem can also be approached from another direction. Suppose that instead of registering the weekly demand, we only note whether the stock is sufficient to meet it. We observe a sequence $X_1, \ldots, X_n$, where $X_i = 1_{Z_i > C}$ is equal to 1 when the demand is higher than the number of supplied products C and equal to 0 if the supply suffices. The variables $X_1, \ldots, X_n$ are independent and have the Bernoulli distribution with parameter p, where p is the probability that there is shortage during an arbitrary week. Since we want to study whether the probability is greater than 10%, we test the null hypothesis $H_0 \colon p \leq 0.10$ against $H_1 \colon p > 0.10$. As test statistic, we take $X = \sum_{i=1}^n X_i$, which is binomially distributed with parameters n and p. Example 4.24 describes how to determine the *p*-value of this test.

Which of the tests above is better? We can judge this using the power function. The power function of the first test, based on the Poisson(θ)-distributed quantities $Z_1, \ldots, Z_n$, is a function of θ. For the second test, based on the Bernoulli-distributed

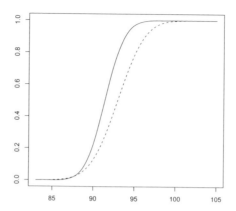

Figure 4.8. Power function as a function of θ based on the Poisson-distributed observations $Z_1, \ldots, Z_n$ (solid curve) and the Bernoulli-distributed observations $X_1, \ldots, X_n$ for $n = 26$, $C = 100$, $\alpha_0 = 0.05$, and $\theta_0 = \max\{\theta : P_\theta(Z > 100) \leq 0.10\} = 88.35$.

variables $X_1, \ldots, X_n$ with distribution parameter p, the power function is, in principle, a function of p. However, for given θ, the probability p can be computed as follows:

$$p = P_\theta(X_i = 1) = P_\theta(Z_i > C) = \sum_{k=C+1}^{\infty} e^{-\theta} \frac{\theta^k}{k!}.$$

We can also use this to compute the power function of the second test as a function of θ.

Figure 4.8 shows the power functions of both tests as a function of θ for the choice $C = 100$, $n = 26$, and $\alpha_0 = 0.05$. We can see that the power function of the first test is greater than that of the second test for values of θ under the alternative hypothesis, that is, for $\theta > \theta_0$. Based on this image, our preference would go out to the test based on the Poisson-distributed random variables $Z_1, \ldots, Z_n$. However, if there is any doubt to the assumption that the weekly demand has a Poisson distribution, then the binomial test is more powerful because this does not make any assumptions on the distribution of the weekly demand. An incorrect assumption in the statistical model may lead to a test of size greater than the desired α_0.

4.5 Statistical Significance

The general set-up of the theory of testing as described above is both rather complicated and astonishingly simple because there are only two possible decisions. In many practical situations, the simplicity is misleading. An effect is called *statistically significant* if the relevant null hypothesis is rejected at the given level. This should be interpreted as follows: *the effect we observed in the data is probably not due to random variation; if we were to repeat the entire experiment, we would probably observe the same effect.* This in no way means that the "effect" is *practically significant*. It is therefore conceivable that the test procedure has correctly shown that the new therapy is better, but that the improvement is negligible. If the existing therapy has probability of success $p = 0.5$ and the new one has probability of success $p = 0.500001$, then we will observe this effect and reject H_0 provided that we have sufficiently many observations, but, practically speaking, it will make little difference which therapy we follow.

Because of this, it is desirable to always supplement a test procedure that leads to rejecting H_0 with an estimation procedure that gives an indication of the size of a possible effect. The context then determines whether this effect is of practical interest.

Another possibility to bridge the discrepancy between statistical and practical significance would be to formulate the null hypothesis differently. We could, for example, test the null hypothesis that the difference $p_2 - p_1$ in probabilities of success for the new and old therapies is at least 0.2, instead of the hypothesis that $p_2 - p_1 > 0$. The value 0.2 could then express the practical significance. In practice, however, we are usually satisfied with determining a qualitative difference and test the hypothesis $H_1 : p_2 - p_1 > 0$.

4.6 Some Standard Tests

In this section, we discuss several tests that are frequently used, other than the Gauss test and binomial test. Most of these tests can be understood intuitively.

For a given test problem, the general idea is to find a test statistic that is "reasonable" (often based on a good estimator for the parameter) and for which we can easily compute a critical value or p-value. For the latter, the probability distribution of the test statistic under the ("boundary" of the) null hypothesis must be either given in a table or computable. Often, however, the probability distribution under the null hypothesis does not belong to the usual list of computed distributions in probability theory. We can then introduce a new standard probability distribution and produce tables. An alternative is to approximate the probability distribution "on-the-fly" using stochastic simulation. Let us discuss examples of both approaches.

We begin this section with a discussion of the two most important random probability distributions, the chi-square and t-distributions. These families of probability distributions are both related to the normal distribution and occur both when testing the parameters of the normal distribution and when carrying out approximations in large samples.

4.6.1 Chi-Square and t-Distribution

The chi-square and t-distributions are continuous probability distributions whose densities are given by relatively simple expressions. For our purpose, however, the following structural definitions of these probability distributions are more appealing.

Definition 4.27 Chi-square distribution

A random variable W has a chi-square distribution with n degrees of freedom, denoted by χ_n^2, if W has the same distribution as $\sum_{i=1}^{n} Z_i^2$ for $Z_1, \ldots, Z_n$ a sample from the $N(0, 1)$-distribution.

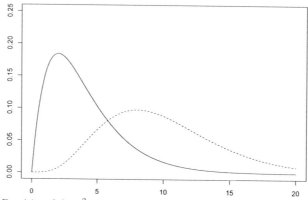

Figure 4.9. Densities of the χ^2-distributions with 4 (solid) and 10 (dashed) degrees of freedom.

Definition 4.28 Student's t-distribution

A random variable T has a t-distribution or Student's t-distribution with n degrees of freedom, denoted by t_n, if T has the same distribution as

$$\frac{Z}{\sqrt{W/n}},$$

for Z and W two independent random variables with the $N(0, 1)$- and χ_n^2-distributions, respectively.

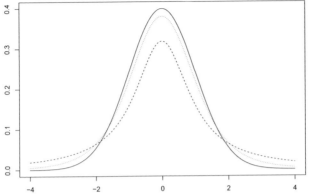

Figure 4.10. Densities of the t-distributions with 1 (dashed), 5 (dotted), and ∞ (solid) degrees of freedom.

Using standard techniques from probability theory, we can derive formulas for the densities of chi-square and t-distributions. These expressions were used in "classical" times to produce tables of the distribution functions. More recently, they are the basis for the standard algorithms in statistical software. We will consider the tables and software as given, and not discuss the exact form of the densities. Figures 4.9 and 4.10 give a qualitative idea of the densities.

The following theorem shows why the chi-square and t-distributions are important.

Theorem 4.29

If $X_1, \ldots, X_n$ is a sample from the $N(\mu, \sigma^2)$-distribution, then
 (i) $\overline{X}$ is $N(\mu, \sigma^2/n)$-distributed;
 (ii) $(n-1)S_X^2/\sigma^2$ is χ_{n-1}^2-distributed;
 (iii) $\overline{X}$ and S_X^2 are independent;
 (iv) $\sqrt{n}(\overline{X} - \mu)/S_X$ has the t_{n-1}-distribution.

Proof. Statement (i) is known from probability theory: the sum of independent normally distributed random variables is again normally distributed. For the proofs of statements (ii) and (iii), we may, without loss of generality, assume $\mu = 0$ and $\sigma^2 = 1$.

The joint density of the random vector $X = (X_1, \ldots, X_n)^T$ is then equal to

$$(x_1, \ldots, x_n) \mapsto \prod_{i=1}^{n} \frac{1}{\sqrt{2\pi}} e^{-\frac{1}{2}x_i^2} = \frac{1}{(2\pi)^{n/2}} e^{-\frac{1}{2}\|x\|^2},$$

where $\|x\|^2 = \sum_{i=1}^{n} x_i^2$ is the square of the Euclidean length of x. Define the vector $f_1 = (1/\sqrt{n}, \ldots, 1/\sqrt{n}) \in \mathbb{R}^n$ with $\|f_1\|^2 = 1$ and extend f_1 arbitrarily to an orthonormal basis $\{f_1, \ldots, f_n\}$ of $\mathbb{R}^n$. Let O be the $(n \times n)$-matrix with rows

129

$f_1, \ldots, f_n$. It immediately follows from the definition that $OO^T = I$ (with I the identity matrix), so that $O^T = O^{-1}$ and O is an orthogonal matrix, $OO^T = O^T O = I$, and consequently $\|Ox\|^2 = x^T O^T Ox = \|x\|^2$ for all x. Define the random vector $Y = OX$. Then

$$Y_1 = f_1 X = \sqrt{n}\, \overline{X},$$

$$\sum_{i=2}^{n} Y_i^2 = \|Y\|^2 - Y_1^2 = \|X\|^2 - n\overline{X}^2 = \sum_{i=1}^{n}(X_i - \overline{X})^2.$$

Statements (ii) and (iii) consequently follow if we can prove that $Y_1, \ldots, Y_n$ are independent and $N(0, 1)$-distributed.

The distribution function of Y is given by

$$P(Y \leq y) = \int \cdots \int_{x:Ox \leq y} \frac{1}{(2\pi)^{n/2}} e^{-\frac{1}{2}\|x\|^2}\, dx_1 \cdots dx_n$$

$$= \int \cdots \int_{u:u \leq y} \frac{1}{(2\pi)^{n/2}} e^{-\frac{1}{2}\|u\|^2}\, du_1 \cdots du_n,$$

where we use the substitution $Ox = u$. Then $\|x\| = \|Ox\| = \|u\|$, and the Jacobian of the transformation $Ox = u$ is equal to $\det O = 1$. It immediately follows from the last expression that Y has the same joint density as X; that is, $Y_1, \ldots, Y_n$ are independent and normally distributed with expectation 0 and variance 1.

For the proof of statement (iv), we write

$$\sqrt{n}\frac{\overline{X} - \mu}{S_X} = \frac{\sqrt{n}(\overline{X} - \mu)/\sigma}{\sqrt{\frac{(n-1)S_X^2/\sigma^2}{(n-1)}}}.$$

By statement (iii), the numerator and denominator are independent of each other, and by statements (i) and (ii) they have, respectively, an $N(0, 1)$-distribution and the square root of a χ_{n-1}^2-distribution divided by $n - 1$. By Definition 4.28, the quotient then has the t_{n-1}-distribution. ∎

The statements of Theorem 4.29 are interesting. In particular, the independence of the sample mean and sample variance for the same data is surprising. Figure 4.11 illustrates that this property depends on the distribution of the observations: the normal distribution has this property, while the exponential distribution does not! For applications, the implication of statement (iv) that the distribution of $\sqrt{n}(\overline{X} - \mu)/S_X$ does not depend on the parameter σ^2 is most important for, among other things, setting up a test. It is nice to know that this distribution is a t-distribution, so that we can refer to standard functions or tables of this distribution. This is, however, less essential, because we can also approximate the distribution using stochastic simulation.

Because the form of the density of Student's t-distribution is known explicitly (determined by W. Gosset, who published his results under the pseudonym "Student"), simulation is unnecessary. Every statistical package contains functions to compute the distribution function and quantiles of chi-square and t-distributions numerically.

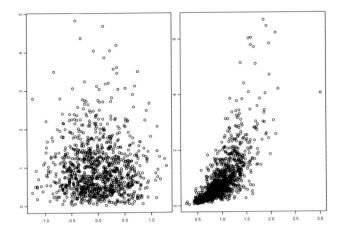

Figure 4.11. Scatter plot of the sample mean (x-axis) against the sample variance (y-axis) for 1000 samples of size 5 from the standard normal distribution (on the left) and 1000 samples of size 10 from the exponential distribution (on the right). On the left, the two coordinates are independent; on the right, there is a positive correlation.

4.6.2 One-Sample Tests

Given a sample $X_1, \ldots, X_n$, we often want to test whether the location of the marginal distribution of the sample is to the left or to the right of a certain value. For the "location," we can take, for example, the "expectation" or "median." If we also assume that the sample comes from a normal distribution, then the well-known t-test is the correct test for the problem when the variance is unknown. If the variance were known, we would use the Gauss test (see Example 4.12).

Example 4.30 t-Test

Let $X_1, \ldots, X_n$ be a sample from the $N(\mu, \sigma^2)$-distribution with μ and σ^2 unknown. We want to test $H_0: \mu \leq \mu_0$ against $H_1: \mu > \mu_0$, where μ_0 is a fixed number (for example $\mu_0 = 0$). Formally, the parameter in this case is given by the pair $\theta = (\mu, \sigma^2)$, and the null hypothesis is equal to $\Theta_0 = \{(\mu, \sigma^2): \mu \leq \mu_0, \sigma^2 > 0\}$.

Since the test statistic and the critical region K of the Gauss test from Example 4.12 depend on σ, and this parameter is now unknown, we cannot use that test here. A logical extension of the Gauss test is to replace σ in the definition of the test statistic by an estimator. We use the sample standard deviation S_X. This gives the test statistic

$$T = \sqrt{n} \frac{\overline{X} - \mu_0}{S_X}.$$

We reject the null hypothesis for large values of this variable. Since the substitution of S_X for σ also changes the distribution of this variable, it is no longer normally distributed when $\mu = \mu_0$. It is therefore not immediately clear which critical values we should take. In the next section, we explain, using Theorem 4.29, that this must be the $(1 - \alpha_0)$-quantile of the t-distribution with $n - 1$ degrees of freedom, which we denote by $t_{n-1, 1-\alpha_0}$.

By Convention 4.9, the critical value c_{α_0} for finding a test of significance level α_0 must satisfy

$$\sup_{\mu \leq \mu_0, \sigma^2 > 0} P_{\mu, \sigma^2} \left(\sqrt{n} \frac{\overline{X} - \mu_0}{S_X} \geq c_{\alpha_0} \right) \leq \alpha_0,$$

for α_0 the level of the test. Note that the supremum must be taken over both $\mu \leq \mu_0$ and all possible values of σ^2, that is, over the entire parameter space under the null hypothesis. However, the supremum over μ (for every σ) is taken in the boundary point $\mu = \mu_0$, which is clear intuitively (but not trivial to prove), so that the inequality reduces to

$$\sup_{\sigma^2 > 0} P_{\mu_0, \sigma^2} \left(\sqrt{n} \frac{\overline{X} - \mu_0}{S_X} \geq c_{\alpha_0} \right) \leq \alpha_0.$$

Now, by Theorem 4.29(iv), the distribution of $\sqrt{n}(\overline{X} - \mu_0)/S_X$ under (μ_0, σ^2) does not depend on (μ_0, σ^2) and is equal to the t_{n-1}-distribution. It now follows from the inequality above that $c_{\alpha_0} \geq t_{n-1, 1-\alpha_0}$. To obtain the largest possible power function, in accordance with Convention 4.10, we choose the critical region as large as possible and take $c_{\alpha_0} = t_{n-1, 1-\alpha_0}$. The size α of the test is then exactly equal to the level: $\alpha = \alpha_0$.

The resulting test, called the *t-test* or *Student's t-test*, states: "Reject H_0 if $\sqrt{n}(\overline{X} - \mu_0)/S_X \geq t_{n-1, 1-\alpha_0}$." The corresponding p-value for observed values $\overline{x}$ and s_x is equal to

$$P_{\mu_0, \sigma^2} \left(T \geq \sqrt{n} \frac{\overline{x} - \mu_0}{s_x} \right) = P \left(T_{n-1} \geq \sqrt{n} \frac{\overline{x} - \mu_0}{s_x} \right),$$

where T_{n-1} is a random variable with the t_{n-1}-distribution.

Adjusting the t-test for the test problems $H_0: \mu \geq \mu_0$ and $H_0: \mu = \mu_0$ is analogous to adjusting the Gauss test. It is important for this to know that the t-distribution is symmetric around 0, like the normal distribution, so that $t_{n,\alpha} = -t_{n, 1-\alpha}$.

For small values of n ($n \leq 10$), the t_n-distribution differs greatly from the normal distribution. Using normal quantiles instead of t_{n-1}-quantiles (that is, the Gauss test with σ taken equal to S_X) then leads to a test with a size that is much greater than the desired size. This violates Convention 4.9. For increasing values of n, the t_n-distribution increasingly resembles the standard normal distribution, with convergence to the normal distribution as $n \to \infty$. For $n \geq 20$, the similarity is already good enough that the t-test and Gauss test give practically identical results.

We have introduced the t-test using ad hoc arguments. We can, however, show that the test is uniformly most powerful within the class of all unbiased tests (see Section 6.4.3).

The t-test is the correct test for testing the location when the observations $X_1, \ldots, X_n$ form a sample from the normal distribution. If the last assumption does not hold, then it may be possible and desirable to transform the observations (for example using the logarithmic function) to observations for which the normality assumption is reasonable. An alternative is to use a test that does not require normality. There are many examples of such tests, of which we will discuss only one.

Example 4.31 Sign test

The sign test can be applied under minimal assumptions and is therefore well suited when the observations do not come from the normal distribution. It is a test for the median and not for the expectation, as is the case for the Gauss test and t-test. Suppose that we want to test whether the median μ of the distribution giving the independent observations $X_1, \ldots, X_n$ is greater than a given value μ_0, that is, test $H_0: \mu \leq \mu_0$ against $H_1: \mu > \mu_0$. The test statistic is $T = \#(1 \leq i \leq n: X_i > \mu_0)$; we count how many observations are greater than μ_0 or, equivalently, how many differences $X_i - \mu_0$ are positive. We, in fact, apply the binomial test to the signs (positive or negative) of the differences $X_1 - \mu_0, \ldots, X_n - \mu_0$. The test statistic is binomially distributed with parameters n and $p_\mu = P_\mu(X_i > \mu_0)$. If the median of the distribution of the observations is equal to μ_0, then the parameters are n and $\frac{1}{2}$. If, on the other hand, the distribution of the observations has median μ with $\mu \leq \mu_0$, then the probability p_μ is less than or equal to $1/2$. The null hypothesis $H_0: \mu \leq \mu_0$ can therefore be tested by testing the equivalent null hypothesis $H_0: p_\mu \leq 1/2$ using T. We reject H_0 for large values of T, where the critical value is determined as in Example 4.13. ▭

*Example 4.32 Tests for σ^2

Let $X_1, \ldots, X_n$ be a sample from the $N(\mu, \sigma^2)$-distribution with μ and σ^2 unknown parameters. Consider the problem of testing $H_0: \sigma^2 \leq \sigma_0^2$ against $H_1: \sigma^2 > \sigma_0^2$, where σ_0^2 is a fixed number. Formally, the parameter is given by the pair $\theta = (\mu, \sigma^2)$, and the parameter space under the null hypothesis is equal to $\Theta_0 = \{(\mu, \sigma^2): \mu \in \mathbb{R}, \sigma^2 \leq \sigma_0^2\}$. A reasonable estimator for σ^2 is the sample variance S_X^2. Large values of S_X^2 indicate that the alternative hypothesis might be correct. We therefore reject the null hypothesis for large values of S_X^2.

The probability distribution of $(n-1)S_X^2/\sigma^2$ under (μ, σ^2) does not depend on the parameter (μ, σ^2), and is exactly the chi-square distribution with $n-1$ degrees of freedom (see Theorem 4.29). If we denote the α-quantile of the chi-square distribution with $n-1$ degrees of freedom by $\chi^2_{n-1,\alpha}$, then the obvious choice for a test is: "Reject H_0 when $(n-1)S_X^2/\sigma_0^2 \geq \chi^2_{n-1,1-\alpha_0}$" (with α_0 the level of the test). We can show that the size of this test is α_0, in the same way as in previous examples.

The tests for the other one-sided test problem $H_0\colon \sigma^2 \geq \sigma_0^2$ and the two-sided test problem $H_0\colon \sigma^2 = \sigma_0^2$ are obvious. Note, however, that the chi-square distribution is not symmetric and puts all probability mass on $(0, \infty)$. There is consequently no direct relation between the quantiles $\chi^2_{n-1,\alpha}$ and $\chi^2_{n-1,1-\alpha}$, and we cannot use the absolute value of the test statistic to describe the two-sided test. The two-sided test states: "Reject H_0 when $(n-1)S_X^2/\sigma_0^2 \leq \chi^2_{n-1,\alpha_0/2}$ or $(n-1)S_X^2/\sigma_0^2 \geq \chi^2_{n-1,1-\alpha_0/2}$." We can also carry out these tests with p-values.

4.6.3 Two-Sample Tests

In the *two-sample problem*, we have two samples $X_1, \ldots, X_m$ and $Y_1, \ldots, Y_n$ from possibly different probability distributions, and we are interested in comparing these distributions, for example their locations. Depending on the assumptions, there exist different types of two-sample tests.

We can make an important distinction between tests for *paired* and *independent* observations. In the first case, the two samples come from a sample $(X_1, Y_1), \ldots, (X_n, Y_n)$ of pairs of observations, where the X- and Y-variables in each pair may be related, but the pairs themselves are assumed to be independent. For example, when studying the effectivity of treatment, the X might give the state of a patient before treatment, while the Y gives the state after treatment. Since X_i and Y_i are measurements on the same patient, it is logical that they are stochastically dependent. Indeed, a low value in the first observation indicates that the patient is in poor health, which makes it more probable that the second observation will also be low (with respect to the rest of the population, though possibly higher than the first observation if the treatment is successful).

In the case of repeated measures on the same object or person, the dependence of the measurements is unavoidable. In other applications, the X- and Y-components may be intentionally chosen dependent by the set-up of the experiment. For example, a group of subjects may be paired up using background variables such as age, gender, prior treatment, or case history, so that the two persons in each pair are comparable with respect to these variables. Then, one (arbitrarily chosen) person in each pair receives the drug while the other receives a placebo. A difference in the state of the patient after this treatment gives an indication of the efficacy of the drug. The purpose of pairing the subjects in this approach is to emphasize the effect of the treatment. Indeed, an observed difference within the pair cannot be explained by fluctuations in the background variables but must be due to the treatment (or to an as yet unknown background variable). If we do not pair the observations, then the additional random fluctuation caused by the background variables can mask the effect of the treatment.

Example 4.33 t-Test for paired observations

It seems natural to base a test for comparing the locations of two paired samples $(X_1, Y_1), \ldots, (X_n, Y_n)$ on the differences $Z_i = X_i - Y_i$. The t-test for pairs is the usual t-test applied to the differences $Z_1, \ldots, Z_n$.

To apply the t-test, we assume that the differences $Z_1, \ldots, Z_n$ are independent and $N(\Delta, \sigma^2)$-distributed, where the parameter Δ is equal to the difference $\mathrm{E}X_i - \mathrm{E}Y_i$ of the expectations. Suppose that we want to test the null hypothesis $H_0: \Delta = 0$ that the treatment does not have any effect, or one of the hypotheses $H_0: \Delta \geq 0$ or $H_0: \Delta \leq 0$.

If all X_i and Y_i are independent and have the $N(\mu, \sigma_1^2)$- and $N(\mu - \Delta, \sigma_2^2)$-distributions, respectively, then the differences $Z_i = X_i - Y_i$ have the normal distribution with expectation Δ and variance $\sigma^2 = \sigma_1^2 + \sigma_2^2$, and we can apply the test mentioned above. In many applications, however, the X_i and Y_i will not be independent, because they concern measurements on the same object. Fortunately, even without that assumption, the normality and independence of the differences are reasonable assumptions.

The power function of the t-test depends strongly on the variance σ^2. If the variance is large, then the difference in expectations is difficult to detect, and the power function of the t-test is small. A small variance is favorable and ensures a large power function. This finding makes it clear that it can be wise to intentionally make the samples in the two-sample problem dependent. Indeed, by the rule for the variance of a difference, we have var $Z_i = $ var $Y_i + $ var $X_i - 2\,\mathrm{cov}(X_i, Y_i)$, which is less than var $Y_i + $ var X_i if X_i and Y_i have a positive correlation. An intuitive explanation is that taking differences eliminates random fluctuations that are present in both the X- and Y-components and do not interest us. After eliminating this variation, it is easier to discover a possible difference caused by the treatment.

A correct application of the t-test does require that the differences $Z_1, \ldots, Z_n$ may be viewed as a sample from a normal distribution. ▭

Example 4.34 Two-sample t-test

Let $X_1, \ldots, X_m$ and $Y_1, \ldots, Y_n$ be independent samples from, respectively, the $N(\mu, \sigma^2)$- and the $N(\nu, \sigma^2)$-distributions. We want to carry out a one-sided test for $\mu - \nu > 0$, that is, $H_0: \mu - \nu \leq 0$ against $H_1: \mu - \nu > 0$. An obvious estimator for $\mu - \nu$ is the difference $\overline{X} - \overline{Y}$ of the sample means. Large values of this difference are an indication that H_1 is correct. The distribution of $\overline{X} - \overline{Y}$ is normal with expectation $\mu - \nu$ and variance

$$\mathrm{var}(\overline{X} - \overline{Y}) = \mathrm{var}\,\overline{X} + \mathrm{var}\,\overline{Y} = \frac{\sigma^2}{m} + \frac{\sigma^2}{n} = \sigma^2 \left(\frac{1}{m} + \frac{1}{n}\right)$$

because of the independence of the two samples. Because this distribution depends on the unknown parameter σ^2, we choose not $\overline{X} - \overline{Y}$ as test statistic, but rather the quantity

$$T = \frac{\overline{X} - \overline{Y}}{S_{X,Y}\sqrt{\frac{1}{m} + \frac{1}{n}}},$$

where

$$S_{X,Y}^2 = \frac{1}{m+n-2}\left(\sum_{i=1}^{m}(X_i - \overline{X})^2 + \sum_{j=1}^{n}(Y_j - \overline{Y})^2\right)$$

is an unbiased estimator for σ^2 (the maximum likelihood estimator for σ^2 is equal to $(m+n-2)/(m+n)S^2_{X,Y}$; verify). If $\mu = \nu$, then T has a t-distribution with $m+n-2$ degrees of freedom (see the next paragraph for a deduction). As with the t-test for one sample in Example 4.30, it suffices to consider the distribution of T in the boundary point $\mu = \nu$, which must then be independent of σ^2. We therefore choose the critical value equal to $t_{m+n-2,1-\alpha_0}$, and the test states: "Reject H_0 when $T \geq t_{m+n-2,1-\alpha_0}$."

To see that T has a t-distribution, we write it as

$$T = \frac{(\overline{X} - \overline{Y})/\sqrt{\frac{\sigma^2}{m} + \frac{\sigma^2}{n}}}{\sqrt{\frac{(m+n-2)S^2_{X,Y}/\sigma^2}{m+n-2}}}.$$

When $\mu = \nu$, the numerator of this expression has the $N(0,1)$-distribution. To determine the distribution of the denominator, we note that the sum of two independent chi-square distributed random variables again has a chi-square distribution, with the number of degrees of freedom equal to the sum of the numbers of degrees of freedom of the two terms. Using Theorem 4.29, we see that $(m+n-2)S^2_{X,Y}/\sigma^2$ has a χ^2_{m+n-2}-distribution and is independent of $\overline{X} - \overline{Y}$. The numerator and denominator of the test statistic are therefore independent. That T has distribution t_{m+n-2} when $\mu = \nu$ now follows from the definition of the t-distribution.

Tests for other one-sided and two-sided problems, and the corresponding p-values, can be deduced in a way similar to that used for the one-sample problem.

⎯⎯

We call the test in Example 4.34 the t-test (or Student's t-test) for two samples. This test differs essentially from the one-sample test for differences from Example 4.33 because in that case, pairs (X_i, Y_i) were defined in a natural way, which is not the case here. If the coordinates X_i and Y_i in a pair (X_i, Y_i) are dependent, we may not use the two-sample t-test, or at least not with critical value $t_{m+n-2,1-\alpha_0}$, because in that case, there is no guarantee that the size is less than or equal to α_0. If, however, the pairs $(X_1, Y_1), \ldots, (X_n, Y_n)$ and the coordinates X_i and Y_i in each pair are independent and normally distributed, then both the t-test for pairs (Example 4.33) and the two-sample t-test (Example 4.34) have size α_0 and are admissible. The two-sample t-test is then preferable because of its larger power function. The intuitive reason is that in the t-test for pairs, the unknown parameter σ^2 is estimated using n independent observations (the differences $Z_i = X_i - Y_i$, of which $n - 1$ are "free," that is, with $n - 1$ degrees of freedom), while $S^2_{X,Y}$ is based on $2n$ independent observations (with $2n - 2$ degrees of freedom). The latter is obviously better.

In Example 4.34, we assumed that the variance σ^2 is the same for both samples, but in many practical problems, this is uncertain or not true. A more general problem is obtained by assuming that the two samples come from the $N(\mu, \sigma^2)$- and $N(\nu, \tau^2)$-distributions. We want to test the same null hypothesis $H_0 : \mu \leq \nu$, but now with unknown σ^2 and τ^2. This is the well-known *Behrens–Fisher problem*. Unlike in the case $\sigma^2 = \tau^2$, where the test we just discussed is the uniformly most powerful test among the unbiased tests, there is no absolutely best test in the situation of the

Behrens–Fisher problem (whence the word "problem"). There are several reasonable tests (for which we refer to other textbooks).

If we ignore the possible inequality of σ^2 and τ^2 and apply the two-sample t-test from Example 4.34, then the true size of the test may differ greatly from the desired size (called the *nominal size* in this context). Table 4.1 provides an impression of this. The effect of different variances is relatively small when m and n are approximately equal and not too small. (We can prove that the size converges to α_0 as $m = n \to \infty$ for every σ^2/τ^2!) This leads to the advice to choose samples of equal size whenever possible. This is also wise when $\sigma^2 = \tau^2$, because the power function of the two-sample t-test is maximal when $m = n$ (for fixed $m + n$).

	σ^2/τ^2	0.2	0.5	1	2	3
m	n					
5	3	0.100	0.072	0.050	0.038	0.031
15	5	0.180	0.098	0.050	0.025	0.008
7	7	0.063	0.058	0.050	0.058	0.063

Table 4.1. True size of the two-sided two-sample t-test for different variances and nominal level 0.05.

Example 4.35 Asymptotic t-test

A correct application of the two-sample t-test from Example 4.34 requires that the two samples be normally distributed with equal variances. If the two samples are both sufficiently large, then neither the normality nor the assumption that the variances are equal is essential, provided that the test be adjusted as follows. As test statistic, we choose

$$T = \frac{\overline{X} - \overline{Y}}{\sqrt{\frac{S_X^2}{m} + \frac{S_Y^2}{n}}}.$$

This variable differs from the test statistic in Example 4.34 by the use of a different estimator for the standard deviation in the denominator.

Using the central limit theorem, Theorem A.28, we can show that under the hypothesis $\mu = \nu$ of equal expectations for the two samples, the variable $T = T_{m,n}$ converges in distribution to a standard normal distribution as $m, n \to \infty$, provided that the variances of the two samples exist and are finite. For large values of m and n, we can therefore test the null hypothesis $H_0: \mu \leq \nu$ using the test that states: "Reject H_0 when $T \geq \xi_{1-\alpha_0}$."

The size of this test converges to the level α_0 as $m, n \to \infty$, for every pair of underlying distributions with finite variances. For distributions that are not too asymmetric, we can already use this result for $m = n = 20$.

It is not always reasonable to assume that the data come from normal distributions. If there are good reasons to assume a different parametric model, for example exponential distributions, then this will in general lead to a different test, because in that case the t-test does not have the right level and may have an unnecessarily small power function. The general methods for constructing a test, like the likelihood ratio test from Section 4.7, suggest which test is reasonable.

It is also possible to find correct tests that require hardly any assumptions on the distribution. So-called *distribution-free tests* work for very broad classes of distributions. The sign test from Example 4.31 belongs to this group. Below, we discuss an example of a distribution-free two-sample test.

Example 4.36 Wilcoxon signed-rank test

Given samples $X_1, \ldots, X_m, Y_1, \ldots, Y_n$, we define the *ranks* $R_1, \ldots, R_m$ of the first sample in the total sample as the positions (or ranks) of $X_1, \ldots, X_m$ after the data $X_1, \ldots, X_m, Y_1, \ldots, Y_n$ have been sorted by size. (For example, if X_1 is the fourth smallest observation, then we set $R_1 = 4$; if X_2 is the largest, then $R_2 = m + n$, etc.) The test statistic of the *Wilcoxon test* (also called the *Mann–Whitney U test*) is $W = \sum_{i=1}^{m} R_i$. Large values of W indicate that $X_1, \ldots, X_m$ are relatively large with respect to $Y_1, \ldots, Y_n$. This leads to rejecting the null hypothesis H_0 that the two samples are identically distributed for the alternative hypothesis that the first sample comes from a "stochastically larger distribution" for large values of W. Of course, we can also do a one-sided test in the other direction and do two-sided tests.

Under the null hypothesis, $X_1, \ldots, X_m, Y_1, \ldots, Y_n$ can be viewed as a sample of size $m + n$ from a fixed (unknown) distribution. The ranks $R_1, \ldots, R_m$ can then be viewed as an arbitrary selection of m numbers out of the numbers $\{1, 2, \ldots, m + n\}$. (For simplicity, we assume that the observations are continuously distributed, so that the ranks are well defined.) The distribution of the Wilcoxon variable under the null hypothesis is therefore independent of this distribution, and can be determined using combinatorial arguments. This distribution has been tabulated and is available through statistical computer packages.

4.6.4 Goodness-of-Fit Tests

Tests to check whether the distribution of an observation belongs to a certain family are called *goodness-of-fit tests*.

In Chapter 2, we saw how the distribution of a sample could also be evaluated graphically, for example using a QQ-plot. It is not our intention to replace these graphical methods by formal tests; rather, we view the tests as a supplement. The formal set-up of testing is an advantage because of the clarity, but has the disadvantage of only giving a yes/no answer, without giving insight into the deviation from normality when the answer is "no." On the other hand, the method of testing is well adapted to confirming or refuting an alleged deviation in a QQ-plot objectively.

Example 4.37 Application: Black–Scholes model

The Black–Scholes model for log returns on the value of shares (see Example 2.17) says that these log returns can be viewed as independent samples from a normal distribution. The distribution of the log returns is important both for "risk management" and for the prices of derivatives (such as options). If we assume normality, but in reality, the log returns have a distribution with thicker tails (many extreme values), then someone who owns these shares runs a higher risk than was factored in, and the option price will not be realistic. This explains the interest in testing the normality assumption. Can we view the log returns as samples from a normal distribution, or not?

In addition to the marginal distribution of a sample, we can also study other aspects with the help of a test. In the case of log returns, for example, it could be interesting to study the time dependence.

This category of tests does not fit well in the general philosophy of testing because with goodness-of-fit tests we generally prefer not to reject the null hypothesis. The null hypothesis says, for example, that the data can be viewed as a sample from a normal distribution, and confirming this null hypothesis would provide the most information. However, the general set-up of testing does not give us this possibility: the only possible strong conclusion is that the null hypothesis is incorrect; in the other case, we do not obtain a strong statement. One could think that interchanging the null and alternative hypotheses would solve the problem, for when we then reject the null hypothesis, we would have the strong conclusion that the data comes from a normal distribution. However, in practice, this null hypothesis will never be rejected. In that case, the null hypothesis contains all nonnormal distributions. Every normal distribution in the alternative hypothesis can be approximated arbitrarily closely by a nonnormal distribution under the null hypothesis. It is consequently impossible to make a clear distinction between the null and alternative hypotheses and draw the strong conclusion. We therefore choose for the first-mentioned null hypothesis that the distribution from which the observations come is a normal distribution.

Following this course of action, it is wise to interpret the results of goodness-of-fit tests pragmatically. If, for example, the null hypothesis of normality is not rejected, then we view this as an indication that using the normal distribution is not unreasonable, without interpreting it as sufficient proof that we have normality. It is simply impossible to show that a given distribution is correct.

Example 4.38 Kolmogorov–Smirnov test

Let $X_1, \ldots, X_n$ be a sample from an unknown distribution F. We want to test the null hypothesis $H_0: F = F_0$ that this distribution is equal to a given reference distribution F_0 against the alternative $H_1: F \neq F_0$ that this is not the case. The distribution F_0 could, for example, be the standard normal distribution.

The *Kolmogorov–Smirnov test* is based on the *empirical distribution function* $\mathbb{F}_n$ of $X_1, \ldots, X_n$, which is defined as

$$\mathbb{F}_n(x) = \frac{1}{n} \#(X_i \leq x) = \frac{1}{n} \sum_{i=1}^{n} 1_{\{X_i \leq x\}}$$

(see Figure 4.12). The value $\mathbb{F}_n(x)$ is equal to the number of observations that are less than or equal to x, divided by n. By the law of large numbers, we have $\mathbb{F}_n(x) \overset{P}{\to} \mathrm{E}1_{\{X \leq x\}} = F(x)$ as $n \to \infty$. For sufficiently large values of n, the function $\mathbb{F}_n$ must therefore be close to the true distribution, so close to F_0 if H_0 is true. The Kolmogorov–Smirnov statistic is the maximal distance between $\mathbb{F}_n$ and F_0,

$$T = \sup_{x \in \mathbb{R}} \left| \mathbb{F}_n(x) - F_0(x) \right|.$$

We reject $H_0 \colon F = F_0$ for large values of T. We can deduce the critical value for the test from the probability distribution of T under H_0. This does not have a special name, but it has been tabulated and is available in statistical computer programs. It is good to know that the distribution is the same for every continuous distribution function F_0, so that one table suffices. For large values of n, we can also use the limit result

$$\lim_{n \to \infty} P_{F_0} \left(\sup_{x \in \mathbb{R}} \left| \mathbb{F}_n(x) - F_0(x) \right| > z/\sqrt{n} \right) = 2 \sum_{j=1}^{\infty} (-1)^{j+1} e^{-j^2 z^2}.$$

The sequence on the right can easily be computed numerically for given z. Consequently, this equality is, in particular, useful for determining p-values.

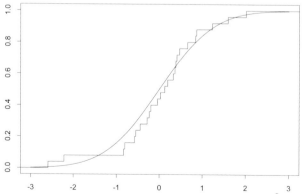

Figure 4.12. The empirical distribution function of a sample of size 25 from the $N(0, 1)$-distribution and the actual distribution function.

In many practical cases, the problem we just discussed is too simple. We often do not want to test a simple null hypothesis $H_0\colon F = F_0$, but rather a hypothesis of the form $H_0\colon F \in \{F_\theta\colon \theta \in \Theta\}$ for a given statistical model $\{F_\theta\colon \theta \in \Theta\}$. For example, tests whether the observations are "normally distributed" correspond to the choices $\theta = (\mu, \sigma^2) \in \mathbb{R} \times (0, \infty)$ and $F_{\mu,\sigma^2} = N(\mu, \sigma^2)$. An extension of the Kolmogorov–Smirnov test statistic is

$$T^* = \sup_{x \in \mathbb{R}} |\mathbb{F}_n(x) - F_{\hat{\theta}}(x)|,$$

for an estimator $\hat{\theta}$ of θ. We again reject for large values of T^*. Because of the substitution of $\hat{\theta}$, however, the distribution of T^* is not equal to that of T. In general, this distribution depends on the model we are testing, on the estimator $\hat{\theta}$ that is used, and even on the true parameter θ. The distribution has been tabulated for some special cases. In other cases, we use approximations or determine critical values using computer simulations.

Consider, for example, the application to testing normality. It seems natural to estimate the unknown parameter $\theta = (\mu, \sigma^2)$ using the sample mean and the sample variance. We reject the null hypothesis of normality for large values of the statistic

$$(4.4) \qquad T^* = \sup_{x \in \mathbb{R}} \left| \mathbb{F}_n(x) - \Phi\left(\frac{x - \overline{X}}{S_X}\right) \right|.$$

To determine the critical value for the test, or a p-value, we need to know the distribution of this statistic under the assumption that the null hypothesis is correct. Although the null hypothesis is composite, we can show that the distribution of T^* is the same under each element of the null hypothesis (Exercise 4.46). It is not easy to derive an analytic expression for this distribution, but we can easily approximate it using a simulation. We simulate a sample from the normal distribution of the same size as the data a great number of times, for example 1000 times, and compute the value of the Kolmogorov–Smirnov statistic T^* for each of the 1000 samples. As an approximation for the p-value, we then take the proportion of the 1000 values that is greater than the value of the statistic on the actual data.

Example 4.39 Chi-square test

Let $X_1, \ldots, X_n$ be a sample from an unknown distribution F. An alternative for the Kolmogorov–Smirnov test for a simple null hypothesis, $H_0\colon F = F_0$ against $H_1\colon F \neq F_0$, is the chi-square test for independence. We partition the codomain of X_1 into a number of contiguous intervals $I_1, I_2, \ldots, I_k$. The number of observations in the sample in each interval, denoted by N_j for $j = 1, \ldots, k$, is the random variable

$$N_j = \#(1 \le i \le n\colon x_i \in I_j).$$

Under the null hypothesis, the probability $p_j := \mathrm{P}_{H_0}(X_1 \in I_j)$ that an observation X_i lies in the interval I_j follows from the distribution F_0 for $j = 1, \ldots, k$, and the expectation of the number of observations in the interval I_j is equal to np_j. The test statistic for the chi-square tests gives a normalized measure of the difference between the realized number of observations and the expected number of observations in the intervals:

$$ X^2 = \sum_{j=1}^{k} \frac{(N_j - np_j)^2}{np_j}. $$

Under the null hypothesis, for fixed k, X^2 has approximately a chi-square distribution with $k - 1$ degrees of freedom. This approximation is reliable for sufficiently large values of n. As a rule of thumb, we use that the expected number of observations in each interval under the null hypothesis, np_j for $j = 1, \ldots, k$, should be at least 5. Chi-square tests are also used in other situations. In Example 4.48, we discuss another application of a chi-square test.

Example 4.40 Autocorrelations

In Section 2.3.1, we define the sample autocorrelation coefficient of order $h \in \mathbb{N}$ for a given sample $X_1, \ldots, X_n$ as

$$ \hat{\rho}_{X,n}(h) = \frac{\sum_{i=1}^{n-h}(X_{i+h} - \overline{X}_n)(X_i - \overline{X}_n)}{(n-h)S_{X,n}^2}. $$

Here, we write $\hat{\rho}_{X,n}(h)$, $\overline{X}_n$, and $S_{X,n}^2$ instead of $\hat{\rho}_X(h)$, $\overline{X}$, and S_X^2 to emphasize the dependence of these random variables on n. The sample autocorrelation coefficient of order h is a measure for the linear correlation between a variable X_t and a variable X_{t+h} measured h "points of time" later. If the sample autocorrelation coefficient takes on values close to 0, this is an indication for (linear) time dependence.

Suppose that we want to test the null hypothesis that $X_1, \ldots, X_n$ are independent and identically distributed; then we could choose the sample autocorrelation coefficients $\hat{\rho}_{X,n}(h)$ as test statistics and reject the null hypothesis when these coefficients are too far from 0. To specify "too far from 0," we need to know the distribution of the sample autocorrelation coefficients under the assumption of independence, so that we can define a critical region and determine p-values.

Since the sample autocorrelation coefficients are a complicated function of the variables $X_1, \ldots, X_n$ and their distribution moreover depends on the marginal distributions of the X_i, it is not easy to determine the null distribution. For large values of n, however, we can carry out an approximation, based on the following limit theorem. *If $X_1, \ldots, X_n$ is a sample from a distribution with finite fourth moment, then for every h, the sequence $\sqrt{n}\hat{\rho}_{X,n}(h)$ converges in distribution to a standard normal distribution as $n \to \infty$. Moreover, the sequences for different values of h are asymptotically independent.* In practice, we consider this mathematical theorem justification for viewing the sample autocorrelation coefficients $\sqrt{n}\rho_{X,n}(1), \sqrt{n}\rho_{X,n}(2), \ldots$ as a sequence of independent standard normal variables for large values of n and under H_0. If this were true, then the test "Reject H_0 when $\sqrt{n}|\rho_{X,n}(h)| \geq \xi_{1-\alpha_0/2}$" would have size α_0. Since this approximation only holds for large values of n, in reality, this test has size approximately α_0.

We can carry out the test for every $h > 0$ (where the normal approximation is only satisfactory when h is relatively small with respect to n). If we carry out the test for k values of h, each time with size α_0, then the size of all tests taken together is approximately equal to $1 - (1-\alpha_0)^k$. (Use the rule $P(\cup_h A_h) = 1 - P(\cap_h A_h^c)$ with A_h the event that the hth test falsely rejects the null hypothesis of independence. Because of the (asymptotic) independence of the tests, we then have $P(\cap_h A_h^c) = \prod_h (1 - P(A_h))$.) For small α_0, the size is then $1 - (1 - \alpha_0)^k \approx k\alpha_0$, so k times as large as the size of each of the tests individually. If we want to achieve an overall size α_0, then we need to carry out the individual tests with size α_0/k. In practice, we are less formal, and make a plot of, for example, the first 20 sample autocorrelation coefficients with horizontal lines at heights $\pm 1.96/\sqrt{n}$ (compare with Figure 2.13). If the observations are independent, then we expect that one of the 20 sample autocorrelation coefficients, and not significantly more, will fall outside the horizontal strip.

If we restrict the null hypothesis to the hypothesis that $X_1, \ldots, X_n$ are independent and normally distributed, then it is possible to determine the distribution of the sample autocorrelation coefficients more precisely.

4.7 Likelihood Ratio Tests

Tests are often set up based on heuristic arguments. A few examples were discussed in the previous section. In this section and the next, we discuss several general methods for finding a test, beginning with the most important one, the likelihood ratio test. In the next section, we discuss score and Wald tests.

Definition 4.41 Likelihood ratio statistic

If p_θ is the probability density of a random vector X, then the likelihood ratio statistic for testing $H_0: \theta \in \Theta_0$ against $H_1: \theta \in \Theta \setminus \Theta_0$ is defined as

$$\lambda(X) = \frac{\sup_{\theta \in \Theta} p_\theta(X)}{\sup_{\theta_0 \in \Theta_0} p_{\theta_0}(X)}.$$

(Define a/b as ∞ if $a > 0 = b$.)

To compute $\lambda(X)$, we maximize the likelihood twice, once with the parameter θ restricted to Θ_0, and once over the full parameter space Θ. Since Θ_0 is a subset of Θ, the likelihood ratio statistic will always be greater than or equal to 1. If we denote the usual maximum likelihood estimator by $\hat{\theta}$ and the maximum likelihood estimator under the assumption that the null hypothesis H_0 is correct by $\hat{\theta}_0$, then we can also write the likelihood ratio statistic as

$$\lambda(X) = \frac{p_{\hat{\theta}}(X)}{p_{\hat{\theta}_0}(X)}.$$

If $\hat{\theta} \in \Theta_0$, then $\hat{\theta}_0 = \hat{\theta}$ and the likelihood ratio statistic is equal to 1. If the numerator $p_{\hat{\theta}}(X)$ of $\lambda(X)$ is greater than the denominator, this is an indication that the space $\Theta \setminus \Theta_0$ contains "more likely" parameters than the null hypothesis space Θ_0. Large values of $\lambda(X)$ therefore give an indication that H_1 is correct. We therefore take a critical region of the form $\{\lambda(x) \geq c_{\alpha_0}\}$. The critical value c_{α_0} and/or p-values can be determined from the distributions of $\lambda(X)$ under every $\theta_0 \in \Theta_0$.

Example 4.42 Normal distribution

The likelihood ratio statistic for testing $H_0: \mu = \mu_0$ against $H_1: \mu \neq \mu_0$ based on a sample $X = (X_1, \ldots, X_n)$ from the $N(\mu, \sigma^2)$-distribution for a known σ^2 is

$$\lambda_n(X_1, \ldots, X_n) = \frac{\prod_{i=1}^{n}(2\pi\sigma^2)^{-1/2}e^{-\frac{1}{2}(X_i - \hat{\mu})^2/\sigma^2}}{\prod_{i=1}^{n}(2\pi\sigma^2)^{-1/2}e^{-\frac{1}{2}(X_i - \mu_0)^2/\sigma^2}}$$

$$= \exp\left(-\frac{1}{2\sigma^2}\sum_{i=1}^{n}(X_i - \hat{\mu})^2 + \frac{1}{2\sigma^2}\sum_{i=1}^{n}(X_i - \mu_0)^2\right),$$

where $\hat{\mu}$ is the maximum likelihood estimator for μ, that is, $\hat{\mu} = \overline{X}$. It can be useful to consider the distribution of $2 \log \lambda_n$ instead of that of λ_n because twice the logarithm of the likelihood ratio statistic is equal to

$$2 \log \lambda_n(X_1, \ldots, X_n) = -\frac{1}{\sigma^2}\sum_{i=1}^{n}(X_i - \overline{X})^2 + \frac{1}{\sigma^2}\sum_{i=1}^{n}(X_i - \mu_0)^2$$

$$= n\left(\frac{\overline{X} - \mu_0}{\sigma}\right)^2.$$

Under H_0, the variable $\sqrt{n}(\overline{X} - \mu_0)/\sigma$ has the $N(0, 1)$-distribution, so that $2 \log \lambda_n$ has a χ_1^2-distribution (see Definition 4.27). The null hypothesis is therefore rejected when $n(\overline{X} - \mu_0)^2/\sigma^2 \geq \chi_{1,1-\alpha_0}^2$. Since $(\xi_{1-\alpha_0/2})^2 = \chi_{1,1-\alpha_0}^2$, the likelihood ratio test above is identical to the Gauss test, where H_0 is rejected when $|\sqrt{n}(\overline{X} - \mu_0)/\sigma| \geq \xi_{1-\alpha_0/2}$ (see Example 4.12).

Generally, determining the distribution of the likelihood ratio statistic for every $\theta_0 \in \Theta_0$ is complicated, and approximations are used. A large sample approximation is often possible when the observation is a vector $X = (X_1, \ldots, X_n)$ consisting of a sample $X_1, \ldots, X_n$ from a distribution with (marginal) probability density p_θ. The likelihood ratio statistic then has the following form:

$$\lambda_n(X) = \frac{\prod_{i=1}^n p_{\hat{\theta}}(X_i)}{\prod_{i=1}^n p_{\hat{\theta}_0}(X_i)}.$$

Denote by $\sqrt{n}(\Theta - \theta_0)$ the set of vectors $\sqrt{n}(\theta - \theta_0)$ as θ runs through Θ, that is, $\sqrt{n}(\Theta - \theta_0) = \{\sqrt{n}(\theta - \theta_0) : \theta \in \Theta\}$. We assume that Θ is a subset of the Euclidean space.

Theorem 4.43

Suppose that the map $\vartheta \mapsto \log p_\vartheta(x)$ is continuous and differentiable for all x, with gradient $\dot{\ell}_\vartheta(x)$ such that $\|\dot{\ell}_\vartheta(x)\| \leq L(x)$ for every ϑ in a neighborhood of a given $\theta_0 \in \Theta_0$, where L is a function with $E_{\theta_0} L^2(X_1) < \infty$. Suppose, furthermore, that for the same θ_0, the sets $\sqrt{n}(\Theta - \theta_0)$ and $\sqrt{n}(\Theta_0 - \theta_0)$ converge to, respectively, k-dimensional and k_0-dimensional linear subspaces as $n \to \infty$. Finally, suppose that the maximum likelihood estimators $\hat{\theta}_0$ and $\hat{\theta}$ under θ_0 converge in probability to θ_0 and that the Fisher information matrix i_ϑ is invertible for all ϑ and depends continuously on ϑ. Then, under the given θ_0, we have

$$2 \log \lambda_n(X_1, \ldots, X_n) \rightsquigarrow \chi^2_{k-k_0} \qquad n \to \infty.$$

A "regularity condition" consisting of the differentiability of the log probability density with respect to the parameter is essential for the result, as it was in Theorem 5.9 on the asymptotic normality of the maximum likelihood estimator. This theorem (or Lemma 5.14) also gives sufficient conditions for the consistency $\hat{\theta}_0 \overset{P}{\to} \theta_0$ and $\hat{\theta} \overset{P}{\to} \theta_0$ as $n \to \infty$. The notation $\overset{P}{\to}$ means "convergence in probability"; see Definition A.24.

In addition to this, Theorem 4.43 assumes that the sequences of sets $\sqrt{n}(\Theta - \theta_0)$ and $\sqrt{n}(\Theta_0 - \theta_0)$ converge to linear subspaces. This condition is usually satisfied if the true parameter θ_0 is not a boundary point of the parameter spaces Θ and Θ_0. If this condition is not satisfied, then the chi-square approach fails. The convergence should be understood in the following sense: a sequence of sets H_n converges to a set H if

  (i) every element $h \in H$ is the limit $h = \lim_{n \to \infty} h_n$ of a sequence h_n with $h_n \in H_n$ for every n;
  (ii) if $h = \lim_{i \to \infty} h_{n_i}$ for given positive integers $n_1 < n_2 < \cdots$ and elements $h_{n_i} \in H_{n_i}$ for every i, then $h \in H$.

In most cases, the limit set H is exactly the set of limits $h = \lim h_n$ of convergent sequences with $h_n \in H_n$ for every n. Below, we give two general examples, and a concrete example to clarify the convergence.

For a proof of the theorem and an extension to boundary points θ_0, we refer to the book "Asymptotic statistics" written by Van der Vaart (Chapter 16, 1998).

We can extend the theorem to nonidentically distributed or dependent observations. Moreover, the assumption that the parameter space is a subspace of the Euclidean space is unnecessary. The statement of the theorem depends only on the "codimension" of the null hypothesis space Θ_0 within Θ (the number $k - k_0$ in the theorem). The theorem can also be extended to testing finite-dimensional parameters in semi-parametric models such as the Cox model in Section 7.6.

The "theorem" proposes to reject the null hypothesis at level α_0 if

$$2 \log \lambda_n (X_1, \ldots, X_n) \geq \chi^2_{k-k_0, 1-\alpha_0}.$$

This critical region is always one-sided, regardless of whether the null hypothesis is one- or two-sided.

Example 4.44 Simple null hypothesis

Let θ be a one-dimensional parameter, and suppose that we want to test the simple null hypothesis $H_0 : \theta = \theta_0$ for a given value θ_0. If θ_0 is an interior point of the parameter space Θ, then the convergence in the theorem holds with $k = 1$ and $k_0 = 0$. Under certain regularity conditions, twice the logarithm of the likelihood ratio statistic is therefore asymptotically chi-square distributed with one degree of freedom.

It is immediately clear that $k_0 = 0$, because $\Theta_0 = \{\theta_0\}$, so that $\sqrt{n}(\Theta_0 - \theta_0) = \{0\}$ for every n. Obviously, the sequence of sets $\{0\}, \{0\}, \{0\}, \ldots$ converges to the zero-dimensional space $\{0\}$.

The assumption that θ_0 is an interior point of Θ means that $\Theta - \theta_0$ contains a (possibly very small) open ball around 0. Then for large n, the set $\sqrt{n}(\Theta - \theta_0)$ contains a very large ball around 0, and we can verify that this implies that the limit of this sequence of sets is the full space $\mathbb{R}$.

Example 4.45 One-dimensional restriction

Let $\theta = (\theta^1, \ldots, \theta^m)$ be an m-dimensional vector, and suppose that we want to test the null hypothesis $H_0 : \theta^1 = c$ that the first coordinate has a certain value. The remaining $m - 1$ coordinates can be chosen freely in the null hypothesis. Once again, assume that a given vector $\theta_0 = (c, \theta_0^2, \ldots, \theta_0^m) \in \Theta_0$ is an interior point of the parameter space Θ. Arguments similar to those used in the previous example make it seem plausible that in this case, the convergence in the theorem holds with given $k = m$ and $k_0 = m - 1$ for the given θ_0. The limit distribution of twice the logarithm of the likelihood ratio statistic for testing the null hypothesis over the one-dimensional parameter is therefore χ_1^2 (under certain conditions).

The more general form of a one-dimensional null hypothesis is $H_0: b^T\theta = c$, with $b \in \mathbb{R}^m$, $c \in \mathbb{R}$, and $b^T\theta$ the inner product of b and θ. (In the previous paragraph, $b = (1, 0, \ldots, 0)$.) In that case, there exists an orthonormal coordinate transformation U on Θ, $\theta \mapsto U\theta =: \tilde{\theta}$, such that $\tilde{\theta}^1 = b^t\theta/\|b\|$. Using this, the null hypothesis $H_0: b^T\theta = c$ becomes equivalent to $H_0: \tilde{\theta}^1 = \tilde{c}$, with $\tilde{c} = c/\|b\|$. Moreover, the value of the likelihood ratio statistic remains unchanged because, using the substitution $\theta \mapsto U^{-1}\tilde{\theta}$, we can write the likelihood as a function of $\tilde{\theta}$ and the maximum likelihood estimator for $\tilde{\theta}$ is equal to $U\hat{\theta}$. It follows that the limit distribution of $2 \log \lambda_n$ for a general one-dimensional restriction is the χ_1^2-distribution (under certain conditions).

The likelihood ratio test is not uniformly most powerful (Definition 6.37). This is not a deficiency of this test, but a consequence of the fact that for many problems, there does not exist a uniformly most powerful test. For different alternative values, a different test is most powerful each time. The likelihood ratio test is "average" for various alternative values, but often not absolutely optimal for any single alternative value.

Example 4.46 Normal distribution, continue from Example 4.42

Let $X_1, \ldots, X_n$ be a sample from the $N(\mu, \sigma^2)$-distribution with known σ^2. In Example 4.42, we deduce that under the null hypothesis $H_0: \mu = \mu_0$, twice the logarithm of the likelihood ratio statistic is exactly χ_1^2-distributed.

The likelihood ratio statistic for testing $H_0: \mu \leq \mu_0$ is more complicated, and we no longer have the convergence of the sets $\sqrt{n}(\Theta_0 - \theta_0)$ in the "theorem" for all $\theta_0 \in \Theta_0$. Moreover, the asymptotic null distribution is not χ^2! To see the first statement, we take $\Theta_0 = (-\infty, \mu_0]$. Then $\sqrt{n}(\Theta_0 - \mu_0) = (-\infty, 0]$ for every n, and this set does not converge to a linear space.

Example 4.47 Comparing two binomial probabilities

Let X and Y be independent with, respectively, the $\text{bin}(m, p_1)$- and $\text{bin}(n, p_2)$-distributions. We want to test the hypothesis $H_0: p_1 = p_2$ against the alternative $H_1: p_1 \neq p_2$. The maximum likelihood estimator for (p_1, p_2) without restrictions is $(\hat{p}_1, \hat{p}_2) = (X/m, Y/n)$, the vector consisting of the two maximum likelihood estimators when we observe only X or Y. Under the null hypothesis that $p = p_1 = p_2$, the likelihood function is

$$p \mapsto \binom{m}{X} p^X (1-p)^{m-X} \binom{n}{Y} p^Y (1-p)^{n-Y}.$$

This is maximized by $\hat{p}_0 = (X+Y)/(m+n)$. The maximum likelihood estimator for (p_1, p_2) under the null hypothesis is therefore $(\hat{p}_0, \hat{p}_0)$. The likelihood ratio statistic can now be computed as

$$\lambda(X, Y) = \frac{(X/m)^X (1 - X/m)^{m-X} (Y/n)^Y (1 - Y/n)^{n-Y}}{\hat{p}_0^{X+Y} (1 - \hat{p}_0)^{m+n-X-Y}}.$$

The "theorem" can be applied and gives a chi-square approximation with $2 - 1 = 1$ degree of freedom, because it is a one-dimensional restriction. We reject H_0 when $2 \log \lambda \geq \chi^2_{1, 1-\alpha_0}$.

Alternatives for this test are Fisher's exact test and the chi-square test; see other textbooks for more information.

---

* **Example 4.48** Multinomial distribution

Let $Y = (Y_1, \ldots, Y_m)$ be multinomially distributed with parameters $(n, p_1, \ldots, p_m)$. We assume that n is known and want to test a hypothesis on the probability vector $p = (p_1, \ldots, p_m)$. The likelihood function is given by

$$p \mapsto \binom{n}{Y_1 \cdots Y_m} p_1^{Y_1} \cdots p_m^{Y_m}.$$

The maximum likelihood estimator for $p_1, \cdots, p_m$ with respect to the natural parameter space $\{p \in \mathbb{R}^m : p_i \geq 0, \sum_{i=1}^m p_i = 1\}$ (the "unit simplex") is equal to $\hat{p}_i = Y_i/n$ for $i = 1, \cdots, m$. The log-likelihood ratio statistic for testing $H_0 : p \in \mathcal{P}_0$ for a given subset $\mathcal{P}_0$ of the unit simplex is therefore

$$\log \lambda(Y) = \log \frac{\binom{n}{Y_1 \cdots Y_m} \prod_{i=1}^m (Y_i/n)^{Y_i}}{\sup_{p \in \mathcal{P}_0} \binom{n}{Y_1 \cdots Y_m} \prod_{i=1}^m p_i^{Y_i}} = \inf_{p \in \mathcal{P}_0} \sum_{i=1}^m Y_i \log \frac{Y_i}{np_i}.$$

Even for the simple null hypothesis $\mathcal{P}_0 = \{p_0\}$, this statistic has a complicated distribution. Since Y can be viewed as the sum of n independent multinomially distributed variables with parameters 1 and p, and the probability density of Y and the joint density of this sample are proportional, the "theorem" stated previously may be applied. The dimension of the parameter space (the k in the theorem) is equal to $m - 1$ (provided that the true parameter p is an interior point of the unit simplex) because $(p_1, \ldots, p_m)$ varies over an $(m - 1)$-dimensional set.

For a simple null hypothesis, the likelihood ratio test is asymptotically equivalent to the chi-square test of Example 4.39. To see this, we rewrite the log-likelihood ratio statistic $\log \lambda(Y)$. The Taylor approximation of $f(y) = y \log(y/y_0)$ with $f'(y) = \log(y/y_0) + 1$ and $f''(y) = 1/y$ in the neighborhood of y_0 give, for large n, the approximation $f(y) \approx (y - y_0) + \frac{1}{2}(y - y_0)^2/y_0$. This approximation applied to every term in the sum of $\log \lambda(Y)$ with $y = Y_i$ and $y_0 = np_i$ gives

$$\log \lambda(Y) = \sum_{i=1}^m Y_i \log \frac{Y_i}{np_i} \approx \sum_{i=1}^m (Y_i - np_i) + \frac{1}{2} \sum_{i=1}^m \frac{(Y_i - np_i)^2}{np_i}$$

$$= \frac{1}{2} \sum_{i=1}^m \frac{(Y_i - np_i)^2}{np_i}.$$

The last equality follows from $\sum_{i=1}^m np_i = n$ and $\sum_{i=1}^m Y_i = n$.

* **Example 4.49** Application: compound Poisson process

In Example 3.22, the maximum likelihood estimator is determined for the two-dimensional parameter (θ, μ) in the distribution for the monthly payment made by a health insurance company. The assumption was made that the expected number of claims and claim sizes are equal all months of the year. However, it has been theorized that there is a difference between the expected size of the claims in the summer and in the winter. We will test this hypothesis using the likelihood ratio test, based on the data of n winter months and m summer months. We assume that the data from different months are independent.

As a model, we assume that the sizes of the claims in the summer and in the winter have exponential distributions with unknown parameters θ_s and θ_w, respectively. The parameter μ from the distribution of the number of claims is taken the same in the summer and in the winter. The parameter is now three-dimensional, $(\mu, \theta_s, \theta_w)$. The null hypothesis reads $H_0 : \theta_s = \theta_w$, and the alternative hypothesis is $H_1 : \theta_s \neq \theta_w$. The maximum likelihood estimator for the parameter can be determined as described in Example 3.22. Under the null hypothesis, the parameter is equal to $(\mu, \theta_0, \theta_0)$. As in Example 3.22, the log-likelihood function can be written as a sum of terms that each depend on one of the parameters. The maximum likelihood estimator for μ follows by maximizing the term that depends only on μ. Since this term appears in both the log-likelihood of the full model and the log-likelihood under the null hypothesis, this term cancels out in the log-likelihood ratio statistic. We therefore from now on disregard the term and the estimator for μ. Under the null hypothesis, θ_0 is estimated by

$$\hat{\theta}_0 = \frac{\sum_{i=1}^{n+m} N_i}{\sum_{i=1}^{n+m} \sum_{j=1}^{N_i} C_{i,j}},$$

where the data of all $n + m$ months has been taken together. Without the restrictions of the null hypothesis, the maximum likelihood estimators for θ_s and θ_w are given by

$$\hat{\theta}_s = \frac{\sum_{i=1}^{n} N_i^s}{\sum_{i=1}^{n} \sum_{j=1}^{N_i^s} C_{i,j}^s} \quad \text{and} \quad \hat{\theta}_w = \frac{\sum_{i=1}^{m} N_i^w}{\sum_{i=1}^{m} \sum_{j=1}^{N_i^w} C_{i,j}^w},$$

where the superscript s and w indicate data for the summer and winter months, respectively. The log-likelihood ratio statistic is given by

$$\log \lambda_{n,m} = \sum_{i=1}^{n} \log\left(\prod_{j=1}^{N_i^s} \frac{\hat{\theta}_z}{\hat{\theta}_0} e^{-(\hat{\theta}_s - \hat{\theta}_0)C_{i,j}^s}\right) + \sum_{i=1}^{m} \log\left(\prod_{j=1}^{N_i^w} \frac{\hat{\theta}_w}{\hat{\theta}_0} e^{-(\hat{\theta}_w - \hat{\theta}_0)C_{i,j}^w}\right).$$

By the previous "theorem," for large values of n and m, we know that under the null hypothesis, the statistic $2 \log \lambda_{n,m}$ has approximately a chi-square distribution with 1 degree of freedom, because we are dealing with a one-dimensional restriction (see Example 4.45).

* 4.8 Score and Wald Tests

Carrying out the likelihood ratio test requires determining the maximum likelihood estimator for the parameter, both under the null hypothesis and for the full model. This can be demanding. The *score test* is an alternative that requires less computation and has about the same quality when we have many observations.

The score function of a statistical model given by the marginal probability density p_θ is defined as the gradient $\dot{\ell}_\theta(x) = \nabla_\theta \log p_\theta(x)$ of the logarithm of the marginal probability density; see Definition 5.8. Lemma 5.10 in Chapter 5 states that, under certain conditions, $E_\theta \dot{\ell}_\theta(X) = 0$ for every parameter θ. If the value $\dot{\ell}_{\theta_0}(x)$ differs considerably from 0, this is an indication that θ_0 is not the true value of the parameter. This gives the principle of the score test: the null hypothesis $H_0 : \theta = \theta_0$ is rejected when the score function $\dot{\ell}_{\theta_0}(x)$ differs considerably from 0.

The question is how to quantify "considerably." We will answer this question only in the case where $X = (X_1, \ldots, X_n)$ is a sample of independent, identically distributed variables. The probability density of X is then of the form $(x_1, \ldots, x_n) \mapsto \prod_{i=1}^n p_\theta(x_i)$, for p_θ the (marginal) density of one observation. The score statistic for $H_0 : \theta = \theta_0$ is then of the form

$$\sum_{i=1}^n \dot{\ell}_{\theta_0}(X_i),$$

where $\dot{\ell}_\theta$ is now the score function for a single observation. The score statistic is a sum of independent, identically distributed random vectors. Under the null hypothesis, $E_{\theta_0} \dot{\ell}_{\theta_0}(X) = 0$, and when n is large, the sum above has approximately a normal distribution by the central limit theorem (Theorem A.28). This theorem implies that under θ_0,

$$\frac{1}{\sqrt{n}} \sum_{i=1}^n \dot{\ell}_{\theta_0}(X_i) \rightsquigarrow N(0, i_{\theta_0}), \qquad i_{\theta_0} = E_{\theta_0} \dot{\ell}_{\theta_0}(X_i) \dot{\ell}_{\theta_0}^T(X_i)$$

as $n \to \infty$. The number i_{θ_0}, or the matrix i_{θ_0} when the parameter has dimension (which we assume to be finite) greater than one, is exactly the Fisher information, which we will also encounter in Chapter 5. When θ is a one-dimensional real-valued parameter, we can choose the following as test statistic:

$$\left| i_{\theta_0}^{-1/2} \frac{1}{\sqrt{n}} \sum_{i=1}^n \dot{\ell}_{\theta_0}(X_i) \right|.$$

We reject the null hypothesis $H_0 : \theta = \theta_0$ when this test statistic is greater than the $(1 - \alpha_0/2)$-quantile of the standard normal distribution.

When we have a k-dimensional parameter, the displayed expression is a vector. We then choose the square of its norm as test statistic and reject the null hypothesis $H_0 : \theta = \theta_0$ when this quantity is greater than the $(1 - \alpha_0)$-quantile of the chi-square distribution with k degrees of freedom. If n is sufficiently large, the size of the test is approximately equal to α_0.

This gives a complete description of the score test for a simple null hypothesis. To test a composite null hypothesis $H_0: \theta \in \Theta_0$ for a given subset $\Theta_0 \subset \Theta$, we cannot use the test in this form, because if the null hypothesis is true, we do not know the true $\theta_0 \in \Theta_0$. The score test can be extended to this case by substituting the maximum likelihood estimator $\hat{\theta}_0$ for the unknown θ_0 under the null hypothesis for θ. We then use the test statistic

(4.5)
$$\left\| i_{\hat{\theta}_0}^{-1/2} \frac{1}{\sqrt{n}} \sum_{i=1}^{n} \dot{\ell}_{\hat{\theta}_0}(X_i) \right\|^2 .$$

Under the same type of regularity conditions as those for the likelihood ratio test (compare with Theorem 4.43), for large n, under the null hypothesis, this statistic has approximately the chi-square distribution with $k - k_0$ degrees of freedom, with the same k and k_0 as in Theorem 4.43. We therefore reject $H_0: \theta \in \Theta_0$ if the statistic in (4.5) is greater than the $(1 - \alpha_0)$-quantile of the chi-square distribution with $k - k_0$ degrees of freedom.

We see that applying the score test for a composite null hypothesis requires determining the maximum likelihood estimator under the null hypothesis. If the parameter θ is partitioned as $\theta = (\theta_1, \theta_2)$ and the null hypothesis space is of the form $\Theta_0 = \{(\theta_1, \theta_2): \theta_1 \in \mathbb{R}^{k_0}, \theta_2 = 0\}$, then this corresponds to determining the maximum likelihood estimator in a lower-dimensional submodel. If we set $\dot{\ell}_\theta = (\dot{\ell}_{\theta,1}, \dot{\ell}_{\theta,2})$, where $\dot{\ell}_{\theta,i}$ is the vector of partial derivatives of the logarithm of the probability density with respect to the coordinates of θ_i, then $\hat{\theta}_0$ will satisfy $\hat{\theta}_0 = (\hat{\theta}_{0,1}, 0)$ for $\hat{\theta}_{0,1}$ determined by the likelihood equation

$$\sum_{i=1}^{n} \dot{\ell}_{\hat{\theta}_{0,1}}(X_i) = 0.$$

This is a system with number of equations equal to the dimension of θ_1 in $\theta = (\theta_1, \theta_2)$. The vector $\sum_{i=1}^{n} \dot{\ell}_{\hat{\theta}_0}(X_i)$ is now of the form $(0, \sum_{i=1}^{n} \dot{\ell}_{\hat{\theta}_0,2}(X_i))$, and the score test statistic (4.5) reduces to

(4.6)
$$\frac{1}{n} \left(\sum_{i=1}^{n} \dot{\ell}_{\hat{\theta}_0,2}(X_i) \right)^T (i_{\hat{\theta}_0}^{-1})_{2,2} \left(\sum_{i=1}^{n} \dot{\ell}_{\hat{\theta}_0,2}(X_i) \right).$$

Here, $(i_{\hat{\theta}_0}^{-1})_{2,2}$ is the relevant submatrix of the matrix $i_{\hat{\theta}_0}^{-1}$. (Note that the submatrix $(A^{-1})_{2,2}$ of an inverse matrix A^{-1} is not the inverse of the submatrix $A_{2,2}$.) We can interpret this quantity as a measure for the success of the maximum likelihood estimator $\hat{\theta}_0 = (\hat{\theta}_{0,1}, 0)$ in reducing the score equation $\sum_{i=1}^{n} \dot{\ell}_\theta(X_i)$ for the full model to 0. Since $\sum_{i=1}^{n} \dot{\ell}_{\hat{\theta}}(X_i) = 0$ for the maximum likelihood estimator $\hat{\theta}$ for the full model, we can also view the score test statistic as a measure for the difference between the maximum likelihood estimators under the null hypothesis and of the full model.

For a null hypothesis of the form $H_0: g(\theta) = 0$ for a given, general function $g: \mathbb{R}^k \to \mathbb{R}^m$, we can sometimes determine the maximum likelihood estimator $\hat{\theta}_0$ under H_0 by using the Lagrange method. This is a general method from mathematical analysis for determining an extremum of a function under an additional constraint. The idea is to determine the stationary points $(\hat{\theta}_0, \hat{\lambda})$ of the function

$$(\theta, \lambda) \mapsto \sum_{i=1}^{n} \log p_\theta(X_i) + \lambda^T g(\theta).$$

This function is the likelihood function plus a vector parameter $\lambda \in \mathbb{R}^m$, the "Lagrange multiplier," times the additional constraint ($g(\theta) - 0 = g(\theta)$ in our situation). By the Lagrange theorem, under certain conditions, the first coordinate $\hat{\theta}_0$ of such a stationary point is the desired maximum likelihood estimator under H_0. Differentiating with respect to θ gives the stationary equation

$$\sum_{i=1}^{n} \ell_{\hat{\theta}_0}(X_i) + \dot{g}(\hat{\theta}_0)^T \hat{\lambda} = 0,$$

where $\dot{g} \in \mathbb{R}^{m \times k}$ is the gradient of g. This shows that the Lagrange multiplier $\hat{\lambda}$ is "proportional" to $\sum_{i=1}^{n} \dot{\ell}_{\hat{\theta}_0}(X_i)$, which is essentially the score test statistic. In particular, when $\theta = (\theta_1, \theta_2)$ and $g(\theta) = \theta_2$, the functional matrix is equal to $(0, I)^T$ and we have $\hat{\lambda} = \sum_{i=1}^{n} \dot{\ell}_{\hat{\theta}_0,2}(X_i)$, which is essentially the test statistic (4.6). This is probably why in econometrics, the score test is known as the *Lagrange multiplier test*.

As noted before, the score test can be viewed as a comparison of the maximum likelihood estimator $\hat{\theta}_0$ under the null hypothesis and the maximum likelihood estimator $\hat{\theta}$ for the full model. The *Wald test* carries out a direct comparison and can be viewed as a third variant of the likelihood ratio test. In the case of a partitioned parameter $\theta = (\theta_1, \theta_2)$ and a null hypothesis of the form $\Theta_0 = \{\theta = (\theta_1, \theta_2): \theta_2 = 0\}$, the Wald test is based on the second component $\hat{\theta}_2$ of the maximum likelihood estimator $\hat{\theta} = (\hat{\theta}_1, \hat{\theta}_2)$ for the full model. If $\hat{\theta}_2$ differs too much from the maximum likelihood estimator under the null hypothesis, which is 0, the null hypothesis is rejected. "Too much" can be specified by referring to the limit distribution of the maximum likelihood estimator.

More generally, the Wald test is based on the difference $\hat{\theta} - \hat{\theta}_0$. If the quadratic form

$$n(\hat{\theta} - \hat{\theta}_0)^T i_{\hat{\theta}_0}(\hat{\theta} - \hat{\theta}_0)$$

is too great, the null hypothesis is rejected. Under the conditions of Theorem 4.43, we can show that as $n \to \infty$, this sequence of *Wald statistics* converges to a chi-square distribution with $k - k_0$ degrees of freedom, so that the correct critical value can be chosen from the χ^2-table.

We can show that, under certain conditions, the likelihood ratio test, the score test, and the Wald test all have approximately the same power function if the number of observations is large. We again restrict ourselves to the case where the observation is a vector $X = (X_1, \ldots, X_n)$ of identically distributed random variables with density p_θ.

Theorem 4.50

Suppose that the conditions of Theorem 4.43 are satisfied and that, moreover, the map $\vartheta \mapsto \log p_\vartheta(x)$ has a second derivative $\ddot{\ell}_\vartheta(x)$ such that $\|\ddot{\ell}_\vartheta(x)\| \le L(x)$ for a function L with $\mathrm{E}_{\theta_0} L^2(X_1) < \infty$. Then, under θ_0, as $n \to \infty$, we have

$$\frac{1}{n}\left(\sum_{i=1}^{n} \ell_{\hat{\theta}_0}(X_i)\right)^T i_{\hat{\theta}_0}^{-1}\left(\sum_{i=1}^{n} \ell_{\hat{\theta}_0}(X_i)\right) - n(\hat{\theta} - \hat{\theta}_0)^T i_{\hat{\theta}_0}(\hat{\theta} - \hat{\theta}_0) \rightsquigarrow 0,$$

$$2\log \lambda_n(X_1, \ldots, X_n) - n(\hat{\theta} - \hat{\theta}_0)^T i_{\hat{\theta}_0}(\hat{\theta} - \hat{\theta}_0) \rightsquigarrow 0,$$

where the symbol "$\rightsquigarrow$" means convergence in distribution. Moreover, the sequence $n(\hat{\theta} - \hat{\theta}_0)^T i_{\hat{\theta}_0}(\hat{\theta} - \hat{\theta}_0)$ converges in distribution to a chi-square distribution with $k - k_0$ degrees of freedom.

* 4.9 Multiple Testing

Daily, all over the world, many statistical tests are carried out, generally of size 5%. Some 1 out of 20 true hypotheses are then falsely rejected. This means that in many statistically supported papers in medical journals, in which a 5% statistically significant result is standard, the claim may be unfounded (we mean 5% of the claims in situations where there is no effect; this is not the same as 5% of the papers). No one seems worried about this.

The situation is different when one researcher carries out a large number of tests simultaneously. If he chooses a level of 5% for every test, then when carrying out, for example, 1000 tests, he should expect at least 50 "significant" results, even when in reality there is nothing significant to be found. Such a situation occurs, for example, in medical image analysis if every pixel is tested to see whether the value of the image deviates from what is normal, and in the analysis of genetic data if a large number of genes are being studied for their influence. In all these situations, the multiple testing is seen as a problem.

If we carry out N tests simultaneously, each of size α, then the probability that one or more of the null hypotheses is falsely rejected is less than or equal to $N\alpha$. A simple way to obtain an overall size of α is therefore to carry out each individual test with size α/N. This is known as the *Bonferroni correction*.

The disadvantage of this simple correction is that the actual size is often much smaller than the desired α. (The correction is very *conservative*.) To gain more insight, we formalize the test problem. Suppose that we want to test the N null hypotheses $H_0^j : \theta \in \Theta_0^j$ for $j = 1, \ldots, N$, where $\Theta_0^1, \ldots, \Theta_0^N$ are given subsets of the parameter set that describe the probability distribution of the observation X. To test H_0^j, we have a test with critical region K^j, and, in a multiple test procedure, we decide to reject the null hypotheses H_0^j for which $X \in K^j$. If the true parameter θ_0 belongs to Θ_0^j and $X \in K^j$, then we make a type I error with respect to the jth hypothesis. In reality, every combination of correct and false null hypotheses is possible. If the hypotheses H_0^j for every j in a given set $J \subset \{1, \ldots, N\}$ are correct, and the other null hypotheses are incorrect, then a meaningful definition of the *overall size* is

$$\sup_{\theta \in \cap_{j \in J} \Theta_0^j} \mathrm{P}_\theta \big(X \in \cup_{j \in J} K^j \big).$$

This is the maximal probability that we reject at least one of the correct hypotheses. This expression is less than

$$\sup_{\theta \in \cap_{j \in J} \Theta_0^j} \sum_{j \in J} \mathrm{P}_\theta \big(X \in K^j \big) \leq \sum_{j \in J} \sup_{\theta \in \Theta_0^j} \mathrm{P}_\theta \big(X \in K^j \big).$$

The suprema in the sum on the right-hand side are exactly the sizes of the individual tests with critical regions K^j for the null hypotheses H_0^j. If all these tests have size less than or equal to α, then the overall size is less than or equal to $\#J \alpha \leq N\alpha$, as we concluded earlier.

   The computation shows why the Bonferroni correction is conservative. First, the upper bound $N\alpha$ corresponds with the situation that all null hypotheses are correct, while in reality possibly only $\#J$ hypotheses are correct. Second, and more importantly, the upper bound is based on the inequality $\mathrm{P}_\theta \big(X \in \cup_j K^j \big) \leq \sum_j \mathrm{P}_\theta (X \in K^j)$, which in many cases is too pessimistic: if the critical regions overlap, then the probability of their union may be much smaller than the sum of their probabilities. "Overlap" often arises because of "stochastic dependence" between the tests. In image analysis, for example, data concerning different pixels are often dependent. Unfortunately, there is no general method for taking such dependence into account when combining tests. The best solution is often to not combine the individual tests, but define a new overall test.

   In some cases, the stochastic independence of the tests is a reasonable assumption. If the critical regions are stochastically independent, then

$$\mathrm{P}_\theta \big(X \in \cup_j K^j \big) = 1 - \mathrm{P}_\theta (X \in \cap_j (K^j)^c) = 1 - \prod_j \big(1 - \mathrm{P}_\theta (X \in K^j) \big).$$

If all tests have size less than α, then for $\theta \in \cap_j \Theta_0^j$, this expression is bounded by $1 - (1 - \alpha)^N$, which is (of course) less than $N\alpha$. For small values of α, however, the difference is very small. If we want an overall size of α_0, the Bonferroni correction suggests to take size α_0/N for each test, while for independent hypotheses the somewhat larger value $1 - (1 - \alpha_0)^{1/N}$ can be taken. As $N \to \infty$, the quotient $\big(1 - (1 - \alpha_0)^{1/N}\big)/(\alpha_0/N)$ of these choices increases to $-\log(1 - \alpha_0)/\alpha_0$. For $\alpha_0 = 0.05$, the limit is approximately 1.025866, and the Bonferroni correction is only 2.6% greater.

If we carry out many tests simultaneously (for example, $N \approx 1000$ or greater), then controlling the size may not be very meaningful. Because the overall size is defined as the probability of *at least* one type I error, it is connected to preventing *all* type I errors. This extreme aim will usually lead to a very small power function, with possibly the result that no hypothesis can be rejected. Another approach is to accept that a small number of null hypotheses will be falsely rejected if in that case at least a reasonable number of null hypotheses are correctly rejected.

Definition 4.51 False discovery rate

The false discovery rate (FDR) is the expected proportion of falsely rejected null hypothesis among the rejected hypotheses,

$$FDR(\theta) = E_\theta \frac{\#\{j : X \in K^j, \theta \in \Theta_0^j\}}{\#\{j : X \in K^j\}}.$$

An FDR of at most 5% can be a reasonable criterion. The following procedure, derived by *Benjamini and Hochberg*, is often applied to control the FDR. The procedure is formulated in terms of the p-values P_j of the N tests.

(i) Order the p-values according to size: $P_{(1)} \leq P_{(2)} \leq \cdots \leq P_{(N)}$, and let H_0^j be the hypothesis that corresponds with the jth order statistic $P_{(j)}$.

(ii) Reject all null hypotheses H_0^j with $N P_{(j)} \leq j\alpha$.

(iii) Also reject all null hypotheses with p-value less than that of one of the null hypotheses rejected in step (ii).

It is clear that in general, this procedure will reject more null hypotheses than the Bonferroni method. In terms of p-values, the Bonferroni method corresponds to rejecting the hypotheses H_0^j with $N P_j \leq \alpha$, whereas the Benjamini–Hochberg method uses an extra factor j in the equation $N P_{(j)} \leq j\alpha$ in step (ii). Under certain circumstances, however, this does not negatively influence the FDR. In particular, we can prove that

(4.7)
$$FDR(\theta) \leq \frac{\#\{j : \theta \in \Theta_0^j\}}{N} \alpha (1 + \log N),$$

where the factor $1 + \log N$ may be left out when the test statistics of the different tests are independent or have a certain form of positive dependence (see Theorem 4.52). Without the logarithmic term, the right-hand side is certainly less than α; with the logarithmic term, this is not always the case, but the Benjamini–Hochberg procedure applied with the slightly smaller value $\alpha/(1 + \log N)$ certainly gives an $FDR(\theta)$ that is less than α.

The factor $\#\{j : \theta \in \Theta_0^j\}/N$ is the proportion of correct null hypotheses among the N hypotheses. In many applications, this proportion is close to 1. If this is not the case, then the Benjamini–Hochberg is conservative, like the Bonferroni method. If we knew this proportion beforehand, it would be possible to obtain an FDR close to the nominal value α, by using the previous strategy with an adjusted value of α. This

155

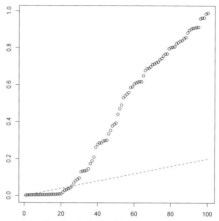

Figure 4.13. Illustration of the Benjamini–Hochberg procedure for multiple tests. The points are ordered p-values $p_{(1)} \leq p_{(2)} \leq \cdots \leq p_{(100)}$ (vertical axis) set out against the numbers $1, 2, \ldots, 100$ (horizontal axis). The dashed curve is the line $x \mapsto 0.20/100x$. The hypotheses corresponding to the p-values left of the intersection point are rejected at $\alpha = 0.20$. (When there are multiple intersection points, the last upcrossing of the p-values can be taken.)

is, however, hardly a realistic situation. There do exist refinements to use the data to determine an estimate for this proportion and correct the value α using this estimate. In the case of independent test statistics, we can, for example, replace α by

$$\alpha \frac{(1-\lambda)N}{\#\{j: P_j > \lambda\} + 1}, \qquad \lambda \in (0, 1).$$

The additional factor can be seen as an estimator for the inverse of the factor $\#\{j: \theta \in \Theta_0^j\}/N$, for every fixed λ, for example $\lambda = \alpha/(1+\alpha)$. Unfortunately, this "adaptive" extension of the Benjamini–Hochberg procedure seems to function less well when the tests are not independent.

For the next theorem, we assume that $P_1, P_2, \ldots, P_N$ are random variables with values in $[0, 1]$ such that the distribution of P_j under a parameter $\theta \in \Theta_0^j$ (for which H_0^j is correct) is stochastically greater than or equal to the uniform distribution; that is, $P_\theta(P_j \leq x) \leq x$ for every $x \in [0, 1]$. This last assumption implies that P_j is a p-value for the null hypothesis $H_0^j: \theta \in \Theta_0^j$, as the test with critical region $\{P_j \leq \alpha\}$ has size $P_\theta(P_j \leq \alpha)$ and for a valid p-value, this is less than or equal to the nominal value α, for every α.

Theorem 4.52 Benjamini–Hochberg

If the distribution of P_j is stochastically greater than or equal to the uniform distribution under every $\theta \in \Theta_0^j$, then (4.7) holds. If, moreover, $P_1, \ldots, P_N$ are independent or the function $x \mapsto P_\theta(K(P_1, \ldots, P_N) \geq y | P_j = x)$ is nonincreasing for every $\theta \in \Theta_0^j$, $y \in \mathbb{N}$, and coordinate-wise decreasing function $K: [0,1]^N \to \mathbb{N}$, then

(4.8) $$FDR(\theta) \leq \frac{\#\{j: \theta \in \Theta_0^j\}}{N}\alpha.$$

Proof. Let $P = (P_1, \ldots, P_N)$ be the vector of the p-values and $K(P) = \max\{j: NP_{(j)} \leq j\alpha\}$. It follows from the definition of the Benjamini–Hochberg procedure that H_0^j is rejected if and only if $j \leq K(P)$. An equivalent statement is that H_0^j is rejected if and only if

$$NP_j \leq K(P)\alpha.$$

The FDR can therefore be written as

$$\mathrm{E}_\theta \frac{\#\{j: P_j \leq K(P)\alpha/N, \theta \in \Theta_0^j\}}{K(P)} = \sum_{j:\theta \in \Theta_0^j} \mathrm{E}_\theta\left(\frac{1\{P_j \leq K(P)\alpha/N\}}{K(P)}\right).$$

The sum is less than or equal to the number of terms times the largest term. To prove the different statements, it therefore suffices to bound the expectation by $(\alpha/N)(1 + \log N)$ for (4.7) and by α/N for (4.8). Since the expectation is taken under the assumption $\theta \in \Theta_0^j$, the quantity P_j is stochastically greater than a uniformly distributed variable.

The inequality necessary for (4.7) therefore immediately follows from the first assertion of Lemma 4.53. Moreover, by the definition, K is a coordinate-wise decreasing function of $P_1, \ldots, P_N$. If $P_1, \ldots, P_N$ are independent, then it follows that $x \mapsto P_\theta(K(P) \geq y | P_j = x)$ is decreasing in x. When $P_1, \ldots, P_N$ are not independent, the assumption that the latter is decreasing in x is added to the theorem. Consequently, the inequality necessary for (4.8) also follows from Lemma 4.53. ∎

Lemma 4.53

Let (P, K) be a random vector with values in $[0,1] \times \{1, 2, \ldots, N\}$. If the distribution of P is stochastically greater than the uniform distribution, then for every $c \in (0,1)$,

$$\mathrm{E}\left(\frac{1\{P \leq cK\}}{K}\right) \leq c\left(1 + \log(c^{-1} \wedge N)\right).$$

If the function $x \mapsto P(K \geq y | P = x)$ is nonincreasing for every y, then this inequality also holds without the factor $1 + \log(c^{-1} \wedge N)$.

Proof. Since $\int_K^\infty (1/s^2)\, ds = 1/K$ for every $K > 0$, we can write the left-hand side of the lemma in the form

$$\mathrm{E} \int_K^\infty \frac{1}{s^2}\, ds \mathbf{1}\{P \leq cK\}$$

$$= \int_0^\infty \mathrm{E}\mathbf{1}\{K \leq s, P \leq cK\} \frac{ds}{s^2}$$

$$\leq \int_0^\infty \mathrm{E}\mathbf{1}\{P \leq c\lfloor s\rfloor \wedge cN\} \frac{ds}{s^2} \leq \int_0^\infty \big(c\lfloor s\rfloor \wedge cN \wedge 1\big) \frac{ds}{s^2}.$$

Here, $\lfloor s\rfloor$ is the greatest integer smaller than s, and in the first inequality we use that K takes on its values in $\mathbb{N}$. The last expression can be computed as $c(1/2 + 1/3 + \cdots + 1/D) + 1/D$, for D the smallest integer greater than or equal to $c^{-1} \wedge N$. This expression is bounded by $c\big(1 + \log(c^{-1} \wedge N)\big)$. This proves the first assertion.

For the second assertion, we denote by $u \mapsto Q(u\,|\,x)$ the quantile function of the conditional distribution of K given $P = x$. By assumption, this conditional distribution decreases stochastically as x increases, which implies that the corresponding quantile function also decreases: $Q(u\,|\,x) \geq Q(u\,|\,x')$ if $x \leq x'$, for every $u \in [0,1]$.

Consider a fixed value $u \in (0,1)$. The function $x \mapsto cQ(u\,|\,x) - x$ takes on the value $cQ(u\,|\,0) \geq c1 > 0$ in $x = 0$ and is strictly decreasing on $[0,1]$. Let x^* be the unique point where the function meets the horizontal axis or let $x^* = 1$ if the function is always positive. In both cases, we have $cQ(u\,|\,P) \geq cQ(u\,|\,x^*-) \geq x^*$ if $P < x^*$, where $Q(u\,|\,x^*-) = \lim_{x \uparrow x^*} Q(u\,|\,x)$. So the event $\{P \leq cQ(u\,|\,P)\}$ is contained in the event $\{P \leq x^*\}$. Consequently,

$$\mathrm{E}\left(\frac{\mathbf{1}\{P \leq cQ(u\,|\,P)\}}{Q(u\,|\,P)}\right) \leq \mathrm{E}\left(\frac{\mathbf{1}\{P \leq x^*\}}{Q(u\,|\,P)}\right) \leq \mathrm{E}\left(\frac{\mathbf{1}\{P \leq x^*\}}{x^*/c}\right) \leq c.$$

If U is uniformly distributed on $[0,1]$, then the variable $Q(U\,|\,x)$ is distributed following the conditional distribution of K given $P = x$. This implies that the vector $\big(P, Q(U\,|\,P)\big)$, for U independent of P, follows the distribution of (P, K). If we replace u by U on the left-hand side in the display, we therefore obtain exactly the left-hand side of the lemma. ∎

4.10 Summary

Let X be an observation with distribution P_θ that depends on an unknown parameter $\theta \in \Theta$. A *test* for the null hypothesis $H_0: \theta \in \Theta_0$ against the alternative hypothesis $H_1: \theta \in \Theta_1 = \Theta \setminus \Theta_0$ is defined by a *critical region* K: if $X \in K$, then H_0 is rejected; if $X \notin K$, then H_0 is not rejected.

The following concepts are important for tests:
- The critical region K is often described using a low-dimensional *test statistic* $T = T(X)$ for which $K = \{x: T(x) \in K_T\}$. The region K_T is often also called the critical region (for simplicity).
- The *(statistical) power function* of a test with critical region K is the function $\theta \mapsto \pi(\theta; K) = P_\theta(X \in K) = P_\theta(T(X) \in K_T)$.
- The *size* of a test with critical region K is $\alpha = \sup_{\theta \in \Theta_0} \pi(\theta; K)$. A test of level α_0 has $\alpha \leq \alpha_0$.
- A *Type I error* is a false positive: H_0 is true but is rejected.
- A *Type II error* is a false negative: H_0 is false but is not rejected.
- An *ideal test* has power 0 for $\theta \in \Theta_0$ and 1 for $\theta \in \Theta_1$. This test does not make any errors, but is not realistic. The probability of a type I error is limited by the size of the test. The probability of a type II error decreases as the sample size increases.

p-Values offer an alternative for a critical region that provides more information:
- For a test $K = \{x: T(x) \leq c_{\alpha_0}\}$, the *p-value* is

$$p = \sup_{\theta \in \Theta_0} P_\theta(T \leq t).$$

If $p \leq \alpha_0$, then H_0 is rejected at size α_0. For a test with $K = \{x: T(x) \geq d_{\alpha_0}\}$, the p-value is defined using $P_\theta(T \geq t)$. In the case of a two-sided critical region, the p-value is

$$2 \min \left(\sup_{\theta \in \Theta_0} P_\theta(T \leq t), \sup_{\theta \in \Theta_0} P_\theta(T \geq t) \right).$$

In addition to Gauss tests, t-tests, and the binomial test, which assume a specific distribution, there also exist the following tests, which can be applied more broadly:
- The *likelihood ratio test* is based on the *likelihood ratio statistic*

$$\lambda(X) = \frac{\sup_{\theta \in \Theta} p_\theta(X)}{\sup_{\theta_0 \in \Theta_0} p_{\theta_0}(X)}.$$

Under certain conditions, under $H_0: \theta = \theta_0$, the statistic $2 \log \lambda_n(X_1 \ldots, X_n)$ based on $X = (X_1, \ldots, X_n)$ asymptotically has a $\chi^2_{k-k_0}$-distribution. For large n, the test that rejects $H_0: \theta = \theta_0$ when $2 \log \lambda_n(X_1, \ldots, X_n) \geq \chi^2_{k-k_0, 1-\alpha_0}$ has approximately size α_0.
- *Nonparametric tests* such as the sign test and the Wilcoxon test require few assumptions and can therefore be applied to a broad class of distributions.

Exercises

1. McRonald advertises quarter pound hamburgers. The Consumers Association wants to research whether these are effectively quarter pounders. They weigh 100 products presented as quarter pound hamburgers. Give a statistical model and describe the test problem.

2. A coffee shop has few clients before 10 A.M. To draw more clients, the owners consider reducing the price of a cup of coffee by 50 cents before 10 A.M. Describe an experiment to evaluate whether such a measure has effect. Give the statistical model and describe the test problem.

3. For each of the following situations, give a statistical model and describe the test problem (null hypothesis, alternative hypothesis).
    (i) A sociologist asks a large group of high school students which academic study they will choose. They expect that a smaller percentage of girls than boys will choose mathematics.
    (ii) A political scientist assumes that there is a correlation between age and voting or not at elections, in particular a negative correlation. He sets 10 age categories and for each category asks 100 persons whether they will vote or not.
    (iii) To measure the effect of a problem session, a group of students is divided randomly into two groups. One group only goes to the lectures, while the other goes to both the lectures and the problem sessions. The observations consist of the exam results of both groups.

4. Traditionally, we assume that there is a linear correlation $y = \alpha + \beta x_1 + \gamma x_2$ between the yield y of an industrial process, the temperature x_1, and the amount of added catalyst x_2. A researcher, however, believes that (within certain boundaries) the temperature does not influence the yield. His colleague does not believe him and wants to use a statistical test to prove that the temperature does play a role. Describe how this question fits into statistical testing (give, among other things, the statistical model and hypotheses).

5. A random number generator is supposed to produce a sequence of numbers $u_1, u_2, \ldots$ that can be viewed as realizations of independent random variables with the uniform distribution on the interval $[0, 1]$. It is impossible to prove that a given generator has this property, but we can try to show, using statistical tests, that the generator does not work properly. Describe the statistical model and the test problem. Also suggest several possible test statistics.

6. The number of clients in a shoe store on Thursdays is approximately normally distributed with expectation 200 and standard deviation 50. By advertising in the local paper that is published on Wednesdays, the store owner hopes to increase the number of clients.
    (i) What conclusion can the store owner draw if the average number of clients on four Thursdays (after the ads appear) is (a) 239 (b) 264? Which assumptions were made?
    (ii) The store owners knows that to cover the costs of the ads, he needs 20 additional clients. Answer the same questions as above with this new aim in mind.

7. According to the packaging, a jar of face cream contains 50 grams of cream. To see whether the manufacturer puts enough cream in each jar, the contents of 100 jars are weighed. The average content turns out to be 49.82 grams. The variance when the jars were filled is supposed to be 1. Give a statistical model and describe the test problem. Use a suitable test to check whether the manufacturer complies with the requirement. Take $\alpha_0 = 0.05$.

8. Let $X_1, \ldots, X_{25}$ be a sample from the $N(\mu, 4)$-distribution. We want to test the null hypothesis $H_0: \mu \leq 0$ against $H_1: \mu > 0$ at level $\alpha_0 = 0.05$. The observed sample mean is 0.63.
   (i) Determine the critical region of a suitable test.
   (ii) Should H_0 be rejected?
   (iii) Determine the power function of the test in $\mu = 1/2$.
   (iv) Determine the p-value of this test.

9. Let $X_1, \ldots, X_{100}$ be independent, $N(\mu, 25)$-distributed random variables. We want to test the the null hypothesis $H_0: \mu = 0$ against $H_1: \mu \neq 0$ at level $\alpha_0 = 0.05$. We find $\bar{x} = -1.67$.
   (i) Use a suitable test to determine whether H_0 should be rejected.
   (ii) Determine the p-value.

10. Let $X_1, \ldots, X_n$ be a sample from the $N(\mu, \sigma^2)$-distribution with μ unknown and $\sigma^2 > 0$ known. Consider the test problem $H_0: \mu \leq \mu_0$ against $H_1: \mu > \mu_0$, where μ_0 is a fixed number. Suppose that, in contrast to Example 4.12, we take $\bar{X}$ as test statistic.
   (i) Show that the critical region $K = \{(x_1, \ldots, x_n): \bar{x} \geq \xi_{1-\alpha_0}\sigma/\sqrt{n} + \mu_0\}$ gives a test of size α_0.
   (ii) Show that the critical region K from the previous part is equal to the critical region based on the test statistic $\sqrt{n}(\bar{X} - \mu_0)/\sigma$ given in Example 4.12.

11. Someone pretends to have telepathic gifts in the sense that if you randomly draw one card from a set with as many red as black cards, he has probability 0.6 of naming the correct color instead of probability 0.5. To test this, we proceed as follows: we let him guess 25 consecutive times, where the drawn card is put back every time. If he guesses correctly at least 17 times, we believe him; otherwise, we do not.
   (i) Reword this problem in terms of null hypothesis, test statistic, alternative hypothesis, critical region.
   (ii) Determine the size of this test.
   (iii) Determine the power function in $p = 0.6$.
   (iv) He guesses correctly 16 times. What is the p-value?
   (v) Do we reject H_0 at level $\alpha_0 = 0.05$? And at level $\alpha_0 = 0.10$?

12. The random variables $X_1, \ldots, X_{25}$ are independent and have the Bernoulli distribution with parameter p. We want to test the null hypothesis $H_0: p \leq 0.6$ against $H_1: p > 0.6$ at level $\alpha_0 = 0.05$. As test statistic, we take $X = \sum X_i$.
   (i) Determine the critical region of the (right one-sided) test.
   (ii) Compute the power function by approximation in $p = 0.6, 0.7, 0.8, 0.9$, and sketch the graph of the power function. (The rule of thumb for the approximation is not satisfied for $p = 0.8$ and $p = 0.9$, but in this exercise and for sketching the graph, we can still use the approximation.)
   (iii) Compute the size of the test.

13. Suppose that in Example 4.11, we choose a test with a critical region of the form $K = \{e, e+1, \ldots, 98\}$.
   (i) Determine e such that $\alpha \leq 0.05$.
   (ii) Compare the power function of this test with that of the test with critical region $\{59, 60, \ldots, 100\}$.

14. According to the polls, during an election, political party A should receive 3.5% of all votes. We think that this is an overestimate. To study this, we ask 250 randomly chosen voters which party they will vote for. We denote by X the number of followers of party A. In our sample, $x = 5$ persons are followers of party A.
    (i) Give a statistical model for this situation.
    (ii) Determine a suitable null hypothesis.
    (iii) Give (an approximation for) the critical region for X at level $\alpha_0 = 0.05$. Test the null hypothesis from the previous part and give your conclusion.
    (iv) Give (an approximation for) the power function in 0.025 corresponding with your answer to part (iii).
    (v) How could we increase the power function in part (iv)?

15. To test the hypothesis $H_0: p \leq 0.5$ that a Bernoulli experiment is unbiased, we carry out a series of n of these experiments, independently from one another, and use the standard test at level 5%. How large must we at least take n for the power function in $p = 0.6$ to be at least 0.9?

16. Let $X_1, \ldots, X_n$ be a sample from the $N(\mu, 4)$-distribution. We want to test the null hypothesis $H_0: \mu \geq 1$ against $H_1: \mu < 1$ at level $\alpha_0 = 0.05$. Since in this case, it is very important to actually reject H_0 if $\mu = 0$, we want to choose n such that in the Gauss test, the probability of a type II error in $\mu = 0$ is at most 0.1. How large must n at least be?

17. To study whether the majority of the inhabitants of the Netherlands go abroad for the summer holidays, we ask n randomly chosen inhabitants where they are going on vacation next summer. We denote by X the number of persons in our sample who are going on vacation abroad. Based on this data, we want to test the null hypothesis $H_0: p \leq 0.5$ against the alternative hypothesis $H_1: p > 0.5$. How large must n at least be chosen to obtain a power function in $p = 0.6$ of at least 95% at level $\alpha_0 = 0.05$.

18. We want to know what percentage of the pieces in bags of candy are red, that is, the probability that a randomly chosen piece of candy from a randomly chosen bag is red. We take a sample of 30 bags of candy, each with 60 pieces. Let Y_i be the number of red pieces in the ith bag ($i = 1, \ldots, 30$). Assume that $Y_1, \ldots, Y_{30}$ are independent and that Y_i is binomially distributed with parameters 60 and p, with p the proportion of red pieces of candy in the bag.
    (i) Determine the maximum likelihood estimator for p.
    (ii) Of the total 1800 pieces of candy, 342 are red. Test the hypothesis $H_0: p = 0.2$ against $H_1: p \neq 0.2$.

19. Let X be a variable with the bin$(25, p)$-distribution. We want to test $H_0: p \geq 0.4$ against $H_1: p < 0.4$. If we want a power function of at least 0.6 in $p = 0.3$, how large must we at least choose the size of the test? Is this satisfactory?

20. A new vaccine for a virus for which there was no vaccine must be tested. Because the illness is in general not very serious, 1000 volunteers are given the virus. The vaccine is deemed successful if it protects in 90% of the cases.
    (i) Give a statistical model and the corresponding test problem.
    (ii) If the experiment gives a p-value of 0.25, what does that mean?
    (iii) The researchers do not find a p-value of 0.25 sufficiently convincing to recommend the vaccine for regular use; do you agree or disagree with this conclusion?

21. A manufacturer studies the life span of two types of fluorescent tubes, type A and type B. In an office building, a large number of fluorescent tubes are placed in pair. Each pair consists of a tube of type A and a tube of type B. The tubes in each pair are switched on and off at the same time. The manufacturer wants to study which type of fluorescent tube has the longest life span. The observations are $(X_1, Y_1), \ldots, (X_n, Y_n)$, where for the ith pair, X_i is the life span of the fluorescent tube of type A and Y_i is that of the tube of type B. Two statisticians analyze the observed values using a statistical test.
    (i) The first statistician defines W_i as equal to 1 if $X_i \geq Y_i$ and 0 otherwise. Then $W_1, \ldots, W_n$ are independent and Bernoulli-distributed with parameter $p = P_p(X_i \geq Y_i) = P_p(X_i - Y_i \geq 0)$. He tests the null hypothesis $H_0: p = 1/2$ against $H_1: p \neq 1/2$. Describe a suitable test based on $W_1, \ldots, W_n$; give a test statistic and an (approximate) critical region. Explain how you have arrived at the critical region. Take α_0 as the size of the test.
    (ii) The second statistician looks at the differences $Z_i = X_i - Y_i$, $i = 1, 2, \ldots, n$ and assumes that $Z_1, \ldots, Z_n$ are independent and normally distributed with unknown expectation μ and unknown variance σ^2. He tests the null hypothesis $H_0: \mu = 0$ against $H_1: \mu \neq 0$. Describe a suitable test for the given problem based on the observed differences; give a test statistic and an (approximate) critical region (a different test than the one for the first statistician). Explain how you have arrived at the critical region. Take α_0 as the size of the test.
    (iii) Assume that the differences $Z_1 \ldots, Z_n$ are normally distributed with expectation μ and variance σ^2. Show that the assumptions of the two statisticians are equivalent under the assumption of normality.
    (iv) Both statisticians have carried out their tests. The first statistician does not reject the null hypothesis; the second does. Is this possible, or has one of the two made a mistake? Explain.

22. Let $X_1, \ldots, X_n$ be a sample from the distribution with probability density $p_\theta(x) = e^{-x+\theta} 1_{x \geq \theta}$. We want to test the null hypothesis $H_0: \theta \geq 0$ against $H_1: \theta < 0$ at level $\alpha_0 = 0.1$. We choose $X_{(1)}$ as test statistic. Construct the critical region for the suitable (one-sided) test.

23. Let X be a random variable with a Poisson distribution with unknown parameter θ. Based on X, we want to test the null hypothesis $H_0: \theta \neq 5$ against $H_1: \theta = 5$. Show that the power function of each test in $\theta = 5$ is not greater than the size. Can a meaningful test be set up for this problem?

24. Let T be a test statistic with a continuous distribution function F_0 under H_0. Then $1 - F_0(t)$ is the p-value of a test that rejects H_0 for large values of t.
    (i) Show that under H_0, the p-value $1 - F_0(T)$ is uniformly distributed on $[0, 1]$.
    (ii) Is the distribution of this variable for a good test under the alternative hypothesis stochastically "greater" of "smaller" than the uniform distribution? (Stochastically greater means that realizations are, in general, greater; more precisely: the distribution function is smaller.)

25.
    (i) Show that the X_2^2-distribution is equal to the exponential distribution with parameter $1/2$.
    (ii) What is therefore the relation between the X_{2n}^2-distribution and a gamma distribution?

26. Show that the expectation and variance of the X_n^2-distribution are equal to n and $2n$, respectively.

27. Consider the estimators $T_c = cS_X^2$ for the variance of a sample $X_1, \ldots, X_n$ from the $N(\mu, \sigma^2)$-distribution. Use Theorem 4.29 and the previous exercise to compute the expected square error of T_c. For what c is this minimal?

28. Determine the distribution of the sum of two independent chi-square-distributed quantities.

29. (*F*-test.) A random variable T has the *F-distribution* with m and n degrees of freedom, denoted by $F_{m,n}$, if T has the same distribution as $(U/m)/(V/n)$ for independent random variables U and V with, respectively, the χ_m^2- and χ_n^2-distributions. Use the critical values from the F-distribution to construct a test for the problem $H_0: \sigma^2/\tau^2 \leq 1$ against $H_1: \sigma^2/\tau^2 > 1$ based on two independent samples $X_1, \ldots, X_m$ and $Y_1, \ldots, Y_n$ from, respectively, the $N(\mu, \sigma^2)$- and $N(\nu, \tau^2)$-distributions (for unknown μ and ν).

30. Based on two independent samples $X_1, \ldots, X_{25}$ and $Y_1, \ldots, Y_{16}$ from the $N(\mu, \sigma^2)$- and $N(\nu, \tau^2)$-distributions, respectively, we want to test $H_0: \sigma^2 \geq 2\tau^2$ against $H_1: \sigma^2 < 2\tau^2$ with unknown μ and ν, and $\alpha_0 = 0.01$.
    (i) What is the conclusion if we find the sums of squares $s_x^2 = 46.7$ and $s_y^2 = 45.1$?
    (ii) Determine the corresponding p-value.

31. Let $X_1, \ldots, X_n$ be a sample from the $N(\mu, \sigma^2)$-distribution, where $\mu \in \mathbb{R}$ and $\sigma^2 > 0$ are unknown.
    (i) Prove that the test "Reject $H_0: \sigma^2 \leq \sigma_0^2$ when $(n-1)S_X^2/\sigma_0^2 \geq \chi_{n-1,1-\alpha}^2$" (described in Example 4.32) has size α.
    (ii) The power function of this test is a function of (μ, σ). Express this function in the distribution function of the chi-square distribution.
    (iii) Sketch the graph of this function.

32. Let $X_1, \ldots, X_n$ be a sample from the $N(\mu, \sigma^2)$-distribution, where μ is known. How could you use the known value of μ to construct a test for $H_0: \sigma^2 = \sigma_0^2$ against $H_0: \sigma^2 \neq \sigma_0^2$? Do you expect this test to have a greater power function than the test from Example 4.32?

33. Show that a t-distribution is symmetric around the origin.

34. A chemical process should produce at least 800 metric tons of chemicals a day. The daily production of a certain week is 785, 805, 790, 793, and 802 metric tons. Do these data give a reason to conclude that there is something wrong with the process? Take $\alpha_0 = 0.05$. Which assumptions were made?

35. The average birth weight of boys in the Netherlands is 3605 grams. A number of midwives want to study whether the expected birth weight of boys in their practice deviates from this. The average birth weight of the 20 most recently born boys in the practice is 3585 grams, and the sample standard deviation is 253 grams.
    (i) Set up a test for the problem described above. Give the null and alternative hypotheses. Give the test statistic, the distribution of the test statistic under the null hypothesis, and the critical region. Take level $\alpha_0 = 0.05$. What is your conclusion?
    (ii) Test the null hypothesis from part (i) again, now based on an (approximate) p-value.

36. In an experiment, the blood pressure of 32 patients with hypertension is measured after they have taken a blood-pressure-lowering drug A. In a second experiment, the blood pressure of 20 patients with hypertension is measured after they have taken B, another blood-pressure-lowering drug. Denote the blood pressure values in the two experiments by $X_1, \ldots, X_{32}$ and $Y_1, \ldots, Y_{20}$. The measured outcomes are $\bar{x} = 163$, $\bar{y} = 158$, $s_X = 7.8$, and $s_Y = 9.0$.
    (i) Use a suitable test to determine which of the two drugs works best. Take a level of 5%.
    (ii) Determine the (approximate) p-value.

37. Ten sweaters are cut in half. One side is washed with product A, the other half with product B. After washing, the sweaters are measured. We find the following lengths:

sweater	1	2	3	4	5	6	7	8	9	10
product A	61, 2	58, 3	56, 7	59, 1	62, 7	61, 3	57, 8	55, 7	61, 8	60, 7
product B	61, 5	58, 2	59, 0	58, 6	62, 4	61, 2	55, 0	55, 0	61, 4	61, 0

Do the sweaters shrink less with product A or with product B? Construct a suitable test and state your conclusion. Take $\alpha_0 = 0.05$. Which assumptions were made?

38. Mister Young has a cab company with 12 cabs. He plans to buy 6 new tires of brand A and 6 new tires of brand B for the back wheels of the cabs. After every 500 km, he will check the wear on the tires. He can either
  (1) put a single new back tire on each of the 12 cabs, or
  (2) put a new back tire of each brand on 6 cabs.
Which of the two methods is preferable statistically? Why?

39. Mister Young from the previous exercise records the following numbers of driven kilometers when the 12 tires are worn:

km with brand A	51000	50500	61500	59000	64000	59000
km with brand B	55000	49500	62500	61500	65500	60000.

  (i) If the results are obtained using method (1), can he see the difference between the brands A and B? Take $\alpha_0 = 0.10$.
  (ii) Same question if the results are obtained using method (2) (where the vertical columns show the 6 cabs).
  (iii) Is it obvious that the two methods should give approximately the same numbers (as we have assumed in this exercise for the sake of convenience)?
  (iv) Is it reasonable to assume that the number of kilometers is exactly normally distributed? And approximately?

40. The content of a sunscreen manufacturer's tubes is checked. The tubes say that the content is equal to 150 grams. The inspector suspects that the manufacturer does not put enough sunscreen in the tubes. On inspection, the following content (in grams) is measured: 150.10, 149.55, 150.00, 149.65, 149.35, 150.15, 149.75, 150.00, 149.65, 150.20, 149.20, 149.95.
  (i) Check with a suitable test whether the inspector's suspicion is correct. Take level $\alpha_0 = 0.05$.
The manufacturer receives a warning from the inspector and claims to have adjusted the filling machines. At the next inspection, the following content (in grams) is measured: 149.85, 150.15, 150.05, 149.90, 150.30, 150.05, 149.95, 149.75, 149.95, 150.10.
  (ii) Set up suitable null and alternative hypotheses to check the manufacturer's claim that the expected weight at the second inspection is higher than that at the first. Carry out the test at level $\alpha_0 = 0.05$.

41. A chemical process should produce 10 metric tons of waste material an hour. Inspectors think that the amount of waste material is too high. The production process is therefore followed during 16 hours, and the amount of waste material produced each hour is recorded. Suppose that the amounts of waste material in these 16 hours, denoted by $X_1, \ldots, X_{16}$, are independent and normally distributed with unknown expectation μ and known variance 1. The sample mean is $\bar{x} = 10.5$.

(i) Give suitable null and alternative hypotheses for the situation describe above. Test the null hypothesis at level $\alpha_0 = 0.05$. Give the test statistic, the critical region, and a conclusion in words.

(ii) How many hours should one measure to have at least a 0.80 probability of discovering a deviation if the true value μ is equal to 10.4 metric tons?

42. Determining the isolating properties of oil can be done by filling a glass tube containing two poles and applying voltage to the two poles, letting it increase until a spark breaks through the isolation. We can repeat this determination of the breakthrough voltage as often as we want. In an experiment described by Youden and Cameron, two determinations are carried out each time ("duplo determinations"). If we denote the breakthrough voltage in the first determination by X and that in the second by Y, then it is reasonable to assume that X and Y have the same distribution (although it is, in general, different for different types of oil). This is, however, in no way certain, since a spark traveling through oil can leave behind ions, which can influence the outcome of the second determination. We want to check whether such an influence is present. In the experiment, we used 10 oil samples (each of a different type of oil); the tube was filled twice from each sample, and two determinations were carried out for each filling of the tube. The outcomes are given below:

oil sample	1st filling		2nd filling	
1	16	12	17	14
2	11	10	12	10
3	14	14	15	14
4	19	17	18	19
5	23	20	21	19
6	13	15	14	14
7	16	15	16	14
8	20	19	19	20
9	15	11	16	13
10	14	12	13	15

Test the null hypothesis that there is no systematic difference between the duplo determinations against the alternative hypothesis that there is one, at level $\alpha_0 = 0.01$, under the assumption that all breakthrough voltages are independent and normally distributed with the same (unknown) variance. Indicate the approximate size of the p-value.

43. To study whether toxic material has been released during a large fire, soil samples have been taken at different locations near the site of the fire. The presence of heavy metals is measured in these samples. For comparison, a number of samples are taken at a safe distance from the fire where the soil has the same type of composition. We want to test whether the concentration in the soil near the fire is higher than the concentration at a safe distance. At both places, 10 samples were taken. The concentrations in the 10 samples near the fire are denoted by $X_1, \ldots, X_{10}$, the concentrations at a safe distance are denoted by $Y_1, \ldots, Y_{10}$. The resulting sample means are $\bar{x} = 101.5, \bar{y} = 99.2$, and the sample variances are $s_x^2 = 5.1$ and $s_y^2 = 5.2$.

(i) Give a statistical model and reformulate the problem outlined above as a test problem. Describe the standard test. Give the test statistic and the critical region. Use level 5%. Carry out the test. What is your conclusion?

Another researcher argues as follows. Since the ground near the site of the fire is clay, heavy metals will not quickly descend to lower soil layers. He therefore wants to study whether the concentration in the top layer is higher than that in the bottom layer; a rise in the top layer may be caused by the fire. For each sample taken near the fire, he determines the concentration in the top and bottom layers.

(ii) Write the problem outlined as a test problem. Give the test statistic and explain how we can determine the p-value.

(iii) Suppose that the p-value is equal to 0.042. What is the conclusion if we take the size α_0 equal to 0.05?

(iv) Which of the two research methods is preferable?

44. Show that a probability density for the t_n-distribution is given by

$$f(x) = \frac{\Gamma\big((n+1)/2\big)}{\Gamma(n/2)} \frac{1}{\sqrt{n\pi}} \left(1 + \frac{t^2}{n}\right)^{-(n+1)/2}.$$

45. Suppose that we have observations $x_1 = 0.5$, $x_2 = 0.75$, and $x_3 = 1/3$. Determine the value of the Kolmogorov–Smirnov statistic for testing whether x_1, x_2, x_3 are realizations of independent U[0, 1]-variables.

46. Make the dependence of the Kolmogorov–Smirnov statistic (4.4) on the observations $X_1, \ldots, X_n$ visible by writing the statistic in the form $T^*(X_1, \ldots, X_n)$. Define $Z_i = (X_i - \mu)/\sigma$. Show that $T^*(X_1, \ldots, X_n) = T^*(Z_1, \ldots, Z_n)$. Deduce that the distribution of the Kolmogorov–Smirnov statistic is the same for every element of the null hypothesis that the observations are normally distributed.

47. Let $X_1, \ldots, X_n$ be a sample from the distribution with probability density $p_\theta(x) = e^{-x+\theta} 1_{x \geq \theta}$.

(i) Determine the likelihood ratio statistic λ_n for testing $H_0: \theta \leq 0$ against $H_1: \theta > 0$.

(ii) Determine the limit distribution of $2 \log \lambda_n$.

48. Let $X_1, \ldots, X_n$ be a sample from the uniform distribution on $[0, \theta]$.

(i) Determine the likelihood ratio statistic λ_n for testing $H_0: \theta \leq \theta_0$ against $H_1: \theta > \theta_0$.

(ii) Determine the likelihood ratio statistic λ_n for testing $H_0: \theta = \theta_0$ against $H_1: \theta \neq \theta_0$.

49. Let $X_1, \ldots, X_n$ be a sample from the Poisson distribution with unknown parameter θ.

(i) Determine the likelihood ratio statistic λ_n for testing $H_0: \theta = \theta_0$ against $H_1: \theta \neq \theta_0$.

(ii) What limit distribution does $2 \log \lambda_n$ have as $n \to \infty$?

50. Let $X_1, \ldots, X_n$ be a sample from the distribution with probability density $p_\theta(x) = 2\theta x e^{-\theta x^2} 1_{(0,\infty)}(x)$, where $\theta > 0$ is an unknown parameter.

(i) Determine the likelihood ratio statistic λ_n for testing $H_0: \theta = \theta_0$ against $H_1: \theta \neq \theta_0$.

(ii) Give the critical region for the likelihood ratio test at level α_0.

51. Let $X_1, \ldots, X_n$ be a sample from the $N(\mu, \sigma^2)$-distribution. We want to test the null hypothesis $H_0: \sigma^2 = \sigma_0^2$ against $H_1: \sigma^2 \neq \sigma_0^2$ at level α_0 (both μ and σ^2 are unknown). Show that the likelihood ratio test rejects H_0 when $(n-1)S_X^2/\sigma_0^2 \notin [c_1, c_2]$, where c_1 and c_2 satisfy

(i) $P\big(\chi_{n-1}^2 \in [c_1, c_2]\big) = 1 - \alpha_0$.

(ii) $c_1 - c_2 = n \log(c_1/c_2)$.

Note that this test differs somewhat from the test in Example 4.32, but not much for large n.

52. (Score test.) Let $X_1, \ldots, X_n$ be a sample from a distribution with the probability density p_θ indexed by a parameter $\theta \in \Theta \subset \mathbb{R}$. To test the null hypothesis $H_0 : \theta = \theta_0$, we consider the test statistic $T_n = 1/n \sum_{i=1}^{n} \dot{\ell}_{\theta_0}(X_i)$, for $\dot{\ell}_\theta$ the score function for p_θ.

  (i) Determine T_n for testing $H_0 : \theta = 1$ based on a sample from the $N(0, \theta^2)$-distribution.
  (ii) Determine a critical region for a test that rejects H_0 for large values of $|T_n|$ and that has approximately size α for large n.
  (iii) Show that the power function of the test converges to 1 as $n \to \infty$ for every $\theta \neq \theta_0$ such that $\mathrm{E}_\theta \dot{\ell}_{\theta_0}(X_1) \neq 0$.
  (iv) Verify that $\mathrm{E}_\theta \dot{\ell}_{\theta_0}(X_1) \neq 0$ for every $\theta \neq \theta_0$.

In the 1970s, Black and Scholes introduced an economic theory for the pricing of options on shares or other tradable "assets." After Black's death, Scholes received the Nobel prize for this work, together with Merton. Even today, the model is the basis for pricing so-called "financial derivatives," financial products that are derived from underlying products such as shares. Below, we will study certain characteristics of this model statistically.

The top image of Figure 4.14 shows the value of a share of Hewlett Packard at the New York stock exchange plotted against the time, in the period 1984–1991. The values A_t of the share at closing time on consecutive exchange days ($i = 1, 2, \ldots, 2000$) are plotted; in the graph, these values have been interpolated linearly.[†] According to the Black–Scholes model, the share price follows a "geometric Brownian motion." This corresponds to the log returns, defined by

$$X_t = \log \frac{A_t}{A_{t-1}},$$

forming a sequence $X_1, X_2, \ldots$ of independent, $N(\mu, \sigma^2)$-distributed random variables. In other words, the logarithms of the relative changes in the share price form an unpredictable noise with a normal distribution. The log returns are shown in the lower image in Figure 4.14; they have also been interpolated linearly. We will study this assumption of the Black–Scholes model in several ways.

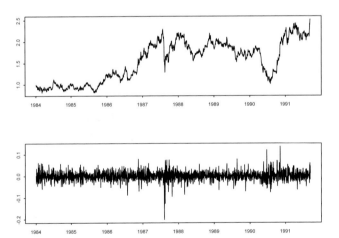

Figure 4.14. Price and log return of a share of Hewlett Packard at the New York stock exchange; initial value set equal to 1.

[†] The data can be found on the book's webpage at http://www.aup.nl under hpprices (share prices) and hplogreturns (log returns).

If the Black–Scholes model holds, then the sample mean $\overline{X}$ and the sample variance S_X^2 are good estimators for the parameters μ and σ^2 of the normal distribution of the log returns. The corresponding estimates, computed both over the full period and over four quarters, are

period	'84–'91	'84-'85	'86–'87	'88–'89	'90–'91
$\hat{\mu}$	0.000463	0.000164	0.001111	-0.000132	0.000710
$\hat{\sigma}$	0.022673	0.020514	0.026304	0.019102	0.024100

The estimate $\hat{\mu} \approx 0.00046$ over the full period means that on average, the value has increased between 1984 and 1991. If, for a moment, we ignore the stochastic fluctuations (not a good idea, see below!), then $A_t \approx A_{t-1}e^{0.000463} \approx A_{t-1}1.000463$. The average increase per day is then almost 0.05%. Yearly (250 stock market days), we have $A_t \approx A_{t-1}e^{0.000463} \approx \ldots \approx A_{t-250}(e^{0.000463})^{250} \approx A_{t-250}1.12$, which gives an average yearly increase of 12%. However, this increase is not uniformly distributed over the full period. In the third quarter '88–'89, the average of the log returns is negative ($\hat{\mu} = -0.000132$).

Using a statistical test, we can study whether such a decrease is compatible with the Black–Scholes model. In the Black–Scholes model, the observations in the four periods form four independent samples from the same normal distribution. We can, for example, test whether the log returns in the second quarter have the same expectation as the log returns in the third quarter, under the assumption that the log returns in the two quarters are samples from the normal distribution with expectation μ and ν, respectively, and variance σ^2. (We chose to study precisely these two quarters after computing the expected values of μ. This means that we in fact use the data twice—to decide what to test and to carry out the test—which makes the interpretation of p-values and sizes suspicious. It would have been better to compare all four periods, but this requires a more complicated test or a comparison of all pairs.) We use the t-test for independent samples. The estimated variance is $\hat{\sigma}^2 = \frac{1}{2}(0.026304^2 + 0.019102^2) \approx 0.00528$, and the t-statistic has value $\sqrt{250}(0.001111 - (-0.000132))/\sqrt{0.00528} \approx 0.27$. For a t-distribution with 998 degrees of freedom, this corresponds to a right p-value of approximately 39%. Despite the practically significant difference in sign in the estimates of μ in the second period, this test therefore does not lead us to doubt the Black–Scholes model. The observed difference in the estimates can be amply explained by the fluctuations of the share prices over time.

In the Black–Scholes model, these fluctuations are measured through the value of the parameter σ^2, which in this context is called the volatility of the share prices. It is unwise to not involve these fluctuations in the computations. According to the Black–Scholes model, in one year (250 stock exchange days), we cannot count on a deterministic growth of approximately 12% $((e^{0.000463})^{250} \approx 1.12)$, but rather on a growth that can be determined using the random variable

$$\frac{A_{250}}{A_0} = \frac{A_{249}e^{X_{250}}}{A_0} = \frac{A_{248}e^{X_{249}+X_{250}}}{A_0} = \cdots = e^{\sum_{t=1}^{250} X_t}.$$

In the Black–Scholes model, the variable $\sum_t X_t$ is normally distributed with expectation 250μ and variance $250\sigma^2$; in other words, it is the random variable $250\mu + \sqrt{250}\sigma Z$, where Z has the standard normal distribution. The distribution of $\exp(\sum_t X_t)$ is called log normal. The expected growth is

$$\mathrm{E}e^{250\mu+\sqrt{250}\sigma Z} = e^{250\mu}\mathrm{E}e^{\sqrt{250}\sigma Z} \approx 1.12e^{250\sigma^2/2} \approx 1.19,$$

where we have substituted the estimates $\hat\mu = 0.000463$ and $\hat\sigma = 0.022673$ for μ and σ. The expected yearly growth in the Black–Scholes model is therefore 19%. It is somewhat surprising that this value is considerably larger than the value 12% we found earlier by ignoring the randomness of the share prices. The form of the Black–Scholes model, where the price is an exponential function of the (normally distributed) log returns, is responsible for this. The expected daily growth is $\exp(\mu + \frac{1}{2}\sigma^2)$ and not $\exp(\mu)$, in accordance with the inequality $\mathrm{E}\exp(X) \geq \exp(\mathrm{E}X)$, which is strict when X is nondegenerate. (To cancel out the apparent contradiction, the reparametrization $(\mu, \sigma^2) \to (\mu - \frac{1}{2}\sigma^2, \sigma^2)$ is often applied, so that the distribution of the log returns is $N(\mu - \frac{1}{2}\sigma^2, \sigma^2)$, and the expected daily growth is $exp(\mu)$.) The estimate of $\mu + \frac{1}{2}\sigma^2$ in the third quarter of the full period is positive, though just barely, so that on closer inspection, the investment does have a positive yield.

The volatility σ also plays a decisive role in the Black–Scholes formula for the price of an option on the HP share. In the dealing rooms and back offices of banks, this price is even expressed with the volatility as unit. The Black–Scholes model is then often deviated from in the sense that the parameter σ is not taken as fixed, but may depend on time. In the four quarter periods, we for example find fluctuations of σ of size 13%. As for the parameter μ, we can test whether these fluctuations are significant. To compare the volatility in the second and third quarters, we compute the F-statistic (see Exercise 4.29) $0.019102^2/0.026304^2$. This leads to a left p-value of approximately $7 * 10^{-12}$ with respect to the F-distribution with 499 and 499 degrees of freedom. This is a strong indication that the volatility is effectively not constant over time.

Up to now, we have not truly tested the basic assumption of the Black–Scholes model that the log returns can be viewed as a sample from a normal distribution. We can, however, argue about both the normality assumption and the assumption that the log returns are independent variables. In fact, almost no one truly believes in the model, although it is applied by default.

We first study the normality of the log returns, under the assumption that the independence holds. In that case, we can test whether the log returns $X_1, \ldots, X_{2000}$ can be viewed as a sample from a normal distribution. Since we have already seen that the volatility σ is not constant over time, we will test the less stringent assumption that the log returns in the third quarter can be seen as a sample from the normal distribution. Figure 4.15 gives a first graphical impression of the distribution of this sample, through a histogram and a QQ-plot. These two graphs lead us to doubt the normality assumption, although the deviation from normality is not very strong. We can study the assumption formally by applying a statistical test such as the Kolmogorov–Smirnov test (see Example 4.38). Figure 4.16 shows

171

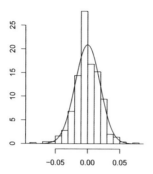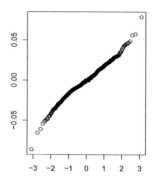

Figure 4.15. Histogram and QQ-plot against the normal distribution of the log returns in the period '88–'89 on the HP-shares. The curve in the histogram is the normal density with parameters equal to the sample mean and sample variance of the log returns.

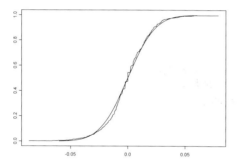

Figure 4.16. Empirical distribution function of the log returns in '88–'89 and distribution function of the normal distribution with parameters equal to the sample mean and sample variance of the log returns.

the empirical distribution function of $X_{1001}, \ldots, X_{1500}$ and the distribution function of the normal distribution with expectation and variance equal to, respectively, the sample mean and the sample variance of this sample. The Kolmogorov–Smirnov statistic is the maximal vertical distance between these two distribution functions and can be shown to equal 0.052. The corresponding critical value can be determined by computing the Kolmogorov–Smirnov statistic for a large number of samples simulated from the normal distribution. From 10 000 simulated samples, the value of the Kolmogorov–Smirnov statistic was greater than 0.052 in 6% of the cases. This means an (approximate) p-value of 6%, so that the null hypothesis of normality is not rejected at size 5%, but only barely.

Finally, we consider the stochastic time-independence of the log returns assumed by Black and Scholes. As first control quantity, we compute the sample autocorrelation coefficients of the log returns. These are shown on the left in Figure 2.13 and do not seem to contradict the independence. The sample autocorrelation coefficients of the squares of the log returns, on the right in Figure 2.13, however, are clearly different from 0. Carrying out the test from Example 4.40 therefore leads us to rejecting the null hypothesis that the log returns are independent, identically distributed random variables.

We can argue about the choice of this method. After all, we had already established that the volatility is not constant over time, so that the null hypothesis that the log returns are identically distributed and independent is not the most relevant hypothesis. We can repeat the analysis for each of the four periods individually. This leads to the same result.

The interesting question is now which dependence between the log returns exists on different days. This is not a simple question, because "dependence" includes many possibilities: all possible denials of "independence," which, by contrast, is uniquely determined. From the different models, the GARCH(1,1) model is seen as the benchmark. This model postulates

$$\sigma_t^2 = \alpha + \theta X_{t-1}^2 + \phi \sigma_{t-1}^2,$$
$$X_t = \sigma_t Z_t.$$

The first equation concerns the propagation of the volatility σ_t. This is not observed directly, but seen as a primary driving process under the log returns. The volatility on day t is a function of the square of the return and volatility on day $t-1$, and increases as these increase ($\phi, \theta \geq 0$). Given the volatility σ_t, the log return at time t is equal to σ_t times a variable Z_t, which is often assumed to be normally distributed and independent of the past $(X_{t-1}, \sigma_{t-1}, X_{t-2}, \ldots)$.

5 Confidence Regions

5.1 Introduction

In Chapter 3, we saw how a parameter θ could be estimated by the value $t = T(x)$ of an estimator T. In the context of this chapter, we will also refer to such estimates as *point estimates*. As a rule, an estimate t differs from the parameter θ to be estimated. Using the confidence regions described in this chapter, we can quantify the possible difference between the estimator T and θ. In many cases, this leads to an *interval estimate* $[L(x), R(x)]$, with the interpretation that θ has a high probability of lying in this interval.

5.2 Interpretation of a Confidence Region

The definition of a confidence region is as follows.

Definition 5.1 Confidence region

Let X be a random variable with a probability distribution that depends on a parameter $\theta \in \Theta$. A map $X \mapsto G_X$ whose codomain is the set of subsets of Θ is a confidence region for θ of confidence level $1 - \alpha$ if

$$P_\theta\big(G_X \ni \theta\big) \geq 1 - \alpha \qquad \text{for all } \theta \in \Theta.$$

In other words, a confidence region is a "stochastic subset" G_X of Θ that has a "high probability" of containing the true parameter θ. Because we do not know beforehand which value of θ is the true value, the condition in the definition holds for all values of θ: under the assumption that θ is the true value, this true value must have probability at least $1 - \alpha$ of being in G_X. After $X = x$ has been observed, the stochastic set G_X changes into a normal, nonstochastic subset G_x of Θ. Generally, α is taken small, for example $\alpha = 0.05$, so that the probability that θ lies in the confidence region is high. As we decrease α, the confidence region will of course grow and therefore give less information on θ, which, however, will then be "more certain." We again have a trade-off between two goals, as we already encountered with tests.

We often say that the probability that the realization G_x contains the true value θ is at least $1 - \alpha$. This probability statement can easily be interpreted incorrectly. In our interpretation, the true value of θ is fixed; the realized confidence region G_x is also not a random variable. Consequently, the true θ either lies in the confidence region G_x or does not. (Unfortunately, we do not know which of the two cases occurs.) The probability statement can be interpreted in the sense that if we, for example, carry out the experiment that gives X independently 100 times and compute the confidence region G_x 100 times, then we may expect that (at least) approximately $100(1 - \alpha)$ of the regions will contain the true θ. This is illustrated in Figure 5.1, which shows 100 independent realizations of a 90% confidence interval for the expectation parameter of the normal distribution. The true value of the parameter is 0 and is contained in 89 of the intervals. In practical situations, we, of course, cannot repeat experiments and can only determine one confidence region. This can be one of the 100α regions that do not contain the true parameter, without our being able to know this!

Because G_X is stochastic and θ deterministic, we have written $G_X \ni \theta$ instead of $\theta \in G_X$. In our notation for probabilities, the random variable is always on the left. For the same reason, some people disapprove of a statement such as "θ has a high probability of lying in G_X." We do not follow this last convention, but again emphasize that confidence regions have a subtle interpretation. In the Bayesian terminology of Section 3.5, on the other hand, the parameter is a random variable. This allows us to see, in that context, the probability statement on the event $\theta \in G_X$ as a statement concerning the random variable θ. The probability of this event can be determined with respect to the posterior distribution. We discuss this approach in Section 5.7.

When θ is a numerical parameter (that is, $\Theta \subset \mathbb{R}$), we typically use *confidence intervals*. These are confidence regions of the form $G_X = [L(X), R(X)]$ for two functions L and R of X. We then also speak of the confidence interval $[L, R]$ for the parameter θ. Sometimes the center of the confidence interval is exactly the used point estimate $T = T(X)$ for θ. We then also write the interval in the form $\theta = T \pm \eta$, with $\eta = \frac{1}{2}(R(X) - L(X))$ half the length of the interval. In other cases, the interval is intentionally chosen asymmetric around the used point estimate, which can be an expression of a "higher precision" upward or downward.

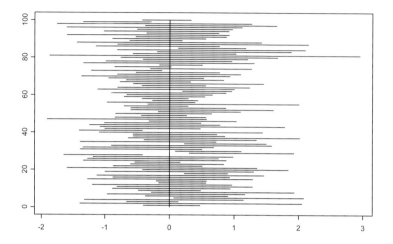

Figure 5.1. 100 realizations of the confidence interval of the expectation of the normal distribution (as computed in Example 5.4) based on 100 independent samples of size 5.

Example 5.2 Normal distribution

Let $X = (X_1, \ldots, X_n)$ be a sample from the normal $N(\mu, \sigma^2)$-distribution with unknown $\mu \in \mathbb{R}$ and known variance σ^2. Then

$$G_X = \left[\overline{X} - \frac{\sigma}{\sqrt{n}} \xi_{1-\alpha/2}, \overline{X} + \frac{\sigma}{\sqrt{n}} \xi_{1-\alpha/2} \right]$$

is a confidence interval for μ of confidence level $1 - \alpha$. We can see this as follows. The sample mean $\overline{X}$, the natural estimator for μ, has the $N(\mu, \sigma^2/n)$-distribution, and therefore $\sqrt{n}(\overline{X} - \mu)/\sigma$ has the standard normal distribution. We then have

$$P_\mu \left(\xi_{\alpha/2} \leq \sqrt{n} \frac{\overline{X} - \mu}{\sigma} \leq \xi_{1-\alpha/2} \right) = 1 - \alpha,$$

where ξ_α is the α-quantile of the standard normal distribution. We can rewrite this in the form

$$P_\mu \left(\overline{X} - \frac{\sigma}{\sqrt{n}} \xi_{1-\alpha/2} \leq \mu \leq \overline{X} + \frac{\sigma}{\sqrt{n}} \xi_{1-\alpha/2} \right) = 1 - \alpha,$$

where we have used that $\xi_{\alpha/2} = -\xi_{1-\alpha/2}$. It follows that $P_\mu(G_X \ni \mu) = 1 - \alpha$ for the G_X mentioned above. This interval is symmetric around the estimator $\overline{X}$ and is often written as

$$\mu = \overline{X} \pm \frac{\sigma}{\sqrt{n}} \xi_{1-\alpha/2}.$$

A realization of this interval contains μ with probability $1 - \alpha$.

The smaller σ and the larger n, the shorter (and therefore more informative) the interval. Note that to cut the interval in half, we need four times as many observations. For larger α, the interval is also shorter, but this goes at the expense of the confidence level.

5.3 Pivots and Near-Pivots

Many confidence regions are constructed using a pivot.

Definition 5.3 Pivot

A *pivot is a function* $T(X, \theta)$ *of the observation and parameter whose probability distribution does not depend on* θ *or any other unknown parameters if the probability distribution of* X *is given by the "true" parameter* θ.

A pivot is therefore not a statistic, because the pivot may depend on both the observation X and the parameter θ. For a pivot $T(X, \theta)$, the probability $P_\theta\big(T(X, \theta) \in B\big)$ is in principle known for every set B. Here, "known" means "independent of θ"; the two occurrences θ in the expression $P_\theta\big(T(X, \theta) \in B\big)$ must therefore cancel each other out. In Example 5.2, we in fact already saw an example of a pivot: $\sqrt{n}(\overline{X} - \mu)/\sigma$, which has the standard normal distribution.

For every set B such that $P_\theta\big(T(X, \theta) \in B\big) \geq 1 - \alpha$, the set

$$\left\{ \theta \in \Theta \colon T(X, \theta) \in B \right\}$$

is a confidence region for θ of confidence level $1 - \alpha$. In general, many sets B exist with this property, and we want to choose a "suitable" candidate from these. Although it seems natural to look for sets for which the volume of the confidence region is small, the choice is not unique. We illustrate this with the following examples.

Example 5.4 Normal distribution

Let $X = (X_1, \ldots, X_n)$ be a sample from the $N(\mu, \sigma^2)$-distribution with $\mu \in \mathbb{R}$ and $\sigma^2 > 0$ unknown. By Theorem 4.29,

$$\sqrt{n}\frac{\overline{X} - \mu}{S_X}$$

has a t_{n-1}-distribution, which does not depend on the parameter (μ, σ^2). This variable is therefore a pivot, and we have

$$P_\mu\left(t_{n-1,\alpha/2} \leq \sqrt{n}\frac{\overline{X} - \mu}{S_X} \leq t_{n-1,1-\alpha/2}\right) = 1 - \alpha.$$

It immediately follows from computations analogous to those in Example 5.2 that

$$\left[\overline{X} - \frac{S_X}{\sqrt{n}}t_{n-1,1-\alpha/2}, \overline{X} + \frac{S_X}{\sqrt{n}}t_{n-1,1-\alpha/2}\right]$$

is a confidence interval for μ of confidence level $1 - \alpha$. Since the interval is symmetric around $\overline{X}$, it can also be written as

$$\mu = \overline{X} \pm \frac{S_X}{\sqrt{n}}t_{n-1,1-\alpha/2}.$$

This interval greatly resembles the interval from the previous example, with σ replaced by S_X and ξ_α replaced by $t_{n-1,\alpha}$. Since the t-distribution has thicker tails than the standard normal distribution, the t-quantiles lie further from 0 than the quantiles of the standard normal distribution, and the interval we found here is in general somewhat longer than in the case where σ is known (although that also depends on the value of S_X). This is the price we have to pay for σ being unknown. As $n \to \infty$, the t_n-distribution increasingly resembles the normal distribution, and S_X converges in probability to σ. Hence the difference between the two intervals disappears as $n \to \infty$.

By the choice of the quantiles, the interval above is symmetric around the maximum likelihood estimator for μ. Nonsymmetric intervals of confidence level $1-\alpha$ can be constructed by choosing other quantiles of the t-distribution:

$$P_\mu\left(t_{n-1,\beta} \le \sqrt{n}\frac{\overline{X} - \mu}{S_X} \le t_{n-1,1-\gamma}\right) = 1 - \alpha$$

for $\beta + \gamma = \alpha$. The confidence interval for μ based on these quantiles is equal to

$$\left[\overline{X} - \frac{S_X}{\sqrt{n}}t_{n-1,1-\gamma}, \overline{X} - \frac{S_X}{\sqrt{n}}t_{n-1,\beta}\right].$$

The shortest confidence interval of confidence level $1 - \alpha$ is obtained by taking $\beta = \gamma = \alpha/2$; this results in the interval given earlier.

Example 5.5 Uniform distribution

If $X = (X_1,\ldots,X_n)$ is a sample from the $U[0,\theta]$-distribution, then the vector $X_1/\theta,\ldots, X_n/\theta$ is a sample from the $U[0,1]$-distribution. Every function of $X_1/\theta,\ldots, X_n/\theta$ is therefore a pivot.

The most interesting pivot is $X_{(n)}/\theta$, since this pivot is based on the maximum likelihood estimator and sufficient quantity $X_{(n)}$ for θ (see Section 6.2 for the definition of a sufficient quantity). We have

$$P_\theta\left(\frac{X_{(n)}}{\theta} \le x\right) = x^n, \qquad 0 \le x \le 1.$$

This leads to several confidence intervals for θ. If c, d with $0 \le c \le d \le 1$ are numbers such that $d^n - c^n = 1 - \alpha$, then

$$1 - \alpha = d^n - c^n = P_\theta\left(c \le \frac{X_{(n)}}{\theta} \le d\right) = P_\theta\left(\frac{X_{(n)}}{d} \le \theta \le \frac{X_{(n)}}{c}\right).$$

The interval $[X_{(n)}/d, X_{(n)}/c]$ is therefore a confidence interval for θ of confidence level $1 - \alpha$.

The choices $c = 0$ and $d = (1 - \alpha)^{1/n}$ lead to the right-open interval $[X_{(n)}(1 - \alpha)^{-1/n}, \infty)$. The choices $c = \alpha^{1/n}$ and $d = 1$ give the interval $[X_{(n)}, X_{(n)}\alpha^{-1/n}]$. Because we are certain that $\theta \ge X_{(n)}$, this interval puts all uncertainty in the upper bound. A reasonable strategy is to choose c and d such that $|1/d - 1/c|$ is minimal and the interval $[X_{(n)}/d, X_{(n)}/c]$ is the shortest possible (see Exercise 5.21). However, all intervals are allowed and have the same interpretation.

Determining confidence regions exactly from pivots is only possible incidentally, simply because there is not always a pivot. For example, it is impossible for the parameter p in the binomial distribution or for the parameter μ in the Poisson distribution. In such a case, we often settle on an approximate confidence region, which can be deduced from a *near-pivot*. When we are dealing with large samples, such near-pivots are usually amply available.

Example 5.6 Binomial distribution

If X is binomially distributed with parameters n and p, then for large n,

$$\frac{X - np}{\sqrt{np(1-p)}}$$

is approximately $N(0, 1)$-distributed, by the central limit theorem; see Section A.7. By approximation, this function of X and p is therefore a pivot. The set

$$\left\{ p: \xi_{\alpha/2} \leq \frac{X - np}{\sqrt{np(1-p)}} \leq \xi_{1-\alpha/2} \right\}$$

is consequently approximately a confidence region for p of confidence level $1-\alpha$. This set is an interval that can be found by solving the quadratic equation $(X - np)^2 \leq \xi_{1-\alpha/2}^2 np(1-p)$. Figure 5.2 shows this interval for certain values of α, n, and p.

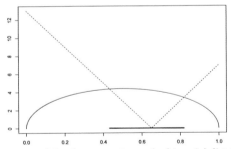

Figure 5.2. Confidence interval for the parameter p of a binomial distribution. The graph shows the functions $p \mapsto |x - np|$ and $p \mapsto 1.96\sqrt{np(1-p)}$ for the case $0 < x < n$ (namely, $x = 13$ and $n = 20$). The confidence interval is the interval on the horizontal axis between the two intersection points.

As long as we are taking approximations, we can also go a step further. By the law of large numbers (Theorem A.26), X/n converges in probability to p as n goes to infinity. Hence the random variable

$$\frac{X - np}{\sqrt{n(X/n)(1 - X/n)}}$$

179

is approximately $N(0,1)$-distributed. (This follows using Slutsky's lemma; see Lemma 5.15.) The approximate confidence interval based on this near-pivot is of the simple form

$$p = \frac{X}{n} \pm \frac{1}{\sqrt{n}} \sqrt{\frac{X}{n}\left(1 - \frac{X}{n}\right)} \xi_{1-\alpha/2}.$$

This interval is often used as an indication of the size when estimating the proportion of elements from a population with a certain characteristic, for example during a poll. It is remarkable that the size of the population does not pay a role in the length of the interval. Only the sample size counts, and to a lesser degree the true proportion. If $p = 1/2$ and $n = 1500$, then the 95% confidence interval is approximately $(X/n) \pm 2\%$. This 2% is probably the value that is meant in newspapers when a deviation of at most 2% is promised in the results of a given poll. The correct interpretation of this margin is that in 95% of the polls, the deviation of the sample proportion to the true proportion is not greater than 2%. Unfortunately, the press often translates this complicated statement into a firm error margin.

As $p \to 0$ or $p \to 1$, the function $p \mapsto \sqrt{p(1-p)}$ converges to 0. The confidence interval is therefore shorter for extreme values of p. The length of the confidence interval is the least favorable for $p = 1/2$.

Example 5.7 Application: counting bacteria

In Example 3.19, we assumed that the number of colony-forming units of bacteria in a centiliter of contaminated water is Poisson-distributed with parameter μ. To estimate μ, the contaminated water was mixed with 100 liters of pure water and divided over 100 Petri dishes. We only observe $Y_1, \ldots, Y_{100}$, with Y_i equal to 1 when a colony of bacteria forms in the ith dish and equal to 0 otherwise. It follows that Y_i has the Bernoulli distribution with probability $p = 1 - e^{-\mu/100}$ for $i = 1, \ldots, 100$. In Example 3.19, p is estimated using the maximum likelihood estimator $\overline{Y}$. Since $\sum_{i=1}^{100} Y_i$ is binomially distributed with parameters 100 and p, it follows from Example 5.6 that

$$P\left(\overline{Y} - \frac{\xi_{1-\alpha/2}}{\sqrt{100}}\sqrt{\overline{Y}(1-\overline{Y})} \le p \le \overline{Y} + \frac{\xi_{1-\alpha/2}}{\sqrt{100}}\sqrt{\overline{Y}(1-\overline{Y})}\right) \approx 1 - \alpha.$$

This confidence interval for p can be used to deduce a confidence interval for μ by substituting $p = 1 - e^{-\mu/100}$. If we write $\hat{\sigma}^2 = \overline{Y}(1-\overline{Y})$, then

$$\left[-100\log\left(1 - \overline{Y} + \frac{\xi_{1-\alpha/2}}{\sqrt{100}}\sqrt{\hat{\sigma}^2}\right), -100\log\left(1 - \overline{Y} - \frac{\xi_{1-\alpha/2}}{\sqrt{100}}\sqrt{\hat{\sigma}^2}\right)\right]$$

is a confidence interval for μ of confidence level $1 - \alpha$, provided $1 - \overline{Y} - \xi_{1-\alpha/2}\sqrt{\hat{\sigma}^2/100} > 0$. If $1 - \overline{Y} - \xi_{1-\alpha/2}\sqrt{\hat{\sigma}^2/100} \le 0$, the upper bound is replaced by infinity.

The near-pivot in Example 5.6 comes from an asymptotic approximation of the distribution of the estimator. Many estimators T_n for a parameter $g(\theta)$ are asymptotically normally distributed in the sense that for certain numbers $\sigma_{n,\theta}$ (often the standard deviation of T_n), under the assumption that θ is the true parameter, we have

$$\frac{T_n - g(\theta)}{\sigma_{n,\theta}} \rightsquigarrow N(0,1)$$

as $n \to \infty$. The arrow $\rightsquigarrow$ is the notation for "convergence in distribution"; see Definition A.25. More precisely, the statement means that

$$\lim_{n\to\infty} P_\theta\left(\frac{T_n - g(\theta)}{\sigma_{n,\theta}} \le x\right) = \Phi(x) \qquad \text{for all } x.$$

An informal interpretation is that for large n, the variable $(T_n - g(\theta))/\sigma_{n,\theta}$ is approximately $N(0,1)$-distributed if θ is the true parameter. Consequently,

$$\frac{T_n - g(\theta)}{\sigma_{n,\theta}}$$

is a *near-pivot*. This is also called the large sample method. This leads to an approximate confidence region for $g(\theta)$ equal to

$$\left\{ g(\theta) \colon T_n - \sigma_{n,\theta}\xi_{1-\alpha/2} \le g(\theta) \le T_n + \sigma_{n,\theta}\xi_{1-\alpha/2} \right\},$$

of confidence level $1 - \alpha$. For convenience, the expression $\sigma_{n,\theta}$ is often replaced by the estimator $\hat{\sigma}_n$, which gives the symmetric interval

$$g(\theta) = T_n \pm \hat{\sigma}_n\xi_{1-\alpha/2}.$$

The expression $\hat{\sigma}_n$ is usually an estimate for the standard deviation of T_n, the *standard error* or *s.e.* of the estimator (or estimate). In many scientific reports, only the estimate and corresponding standard error are mentioned. Provided that the used estimator is approximately normally distributed, we can roughly interpret this information in the sense of a 95% confidence interval of the form $g(\theta) = T_n \pm \xi_{0.975}$ s.e. $= T_n \pm 1.96$ s.e.

Good statistical software gives both a parameter estimate and the standard error of the estimator. For an estimate of a vector-valued parameter, it gives a standard error for every coordinate, and additionally, the estimated covariances between the estimators, in the form of a matrix with the estimated variances of the estimators (the squares of the standard errors) on the diagonal (see Section B.2 for the definition of the covariance matrix).

5.4 Maximum Likelihood Estimators as Near-Pivots

An important special case of the near-pivots discussed in the previous section is that where T_n is the maximum likelihood estimator. Under certain conditions, the maximum likelihood estimator is asymptotically normally distributed. We discuss the simplest case, that of a sample of independent random variables and first restrict ourselves to parameters $\theta \in \Theta \subset \mathbb{R}$.

Definition 5.8 Score function and Fisher information

Let p_θ be the probability density of the observation X_1, and suppose that the function $\theta \mapsto \ell_\theta(x) := \log p_\theta(x)$ is (partially) differentiable for all x. The gradient

$$\dot{\ell}_\theta(x) = \frac{\partial}{\partial \theta} \log p_\theta(x)$$

is then called the score function *of the model. The* Fisher information *for θ in X_1 is defined as the number*

$$i_\theta = \mathrm{var}_\theta\, \dot{\ell}_\theta(X_1).$$

Now, suppose that we have a sample $X_1, \ldots, X_n$ from the distribution with (marginal) probability density p_θ. The log-likelihood function of the model is then equal to $\theta \mapsto \sum_{i=1}^n \ell_\theta(X_i)$ and has derivative $\theta \mapsto \sum_{i=1}^n \dot{\ell}_\theta(X_i)$, the sum of the score functions over the observations. The maximum likelihood estimator $\hat{\theta}_n$ for θ is the point where the log-likelihood takes on its maximum and is therefore a solution of the likelihood equation $\sum_{i=1}^n \dot{\ell}_\theta(X_i) = 0$, unless the maximum of the likelihood is taken on on the boundary of the parameter space.

We call the parameter $\theta \in \Theta$ *identifiable* if no other parameter gives the same probability distribution or, more technically, if the densities p_ϑ and p_θ differ with positive probability: $P_\theta\big(p_\vartheta(X_1) \neq p_\theta(X_1)\big) > 0$ for every $\vartheta \neq \theta$. This natural property will normally be satisfied by a suitable parameterization of the model and is necessary to estimate θ from the observations in a meaningful way. In the following theorem, we assume that the parameter set Θ is compact. Extension to noncompact sets is possible, in general, for example by showing that the maximum likelihood estimator belongs to a compact set with probability converging to 1.

Theorem 5.9

Suppose that Θ is compact and convex and that θ is identifiable, and let $\hat{\theta}_n$ be the maximum likelihood estimator based on a sample of size n from the distribution with (marginal) probability density p_θ. Suppose, furthermore, that the map $\vartheta \mapsto \log p_\vartheta(x)$ is continuously differentiable for all x, with derivative $\dot{\ell}_\vartheta(x)$ such that $|\dot{\ell}_\vartheta(x)| \leq L(x)$ for every $\vartheta \in \Theta$, where L is a function with $\mathrm{E}_\theta L^2(X_1) < \infty$. If θ is an interior point of Θ and the function $\vartheta \mapsto i_\vartheta$ is continuous and positive, then under θ, the sequence $\sqrt{n}(\hat{\theta}_n - \theta)$ converges in distribution to a normal distribution with expectation 0 and variance i_θ^{-1}. Therefore, under θ, as $n \to \infty$, we have

$$\sqrt{n}(\hat{\theta}_n - \theta) \rightsquigarrow N(0, i_\theta^{-1}).$$

A (partial) proof of the theorem is given in Section 5.4.1. If the statement of the theorem is applicable, then for large n, under θ, the random variable

$$\sqrt{ni_\theta}(\hat{\theta} - \theta)$$

is approximately standard normally distributed and therefore a near-pivot. For convenience, we can replace i_θ by an estimator $\widehat{i_\theta}$, and we find for θ the approximated confidence interval

$$\theta = \hat{\theta} \pm \frac{1}{\sqrt{n\widehat{i_\theta}}} \xi_{1-\alpha/2}$$

of confidence level $1 - \alpha$. This interval is called the *Wald interval*. The statement of the theorem is often read to mean that $1/(ni_\theta)$ is an approximation for the variance of $\hat{\theta}$, and its square root is an approximation for the standard error. For $\alpha = 0.05$, the Wald interval is therefore, in fact, an interval of the general form $\theta = \hat{\theta} \pm \xi_{0.975}\text{s.e.} \approx \hat{\theta} \pm 2\text{s.e.}$ (Note that the theorem does not say anything about the convergence of the variance of the maximum likelihood estimators, but the previous interpretation is usually defendable.)

Common estimators for the Fisher information i_θ are the *plug-in estimator* and the *observed information*. The plug-in estimator is $\widehat{i_\theta} = i_{\hat{\theta}}$; that is, the parameter θ in the expression for i_θ is replaced by the maximum likelihood estimator $\hat{\theta}$. The observed information is defined as

$$\widehat{i_\theta} = -\frac{1}{n}\sum_{i=1}^n \ddot{\ell}_{\hat{\theta}}(X_i), \quad \text{with} \quad \ddot{\ell}_\theta(x) = \frac{\partial^2}{\partial\theta^2}\log p_\theta(x).$$

The plug-in estimator requires the (analytic) computation of the Fisher information i_θ, while the observed information follows more easily from the data. The observed information is $-1/n$ times the second-order derivative of the log-likelihood function $\theta \mapsto \sum_{i=1}^n \ell_\theta(X_i)$ evaluated in $\theta = \hat{\theta}$. If necessary, we can use a numerical derivative (difference quotient) instead of an analytic derivative. Graphically, the observed information gives the curvature of the log-likelihood function in the point $\theta = \hat{\theta}$

183

where the log-likelihood is maximal. If the likelihood function has a flat top, then the observed information is small, and the confidence interval for θ is long: the maximum likelihood estimator is then not very precise. (This does not reflect a weakness of this estimation method; it is due to a parameter that is intrinsically difficult to estimate.) The meaningfulness of the observed information as estimator for i_θ is not immediately clear, but follows (for large n) from the following lemma and the law of large number, according to which $n^{-1}\sum_{i=1}^{n}\ddot{\ell}_\theta(X_i) \to \mathrm{E}_\theta\ddot{\ell}_\theta(X_i)$ as $n \to \infty$ with probability 1 if θ is the true parameter.

Lemma 5.10

Suppose that $\theta \mapsto \ell_\theta(x) = \log p_\theta(x)$ is differentiable twice for all x. Then, under certain regularity conditions, we have $\mathrm{E}_\theta\dot{\ell}_\theta(X_1) = 0$ and $\mathrm{E}_\theta\ddot{\ell}_\theta(X_1) = -i_\theta$.

Proof. We write the formulas under the assumption that X_1 is continuously distributed. (For a discrete probability density, we replace the integrals by sums.) Since p_θ is a probability density, we have $1 = \int p_\theta(x)\,dx$ for all θ. Consequently,

$$0 = \frac{\partial}{\partial\theta}\int p_\theta(x)\,dx = \int \frac{\partial}{\partial\theta}p_\theta(x)\,dx = \int \dot{p}_\theta(x)\,dx,$$

with $\dot{p}_\theta(x) = \partial/\partial\theta\, p_\theta(x)$. Interchanging the differentiation (with respect to θ) and the integration (with respect to x) is permitted under certain regularity conditions. Since the score function equals $\dot{\ell}_\theta(x) = \partial/\partial\theta \log p_\theta(x) = \dot{p}_\theta(x)/p_\theta(x)$, we can rewrite the right-hand side as

$$\int \frac{\dot{p}_\theta(x)}{p_\theta(x)}\,p_\theta(x)\,dx = \int \dot{\ell}_\theta(x)\,p_\theta(x)\,dx = \mathrm{E}_\theta\dot{\ell}_\theta(X_1).$$

This completes the proof of the first assertion: $\mathrm{E}_\theta\dot{\ell}_\theta(X_1) = 0$. For the proof of the second assertion, we differentiate $\int p_\theta(x)dx$ twice with respect to θ and find

$$0 = \frac{\partial^2}{\partial\theta^2}\int p_\theta(x)\,dx = \int \ddot{p}_\theta(x)\,dx,$$

with $\ddot{p}_\theta(x) = \partial^2/\partial\theta^2 p_\theta(x)$.

Differentiating the equality $\dot{\ell}_\theta(x) = \dot{p}_\theta(x)/p_\theta(x)$ with respect to θ gives

$$\ddot{\ell}_\theta(x) = \frac{\ddot{p}_\theta(x)}{p_\theta(x)} - \left(\frac{\dot{p}_\theta(x)}{p_\theta(x)}\right)^2 = \frac{\ddot{p}_\theta(x)}{p_\theta(x)} - \dot{\ell}_\theta(x)^2.$$

We multiply this by $p_\theta(x)$ and take the integral with respect to x to find that

$$\mathrm{E}_\theta\ddot{\ell}_\theta(X_1) = \int \ddot{p}_\theta(x)\,dx - \int \dot{\ell}_\theta(x)^2 p_\theta(x)\,dx$$

$$= 0 - \mathrm{E}_\theta\big(\dot{\ell}_\theta(X_1)^2\big) = -\operatorname{var}_\theta \dot{\ell}_\theta(X_1) = -i_\theta,$$

because $\operatorname{var}_\theta \dot{\ell}_\theta(X_1) = \mathrm{E}_\theta\big(\dot{\ell}_\theta(X_1)^2\big) - (\mathrm{E}_\theta\dot{\ell}_\theta(X_1))^2 = \mathrm{E}_\theta\big(\dot{\ell}_\theta(X_1)^2\big)$, by the first assertion. This proves the second assertion. ∎

Example 5.11 Poisson distribution

Let $X = (X_1, \ldots, X_n)$ be a sample from the Poisson(θ)-distribution, where $\theta > 0$ is unknown. The maximum likelihood estimator for θ is $\hat{\theta} = \overline{X}$ provided $\overline{X} > 0$. The score function is equal to

$$\dot{\ell}_\theta(x) = \frac{\partial}{\partial \theta} \log \frac{e^{-\theta}\theta^x}{x!} = \frac{x}{\theta} - 1.$$

The Fisher information is then

$$i_\theta = \operatorname{var}_\theta\left(\frac{X_1}{\theta} - 1\right) = \frac{1}{\theta}.$$

By Lemma 5.10, we would have found the same expression using the equation

$$i_\theta = -\mathrm{E}_\theta \ddot{\ell}_\theta(X_1) = \frac{\mathrm{E}_\theta X_1}{\theta^2} = \frac{1}{\theta}.$$

If we estimate θ by $\overline{X}$, then the plug-in estimator for i_θ is equal to $1/\overline{X}$. The observed information gives the same estimator since

$$-\frac{1}{n}\sum_{i=1}^{n}\ddot{\ell}_{\hat{\theta}}(X_i) = \frac{1}{n}\sum_{i=1}^{n}\frac{X_i}{\hat{\theta}^2} = \frac{\overline{X}}{(\overline{X})^2} = \frac{1}{\overline{X}}.$$

The symmetric approximate confidence interval of confidence level $1 - \alpha$ is then

$$\theta = \overline{X} \pm \frac{\sqrt{\overline{X}}}{\sqrt{n}}\xi_{1-\alpha/2}.$$

We could also have found this interval by a more direct route, by applying the central limit theorem to $\overline{X}$. After all, the sequence $\sqrt{n}(\overline{X} - \theta)/\sqrt{\theta}$ is approximately standard normally distributed (see Example A.30).

Example 5.12 Cauchy distribution

Let $X_1, \ldots, X_n$ be independent variables with probability density

$$p_\theta(x) = \frac{1}{\pi(1 + (x - \theta)^2)}.$$

The log-likelihood equation is

$$\sum_{i=1}^{n}\frac{2(X_i - \theta)}{1 + (X_i - \theta)^2} = 0.$$

This equation cannot be solved explicitly for θ. Therefore, the maximum likelihood estimator cannot be written as an explicit function of $X_1, \ldots, X_n$. However, the estimator can be determined numerically, for example by reading the position of the maximum in a graph of the log-likelihood function; see for example Figure 5.3. The score function is

$$\dot{\ell}_\theta(x) = \frac{2(x - \theta)}{1 + (x - \theta)^2}.$$

185

The Fisher information can be computed with some difficulty if $i_\theta = 1/2$; it is constant as a function of θ, and therefore easily estimated to be $1/2$. The observed information is not exactly equal to $1/2$; it takes on the form

$$\widehat{i_\theta} = \frac{1}{n} \sum_{i=1}^{n} \frac{2 - 2(X_i - \hat\theta)^2}{\left(1 + (X_i - \hat\theta)^2\right)^2},$$

where $\hat\theta$ is the maximum likelihood estimator.

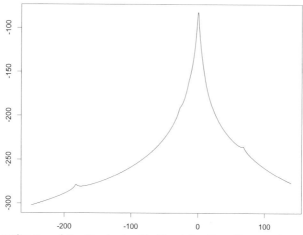

Figure 5.3. A realization of the Cauchy log-likelihood function. The curvature in the top is the observed information.

Example 5.13 Exponential distribution

Let $X = (X_1, \ldots, X_n)$ be a sample from the exponential distribution with unknown parameter λ. The maximum likelihood estimator for λ is $\hat\lambda = 1/\overline{X}$ (see Example 3.12). The score function is equal to

$$\dot\ell_\lambda(x) = \frac{\partial}{\partial\lambda} \log \lambda e^{-\lambda x} = \frac{1}{\lambda} - x,$$

and the Fisher information is

$$i_\lambda = \mathrm{var}_\lambda\left(\frac{1}{\lambda} - X_1\right) = \frac{1}{\lambda^2}.$$

By Lemma 5.10, the Fisher information can also be found using the equation $i_\lambda = -\mathrm{E}_\lambda \ddot\ell_\lambda(X_1) = 1/\lambda^2$. If λ is estimated by the maximum likelihood estimator, then the plug-in estimator for i_λ is equal to $(\overline{X})^2$. The observed information gives the same estimator for i_λ:

$$-\frac{1}{n} \sum_{i=1}^{n} \ddot\ell_{\hat\lambda}(X_i) = \frac{1}{n} \sum_{i=1}^{n} \frac{1}{\hat\lambda^2} = \overline{X}^2.$$

For both estimators for i_λ, we find the symmetric approximate confidence interval

$$\lambda = \frac{1}{\overline{X}} \pm \frac{1}{\sqrt{n}\overline{X}} \xi_{1-\alpha/2}$$

for λ of confidence level $1 - \alpha$.

* 5.4.1 Proof of Theorem 5.9

The proof of Theorem 5.9 consists of two parts: a proof of *consistency* and a proof of *convergence in distribution*. The estimator $\hat{\theta}_n = \hat{\theta}_n(X_1, \ldots, X_n)$ is called *consistent* for θ if $\hat{\theta}_n$ converges in probability to θ, that is, $\hat{\theta}_n \overset{P}{\to} \theta$ under θ as $n \to \infty$. We give a complete proof for the first part, in the form of a lemma, but only prove the second part under stronger conditions than those of Theorem 5.9. For a complete proof of the theorem under weaker conditions than in the theorem, we refer to the book "Asymptotic Statistics" (Van der Vaart (1998)).

The proofs of both parts are based on an analysis of the following stochastic function and its expectation:

$$\mathbb{M}_n(\vartheta) = \frac{1}{n} \sum_{i=1}^{n} \ell_\vartheta(X_i), \qquad M(\vartheta) = \mathrm{E}_\theta \ell_\vartheta(X_1).$$

Note that the argument ϑ of these functions differs from the "true" parameter θ that determines the distribution of the observations and that we use to compute the expectation E_θ. (Under the conditions of Theorem 5.9, the variable $(\ell_\vartheta - \ell_\theta)(X_1)$ has a finite first moment, so that $M(\vartheta) - M(\theta) = \mathrm{E}_\theta(\ell_\vartheta - \ell_\theta)(X_1)$ is always well defined. If this is not the case for $M(\vartheta)$ itself, then we replace ℓ_ϑ in the definitions of $\mathbb{M}_n$ and M and everywhere in the lemma and proof below by $\ell_\vartheta - \ell_\theta$; to simplify the notation, we refrain from doing this.)

Lemma 5.14 Consistency

Suppose that $\Theta \subset \mathbb{R}^k$ is compact and convex and that θ is identifiable. Suppose, moreover, that the map $\vartheta \mapsto \log p_\vartheta(x)$ is continuously differentiable for all x with gradient $\dot{\ell}_\vartheta(x)$ such that $\|\dot{\ell}_\vartheta(x)\| \le L(x)$ for every $\vartheta \in \Theta$, where L is a function with $\mathrm{E}_\theta L^2(X_1) < \infty$. Then $\hat{\theta}_n \overset{P}{\to} \theta$ under θ as $n \to \infty$.

Proof. The proof of the consistency is based on the following two assertions:
(i) The map $\vartheta \mapsto M(\vartheta)$ is continuous with unique absolute maximum in θ.
(ii) The sequence $\Delta_n := \sup_{\vartheta \in \Theta} |\mathbb{M}_n(\vartheta) - M(\vartheta)|$ converges in probability to 0.

Suppose that parts (i)–(ii) hold. It follows from the definition of $\hat{\theta}_n$ that $\mathbb{M}_n(\hat{\theta}_n) \geq \mathbb{M}_n(\theta)$. If we twice replace $\mathbb{M}_n$ in this inequality by M, then it follows from part (ii) that $M(\hat{\theta}_n) \geq M(\theta) - 2\Delta_n$. For any given $\delta > 0$, the closed subset $\{\vartheta \in \Theta : \|\vartheta - \theta\| \geq \delta\}$ of Θ is compact. The continuous function M takes on its maximum in this set, and that maximum is less than its value in θ, where M has a unique absolute maximum. For any given $\delta > 0$, there hence exists an $\varepsilon > 0$ with $M(\vartheta) < M(\theta) - \varepsilon$ for all ϑ with $\|\vartheta - \theta\| \geq \delta$. Inverting this statement gives that $M(\vartheta) \geq M(\theta) - \varepsilon$ implies $\|\vartheta - \theta\| < \delta$. We conclude from $M(\hat{\theta}_n) \geq M(\theta) - 2\Delta_n$ that $\|\hat{\theta}_n - \theta\| < \delta$ as soon as $2\Delta_n \leq \varepsilon$. By part (ii), the latter has probability converging to 1. We therefore have $\|\hat{\theta}_n - \theta\| < \delta$ with probability converging to 1, thus proving the consistency of $\hat{\theta}_n$.

We now need to prove parts (i)–(ii). For the proof of part (i), we apply the mean value theorem to see that for every ϑ_1 and ϑ_2, there exists a value $\tilde{\vartheta}$ between ϑ_1 and ϑ_2 such that $\ell_{\vartheta_1}(x) - \ell_{\vartheta_2}(x) = (\vartheta_1 - \vartheta_2)\dot{\ell}_{\tilde{\vartheta}}(x)$. It follows that

(5.1)
$$\left| \ell_{\vartheta_1}(x) - \ell_{\vartheta_2}(x) \right| \leq \|\vartheta_1 - \vartheta_2\| L(x).$$

If we replace x by X_1 and take the expectation under θ, we find $|M(\vartheta_1) - M(\vartheta_2)| \leq \|\vartheta_1 - \vartheta_2\| \, \mathrm{E}_\theta L(X_1)$, where $\mathrm{E}_\theta L(X_1)$ is finite by assumption. (Note that $|\mathrm{E}Y| \leq \mathrm{E}|Y|$ for every variable Y.) This proves the continuity of M. For the uniqueness of the maximum, we use that $\log x \leq 2(\sqrt{x} - 1)$ for all $x > 0$, so that

$$M(\vartheta) - M(\theta) = \mathrm{E}_\theta \left[\log \frac{p_\vartheta}{p_\theta}(X_1) \right] \leq 2\mathrm{E}_\theta \left[\sqrt{\frac{p_\vartheta}{p_\theta}}(X_1) - 1 \right]$$
$$= 2 \int \sqrt{p_\vartheta}(x)\sqrt{p_\theta}(x)\, dx - 2 = -\int \left(\sqrt{p_\vartheta}(x) - \sqrt{p_\theta}(x) \right)^2 dx.$$

The integral on the right-hand side is strictly positive when $\vartheta \neq \theta$ unless the densities p_ϑ and p_θ are the same, which is excluded by the assumption that the parameter θ is identifiable. This proves part (i).

For the proof of part (ii), we fix a $\delta > 0$. By the assumed compactness, we can cover Θ with finitely many balls of radius δ; denote the centers of the balls by $\vartheta_1, \ldots, \vartheta_k$. For a given $\vartheta \in \Theta$, there then exists a ϑ_j such that $\|\vartheta - \vartheta_j\| < \delta$. Using (5.1), we find that

$$\mathbb{M}_n(\vartheta_j) - \delta \frac{1}{n}\sum_{i=1}^n L(X_i) \leq \mathbb{M}_n(\vartheta) \leq \mathbb{M}_n(\vartheta_j) + \delta \frac{1}{n}\sum_{i=1}^n L(X_i),$$
$$M(\vartheta_j) - \delta \mathrm{E}_\theta L(X_1) \leq M(\vartheta) \leq M(\vartheta_j) + \delta \mathrm{E}_\theta L(X_1).$$

Subtracting the second equation from the first, we find lower and upper bounds for $\mathbb{M}_n(\vartheta) - M(\vartheta)$, and therefore also for the absolute value of this difference. If we then take the supremum over ϑ, we find

$$\sup_\vartheta \left| \mathbb{M}_n(\vartheta) - M(\vartheta) \right| \leq \max_j \left| \mathbb{M}_n(\vartheta_j) - M(\vartheta_j) \right| + \delta \frac{1}{n}\sum_{i=1}^n L(X_i) + \delta \mathrm{E}_\theta L(X_1).$$

It is important that the maximum on the right concerns a finite set of indices j. For every fixed j, we have $\mathbb{M}_n(\vartheta_j) - M(\vartheta_j) \xrightarrow{\mathrm{P}} 0$ as $n \to \infty$, by the law of large numbers (Theorem A.26). The maximum over j therefore converges in probability to 0. Applying the law of large numbers again, we see that $n^{-1}\sum_{i=1}^{n} L(X_i) \xrightarrow{\mathrm{P}} \mathrm{E}_\theta L(X_1)$ as $n \to \infty$. We conclude that the right-hand side of the last display converges in probability to $2\delta\, \mathrm{E}_\theta L(X_1)$ as $n \to \infty$. This is true for every $\delta > 0$. Hence the left-hand side converges in probability to 0. This concludes the proof of part (ii). ∎

We have now proved that $\hat{\theta}_n \xrightarrow{\mathrm{P}} \theta$ as $n \to \infty$, and continue with a proof that $\sqrt{n}(\hat{\theta}_n - \theta)$ converges in distribution to a normal distribution.

Since, by assumption, θ is an interior point of Θ and $\hat{\theta}_n \xrightarrow{\mathrm{P}} \theta$ as $n \to \infty$, we know that $\hat{\theta}_n$ has probability converging to 1 of also being an interior point of Θ. In that case, $\hat{\theta}_n$ satisfies the likelihood equation $\dot{\mathbb{M}}_n(\hat{\theta}_n) = 0$, where the dot means $\partial/\partial\theta$. The mean value theorem gives the existence of a point $\tilde{\theta}_n$ between $\hat{\theta}_n$ and θ such that

$$0 = \dot{\mathbb{M}}_n(\hat{\theta}_n) = \dot{\mathbb{M}}_n(\theta) + (\hat{\theta}_n - \theta)\ddot{\mathbb{M}}_n(\tilde{\theta}_n).$$

We deduce from this that

$$\sqrt{n}(\hat{\theta}_n - \theta) = -\frac{\sqrt{n}\,\dot{\mathbb{M}}_n(\theta)}{\ddot{\mathbb{M}}_n(\tilde{\theta}_n)} = -\frac{n^{-1/2}\sum_{i=1}^{n}\dot{\ell}_\theta(X_i)}{n^{-1}\sum_{i=1}^{n}\ddot{\ell}_{\tilde{\theta}_n}(X_i)}.$$

By Lemma 5.10, we have $\mathrm{E}_\theta\dot{\ell}_\theta(X_1) = 0$; moreover, $\mathrm{var}_\theta\,\dot{\ell}_\theta(X_1)$ is by definition equal to the Fisher information i_θ. By the central limit theorem, the numerator $-n^{-1/2}\sum_{i=1}^{n}\dot{\ell}_\theta(X_i)$ of the fraction on the right-hand side converges in distribution to an $N(0, i_\theta)$-distribution. The denominator of the fraction is an average of the variables $\ddot{\ell}_{\tilde{\theta}_n}(X_i)$. Since $\tilde{\theta}_n$ is stochastic and depends on all observations $X_1, \ldots, X_n$, these variables are not independent, and therefore the law of large numbers cannot be applied as it is. However, $\tilde{\theta}_n \xrightarrow{\mathrm{P}} \theta$ as $n \to \infty$, and below we prove that $n^{-1}\sum_{i=1}^{n}\ddot{\ell}_{\tilde{\theta}_n}(X_i)$ behaves like the average $n^{-1}\sum_{i=1}^{n}\ddot{\ell}_\theta(X_i)$, which does satisfy the law of large numbers, By Lemma 5.10, the limit satisfies $\mathrm{E}_\theta\ddot{\ell}_\theta(X_1) = -i_\theta$, hence we can conclude that $n^{-1}\sum_{i=1}^{n}\ddot{\ell}_{\tilde{\theta}_n}(X_i) \xrightarrow{\mathrm{P}} -i_\theta$ as $n \to \infty$. By Slutsky's lemma, Lemma 5.15, we therefore conclude that $\sqrt{n}(\hat{\theta}_n - \theta)$ converges in distribution to $(1/i_\theta)$ times an $N(0, i_\theta)$-distributed variable, that is, to a variable with the $N(0, i_\theta^{-1})$-distribution.

For a proof that $n^{-1}\sum_{i=1}^{n}(\ddot{\ell}_{\tilde{\theta}_n}(X_i) - \ddot{\ell}_\theta(X_i)) \xrightarrow{\mathrm{P}} 0$ as $n \to \infty$, we now also assume the existence of a *third-order* derivative of $\vartheta \mapsto \ell_\vartheta(x)$ such that $|\dddot{\ell}_\vartheta(x)| \leq K(x)$ for every x and every ϑ in a neighborhood of θ, where K is a function satisfying $\mathrm{E}_\theta K(X_1) < \infty$. Applying the mean value theorem to the second-order derivative then gives $|\ddot{\ell}_\vartheta(x) - \ddot{\ell}_\theta(x)| \leq K(x)|\vartheta - \theta|$ for all x and all ϑ with $|\vartheta - \theta| \leq \varepsilon$ and sufficiently small ε. Consequently, for every $\delta > 0$, we have

(5.2)
$$\mathrm{P}_\theta\left(\left|\frac{1}{n}\sum_{i=1}^{n}(\ddot{\ell}_{\tilde{\theta}_n}(X_i) - \ddot{\ell}_\theta(X_i))\right| > \delta, |\tilde{\theta}_n - \theta| \leq \varepsilon\right)$$

$$\leq \mathrm{P}_\theta\left(\frac{1}{n}\sum_{i=1}^{n}K(X_i)\,|\tilde{\theta}_n - \theta| > \delta\right).$$

By the law of large numbers, the factor $n^{-1}\sum_{i=1}^{n}K(X_i)$ converges in probability to $\mathrm{E}_\theta K(X_1) < \infty$. This implies the existence of a constant M such that $\mathrm{P}_\theta\left(n^{-1}\sum_{i=1}^{n}K(X_i) \leq M\right) \to 1$. Combining this with the inequality $\mathrm{P}_\theta\left(|\tilde{\theta}_n - \theta| > \delta/M\right) \to 0$ shows that the right-hand side of (5.2) converges to 0. The same then holds for the left-hand side. Since $\tilde{\theta}_n \xrightarrow{\mathrm{P}} \theta$ as $n \to \infty$, this remains true when we drop the restriction $|\tilde{\theta}_n - \theta| \leq \varepsilon$.

The only part of the proof of the asymptotic normality of $\sqrt{n}(\hat{\theta}_n - \theta)$ we still need to do in detail consists of the statements of Lemma 5.10. Solidifying the given proof of Lemma 5.10 requires further conditions to justify differentiating under the integral sign. We can also prove the statements in a roundabout way under the existing conditions (see the proof of Theorem 5.39 in Van der Vaart (1998)). It is remarkable that Theorem 5.9 does not assume the existence of the second-order derivative $\ddot{\ell}_\vartheta$, so that the statements of Lemma 5.10 are certainly not *necessary* for a proof; the same holds for the existence of third-order derivatives. We will not discuss this any further.

Lemma 5.15 Slutsky's lemma

Let S_n and T_n *be random variables or vectors with* $S_n \xrightarrow{\mathrm{P}} \sigma$ *for a constant* σ *and* $T_n \rightsquigarrow T$ *as* $n \to \infty$. *Then*
 (i) $S_n + T_n \rightsquigarrow \sigma + T$ *as* $n \to \infty$;
 (ii) *if* $\sigma \neq 0$, *then* $T_n/S_n \rightsquigarrow T/\sigma$ *as* $n \to \infty$.

In part (i), the "constant" σ and T must be vectors of the same length. Part (ii) is true when σ is a scalar but also holds for matrices σ. In the latter case, $\sigma \neq 0$ means that σ is invertible, and dividing by σ means multiplying by its inverse.

Proof. The inequality $\|S_n - \sigma\| \leq \varepsilon$ implies $\sigma - \varepsilon \leq S_n \leq \sigma + \varepsilon$. If $\sigma > 0$, then we can choose ε sufficiently small that $\sigma - \varepsilon > 0$. In that case, the inequality $T_n/S_n \leq x$ implies $T_n \leq x(\sigma + \varepsilon)$, and $T_n \leq x(\sigma - \varepsilon)$ implies $T_n/S_n \leq x$. We conclude that

$$\mathrm{P}\left(T_n \leq x(\sigma - \varepsilon), \|S_n - \sigma\| \leq \varepsilon\right) \leq \mathrm{P}\left(T_n/S_n \leq x, \|S_n - \sigma\| \leq \varepsilon\right)$$
$$\leq \mathrm{P}\left(T_n \leq x(\sigma + \varepsilon), \|S_n - \sigma\| \leq \varepsilon\right).$$

Since $\mathrm{P}\left(\|S_n - \sigma\| > \varepsilon\right) \to 0$, the three probabilities in this equation change at most by a term that converges to 0 if we drop the restriction $\|S_n - \sigma\| \leq \varepsilon$. By applying the convergence $T_n \rightsquigarrow T$ to the first and third probabilities, we conclude that the limit (or lim inf and lim sup) of $\mathrm{P}(T_n/S_n \leq x)$ is asymptotically sandwiched between $\mathrm{P}\left(T \leq x(\sigma - \varepsilon)\right)$ and $\mathrm{P}\left(T \leq x(\sigma + \varepsilon)\right)$, for every x and ε such that $x(\sigma - \varepsilon)$ and $x(\sigma + \varepsilon)$ are continuity points of $x \mapsto \mathrm{P}(T \leq x)$. Since a distribution function can have at most countably many discontinuity points, there exists a sequence $\varepsilon_m \to 0$ such that $x(\sigma - \varepsilon_m)$ and $x(\sigma + \varepsilon_m)$ are continuity points for every m. If x is a continuity point of T/σ, then $x\sigma$ is a continuity point of T, and $\mathrm{P}\left(T \leq x(\sigma - \varepsilon_m)\right)$ and $\mathrm{P}\left(T \leq x(\sigma + \varepsilon_m)\right)$ both converge to $\mathrm{P}(T/\sigma \leq x)$. The sequence $\mathrm{P}(T_n/S_n \leq x)$ then has the same limit.

The proof of part (ii) when $\sigma < 0$ and the proof of part (i) are analogous. ∎

* 5.4.2 Multidimensional Parameters

The above can be extended to the case where the parameter θ is a vector of dimension $k > 1$. The score function is then defined as the gradient

$$\dot{\ell}_\theta(x) = \nabla_\theta \log p_\theta(x) = \left(\frac{\partial}{\partial \theta_1} \ell_\theta(x), \dots, \frac{\partial}{\partial \theta_k} \ell_\theta(x) \right).$$

The Fisher information is generalized to a $(k \times k)$-matrix

$$i_\theta = \left(\operatorname{cov}_\theta \left(\frac{\partial}{\partial \theta_i} \ell_\theta(X_1), \frac{\partial}{\partial \theta_j} \ell_\theta(X_1) \right) \right)_{i,j=1,\dots k}.$$

Theorem 5.9 remains valid, but $\sqrt{n}(\hat\theta - \theta)$ is a random vector and its limit distribution is a *multivariate normal distribution* (see Appendix B). The statement of the "theorem" must be understood in the sense that the near-pivot $(ni_\theta)^{1/2}(\hat\theta - \theta)$ is approximately distributed as a vector $Z = (Z_1, \dots, Z_k)$ of k independent $N(0,1)$-distributed variables.[‡]

The quadratic form

$$(\hat\theta - \theta)^T ni_\theta (\hat\theta - \theta) = \left(\sqrt{n} i_\theta^{1/2} (\hat\theta - \theta) \right)^T \sqrt{n} i_\theta^{1/2} (\hat\theta - \theta)$$

then approximately has the same distribution as $Z^T Z = \sum_{i=1}^k Z_i^2$, that is, a χ_k^2-distribution (see Section 4.6). For $\widehat{i_\theta}$ an estimator for the matrix i_θ, the set

$$\left\{ \theta \colon (\hat\theta - \theta)^T n\widehat{i_\theta} (\hat\theta - \theta) \le \chi_{k,1-\alpha}^2 \right\}$$

is therefore a confidence region of asymptotic confidence level $1 - \alpha$ (for large n). Geometrically, this set is an ellipsoid in the k-dimensional space, because the Fisher information matrix i_θ is positive definite.

Often, we are only interested in a function $g(\theta)$ of a higher-dimensional parameter. Theorem 5.9 can be extended to that case.

Theorem 5.16

Take the situation of Theorem 5.9, but with parameter $\theta \in \Theta \subset \mathbb{R}^k$ and a finite, invertible Fisher information matrix. For a differentiable function $g \colon \Theta \to \mathbb{R}$ with gradient g', under θ, we have

$$\sqrt{n}\big(g(\hat\theta) - g(\theta)\big) \rightsquigarrow N\big(0, g'(\theta) i_\theta^{-1} g'(\theta)^T\big)$$

as $n \to \infty$.

[‡] By $(i_\theta)^{1/2}$, we mean a matrix A of the same dimension as i_θ such that $A^T A = i_\theta$.

Proof. The proof consists of two parts. First, Theorem 5.9 holds, precisely as stated, for multidimensional parameters, where the limit distribution $N(0, i_\theta^{-1})$ is the multivariate normal distribution, with covariance the inverse of the Fisher information matrix. Next, we apply the "delta method" to determine the limit distribution of $\sqrt{n}\big(g(\hat{\theta}) - g(\theta)\big)$. This method corresponds to using the fact that this sequence has the same limit distribution as the first-order Taylor expansion $\sqrt{n}g'(\theta)(\hat{\theta} - \theta)$ at θ. Since $\sqrt{n}(\hat{\theta} - \theta) \rightsquigarrow Z$ as $n \to \infty$, for a normally distributed vector Z with expectation 0 and covariance matrix i_θ^{-1}, we have $\sqrt{n}g'(\theta)(\hat{\theta} - \theta) \rightsquigarrow g'(\theta)Z$ as $n \to \infty$. The random variable $g'(\theta)Z$ has a normal distribution, as in the theorem, because of the properties of the multidimensional normal distribution (see Lemma B.4). A precise justification of the delta method can be found in Chapter 3 of Van der Vaart (1998). ∎

In particular, the first coordinate of $\theta = (\theta_1, \ldots, \theta_k)$ corresponds to the function $g(\theta) = \theta_1$ and gradient $g'(\theta) = (1, 0, \ldots, 0)$. The asymptotic variance of $\sqrt{n}(\hat{\theta}_1 - \theta_1)$ is therefore equal to $(i_\theta^{-1})_{(1,1)}$, the $(1,1)$-element of the inverse matrix i_θ^{-1} (not to be confused with 1 divided by the $(1,1)$-element of i_θ). For θ_1, we use the confidence interval

$$\theta_1 = \hat{\theta}_1 \pm \frac{(\widehat{i_\theta^{-1}})_{(1,1)}^{1/2}}{\sqrt{n}} \xi_{1-\alpha/2}.$$

If $\theta_2, \ldots, \theta_k$ are known and therefore do not need to be estimated, we have a one-dimensional estimation problem. We saw in Theorem 5.9 that in this case the asymptotic variance of $\sqrt{n}(\hat{\theta}_1 - \theta_1)$ is equal to 1 divided by the Fisher information for the one-dimensional estimation problem. This value is equal to $(i_{\theta,(1,1)})^{-1}$, that is, 1 divided by the $(1,1)$-element of the Fisher matrix i_θ in the multidimensional problem above. In general, we have $(i_\theta^{-1})_{(1,1)} \leq (i_{\theta,(1,1)})^{-1}$. This means that when $\theta_2, \ldots, \theta_k$ are unknown, there is loss of information and θ_1 cannot be estimated as precisely, resulting in a greater asymptotic variance and a longer confidence interval for θ_1. In some cases (see Example 5.18), the Fisher information matrix is a diagonal matrix. We then have $(i_{\theta,(1,1)})^{-1} = (i_\theta^{-1})_{(1,1)}$, and not knowing the other parameters does not lead to any loss of information.

Example 5.17 Multinomial distribution

Let $Y = (Y_1, \ldots, Y_m)$ be multinomially distributed with parameters n and $(p_1, \ldots, p_m)$; see Example 4.48. We assume n known and the probabilities $p_1, \ldots, p_m$ unknown. The sum of the probabilities is $\sum_{i=1}^{m} p_i = 1$, and therefore $p_m = 1 - (p_1 + \ldots + p_{m-1})$. Since p_m is fixed whenever $p_1, \ldots, p_{m-1}$ are known, we have a $(m-1)$-dimensional estimation problem. Let $p = (p_1, \ldots, p_{m-1})$ be the vector of the unknown parameters. We want to construct an approximate confidence region for p based on the asymptotic distribution of the maximum likelihood estimator for p. The maximum likelihood estimator for p maximizes the log-likelihood function of the model; this function is given by

$$p \mapsto \log \binom{n}{Y_1 \cdots Y_m} + \sum_{i=1}^{m} Y_i \log p_i.$$

The maximum likelihood estimator for p relative to its parameter space $\{p \in \mathbb{R}^{m-1} : p_i \geq 0, \sum_{i=1}^{m-1} p_i \leq 1\}$ is equal to the vector $(Y_1/n, Y_2/n, \ldots, Y_{m-1}/n)$ (see Exercise 3.15).

In this section, we study the situation where were have a sample of size n. In the multinomial model, we in fact have only one observation $(Y_1, \ldots, Y_m)$, but we can also view this observation as a sum of n independent, identically distributed subobservations X_k for $k = 1, \ldots, n$ with X_k multinomially distributed with parameters 1 and $(p_1, \ldots, p_m)$. We write $X_k = (X_{k,1}, \ldots, X_{k,m})$, so that $\sum_{k=1}^{n} X_k = Y$, where the sum is coordinate-wise. For this model, the maximum likelihood estimators for the parameters $p_1, \ldots, p_m$ are the same as in the multinomial model. This follows from the fact that the log-likelihood functions are equal up to the first term in the log-likelihood of Y, and this term does not depend on the unknown parameters. To illustrate the theory in this section, we assume that we observe the sample $X_1, \ldots, X_n$. The score function of the model is given by the vector

$$\left(\frac{X_{1,1}}{p_1} - \frac{X_{1,m}}{p_m}, \ldots, \frac{X_{1,m-1}}{p_{m-1}} - \frac{X_{1,m}}{p_m} \right).$$

A simple computation gives $\mathrm{var}_p X_{1,i} = p_i(1 - p_i)$ and $\mathrm{cov}_p(X_{1,i}, X_{1,j}) = -p_i p_j$ for $i \neq j$. The (i,j)-element of the Fisher information matrix i_p is therefore given by

$$(i_p)_{i,j} = 1/p_i + 1/p_m \quad \text{for } i = j \text{ and} \qquad (i_p)_{i,j} = 1/p_m \quad \text{for } i \neq j.$$

We can estimate the unknown parameters $p_1, \ldots, p_{m-1}$ in the Fisher information matrix by the maximum likelihood estimators $\hat{p}_1, \ldots, \hat{p}_{m-1}$. The approximate confidence region for p of confidence level $1 - \alpha$ is now equal to

$$\left\{ p : (\hat{p} - p)^T n \widehat{i_p}(\hat{p} - p) \leq \chi^2_{m-1,1-\alpha} \right\},$$

with $\hat{p}$ the maximum likelihood estimator for the parameter p and $\widehat{i_p}$ the estimated Fisher information matrix.

Suppose that we are only interested in estimating p_1. We apply Theorem 5.16, but now with $g(p) = p_1$ and gradient $g'(p) = (1, 0, \ldots, 0)$. It immediately follows that under the assumption that p_1 is the true parameter, as $n \to \infty$, we have

$$\sqrt{n}(\hat{p}_1 - p_1) \rightsquigarrow N(0, (i_p^{-1})_{(1,1)}),$$

with variance equal to the $(1,1)$-element of the inverse Fisher information matrix i_p^{-1}. The (i,j)-element of this matrix is equal to

$$(i_p^{-1})_{(i,j)} = p_i(1 - p_i) \quad \text{for } i = j \text{ and} \qquad (i_p^{-1})_{(i,j)} = -p_i p_j \quad \text{for } i \neq j.$$

Note that the ith diagonal element, $p_i(1 - p_i)$, is equal to $\mathrm{var}_p X_{1,i}$ and the (i,j)-element of i_p^{-1} is equal to $\mathrm{cov}_p(X_{1,i}, X_{1,j})$. In short, $\sqrt{n}(\hat{p}_1 - p_1)$ is asymptotically normally distributed with expectation 0 and variance $p_1(1 - p_1) = \mathrm{var}_p X_{1,1}$. To estimate the variance, we can again replace p_1 by the maximum likelihood estimator. An approximate confidence interval of confidence level $1 - \alpha$ is then equal to

$$p_1 = \hat{p}_1 \pm \frac{\hat{p}_1(1 - \hat{p}_1)}{\sqrt{n}} \xi_{1-\alpha/2}.$$

If we are only interested in estimating the parameter p_1, we could also have reduced the multinomial model to a binomial model with parameters n and p_1. We do not need to estimate the unknown parameters $p_2, \ldots, p_m$ individually; the sum $p_2 + \ldots + p_m = 1 - p_1$ suffices. A simple computation shows that we find the same approximate confidence interval.

Example 5.18 Normal distribution

Let $X = (X_1, \ldots, X_n)$ be a sample from the normal distribution with unknown parameters μ and σ^2. We want to determine a confidence interval for μ.

In Example 5.4, we constructed an exact confidence interval of confidence level $1 - \alpha$ based on the t_{n-1}-distributed random variable $\sqrt{n}(\overline{X} - \mu)/S_X$. This interval is given by

$$\mu = \overline{X} \pm \frac{S_X}{\sqrt{n}} t_{n-1, 1-\alpha/2}.$$

As an alternative, we could also have taken the exact confidence interval from Example 5.2 and replaced the parameter σ^2, which was assumed known, by its estimator S_X^2. We then find an approximate confidence interval of confidence level $1 - \alpha$:

$$\mu = \overline{X} \pm \frac{S_X}{\sqrt{n}} \xi_{1-\alpha/2}.$$

The only difference with the interval from Example 5.4 is the quantiles. For large n, there is hardly any difference between the quantiles of the t_{n-1}-distribution and those of the standard normal distribution, and the intervals will be approximately equal.

When σ^2 is unknown, we can also construct an approximate confidence interval for μ based on the asymptotic distribution of the maximum likelihood estimator for μ. Because σ is unknown, we are dealing with a two-dimensional estimation problem. The score function of the model is given by

$$\dot{\ell}_{(\mu, \sigma^2)}(X_1) = \left(\frac{X_1 - \mu}{\sigma^2}, \ \frac{(X_1 - \mu)^2}{2\sigma^4} - \frac{1}{2\sigma^2} \right)^T = \left(\frac{Z}{\sigma}, \ \frac{Z^2}{2\sigma^2} - \frac{1}{2\sigma^2} \right)^T,$$

where we use the abbreviation $Z = (X_1 - \mu)/\sigma$ and Z has the standard normal distribution. The diagonal elements of the Fisher information matrix are then equal to

$$\text{var}_{(\mu, \sigma)} \left(\frac{Z}{\sigma} \right) = \frac{1}{\sigma^2},$$

$$\text{var}_{(\mu, \sigma)} \left(\frac{Z^2}{2\sigma^2} - \frac{1}{2\sigma^2} \right) = \frac{1}{4\sigma^4} \text{var}_{(\mu, \sigma)} Z^2 = \frac{1}{2\sigma^4},$$

since Z^2 has the χ_1^2-distribution with variance 2. The $(1, 2)$- and the $(2, 1)$-elements of the symmetric Fisher information matrix are equal to

$$\text{cov}_{(\mu, \sigma)} \left(\frac{Z}{\sigma}, \frac{Z^2}{2\sigma^2} \right) = \frac{1}{2\sigma^3} \text{cov}_{(\mu, \sigma)} \left(Z, Z^2 \right) = 0,$$

where the last equality follows from $\mathrm{cov}(Z, Z^2) = EZ^3 - EZ\,EZ^2 = 0$ because the first and third moments of the standard normal distribution are equal to 0. The Fisher information matrix is therefore equal to

$$i_{(\mu,\sigma^2)} = \begin{pmatrix} 1/\sigma^2 & 0 \\ 0 & 1/(2\sigma^4) \end{pmatrix}.$$

Since the Fisher information matrix is a diagonal matrix, we can easily determine its inverse by inverting the diagonal elements:

$$i^{-1}_{(\mu,\sigma^2)} = \begin{pmatrix} \sigma^2 & 0 \\ 0 & 2\sigma^4 \end{pmatrix}.$$

We can again estimate the unknown variance σ^2 using the sample variance S_X^2. We then find the approximate confidence interval for μ using Theorem 5.16,

$$\mu = \overline{X} \pm \frac{S_X}{\sqrt{n}}\xi_{1-\alpha/2}.$$

This is the same approximate interval as at the beginning of this example.

The Fisher information matrix in this example is a diagonal matrix. In this specific case, we have $(i^{-1}_{(\mu,\sigma^2)})_{(1,1)} = (i_{(\mu,\sigma^2),(1,1)})^{-1}$; knowing σ^2 or not does not have any influence on the length of the approximate confidence interval for μ, up to the estimation of σ^2. ▭

5.5 Confidence Regions and Tests

Confidence intervals and tests are closely related. A given set of tests for the problems $H_0: g(\theta) = \tau$ automatically defines a confidence region for $g(\theta)$ and conversely.

Theorem 5.19

Suppose that for every $\tau \in g(\Theta)$, we are given a test of level α for the null hypothesis $H_0: g(\theta) = \tau$ (with a critical region that depends only on τ). Then the set of all values τ that are not rejected in testing is a confidence region for $g(\theta)$ of confidence level $1 - \alpha$.

Conversely, given a confidence region G_X for $g(\theta)$ of confidence level $1 - \alpha$, the critical region $\{x : \tau \notin G_x\}$ gives a test of confidence level $1 - \alpha$ for the null hypothesis $H_0: g(\theta) = \tau$, for all $\tau \in g(\Theta)$.

Proof. For $\tau \in g(\Theta)$, we define the set $\Theta_\tau = \{\theta \in \Theta : g(\theta) = \tau\}$, so that $H_0: g(\theta) = \tau$ is equivalent to $H_0: \theta \in \Theta_\tau$.

In the first part of the theorem, we are given, for every $\tau \in g(\Theta)$, a critical region K_τ of a test of level α for $H_0: \theta \in \Theta_\tau$, and the confidence region we have in mind for $g(\theta)$ is the set $G_X = \{\tau : X \notin K_\tau\}$. That this test has level α means that $P_\theta(X \in K_\tau) \le \alpha$ for all $\theta \in \Theta_\tau$, for every given τ. Since $\tau = g(\theta)$ for every $\theta \in \Theta_\tau$, we therefore also have $P_\theta(X \in K_{g(\theta)}) \le \alpha$ for every $\theta \in \Theta$. It follows from the definition of G_X that $g(\theta) \in G_X$ if and only if $X \notin K_{g(\theta)}$. The first statement of the theorem now follows from $P_\theta(g(\theta) \in G_X) = P_\theta(X \notin K_{g(\theta)}) \ge 1 - \alpha$, for all $\theta \in \Theta$.

In the second part, we are given a confidence region G_X for $g(\theta)$, and the test we have in mind for the null hypothesis $H_0: \theta \in \Theta_\tau$ has critical region $K_\tau = \{x : \tau \notin G_x\}$. That G_X has size α means that $P_\theta(g(\theta) \in G_X) \ge 1 - \alpha$ for all $\theta \in \Theta$. It follows from the definition of K_τ that $X \in K_\tau$ if and only if $\tau \notin G_X$. For $\theta \in \Theta_\tau$, we have $g(\theta) = \tau$, and therefore $P_\theta(X \in K_\tau) = P_\theta(X \in K_{g(\theta)}) = P_\theta(g(\theta) \notin G_X) \le \alpha$. We conclude that the test with critical region K_τ has size α. This proves the second part of the theorem. ∎

At first sight, applying this theorem seems a difficult way to construct a confidence interval: we must test the hypothesis $H_0: g(\theta) = \tau$ for every τ. This can indeed be much work, but in some standard cases, it is quite easy.

Example 5.20 Normal distribution

Let $X = (X_1, \ldots, X_n)$ be a sample from the $N(\mu, \sigma^2)$-distribution with unknown parameters μ and σ^2. The t-test does not reject the null hypothesis $H_0: \mu = \mu_0$ at level α when

$$-t_{n-1,1-\alpha/2} \le \sqrt{n} \frac{\overline{X} - \mu_0}{S_X} \le t_{n-1,1-\alpha/2}.$$

This is equivalent to the inequalities

$$\overline{X} - \frac{S_X}{\sqrt{n}} t_{n-1,1-\alpha/2} \le \mu_0 \le \overline{X} - \frac{S_X}{\sqrt{n}} t_{n-1,\alpha/2}.$$

By Theorem 5.19, the confidence interval of confidence level $1 - \alpha$ for μ is then equal to

$$\mu = \overline{X} \pm \frac{S_X}{\sqrt{n}} t_{n-1,1-\alpha/2}.$$

We had already found this confidence interval in a different way.

* Example 5.21 Exponential distribution

Let $X_1, \ldots, X_n$ be a sample from the exponential distribution with unknown parameter λ. An approximate confidence interval of confidence level $1 - \alpha$ for λ is

$$\lambda = \frac{1}{\overline{X}} \pm \frac{1}{\sqrt{n}\overline{X}} \xi_{1-\alpha/2};$$

see Example 5.13. By Theorem 5.19, the test that rejects the null hypothesis $H_0\colon \lambda = \lambda_0$ when λ_0 is not in this confidence interval is a test for $H_0\colon \lambda = \lambda_0$ against the alternative $H_1\colon \lambda \neq \lambda_0$ of approximate size α. This test corresponds to the Wald test; see Section 4.8.

Example 5.22 Binomial distribution

Let X be binomially distributed with unknown parameter p and known n. We can determine an "exact" confidence interval for p by inverting the exact test for $H_0\colon p = p_0$ discussed in Example 4.24. The best way to do this is to use the test in terms of p-values. The null hypothesis $H_0\colon p = p_0$ is rejected at size α if, for an observed value x,

$$\mathrm{P}_{p_0}(X \geq x) \leq \tfrac{1}{2}\alpha \quad \text{or} \quad \mathrm{P}_{p_0}(X \leq x) \leq \tfrac{1}{2}\alpha.$$

The confidence region for the observed value x is therefore the set

$$\left\{ p\colon \mathrm{P}_p(X \geq x) > \tfrac{1}{2}\alpha \quad \text{and} \quad \mathrm{P}_p(X \leq x) > \tfrac{1}{2}\alpha \right\}.$$

For $x \geq 1$, the function $p \mapsto \mathrm{P}_p(X \geq x)$ is a continuous function of p that is strictly increasing from the value 0 in $p = 0$ to 1 in $p = 1$; see Figure 5.4. Consequently, the set $\{p\colon \mathrm{P}_p(X \geq x) > \tfrac{1}{2}\alpha\}$ is equal to $(p_l, 1]$, where p_l is the solution of the equation

$$\mathrm{P}_{p_l}(X \geq x) = \tfrac{1}{2}\alpha.$$

On the other hand, for $x \leq n - 1$, the function $p \mapsto \mathrm{P}_p(X \leq x)$ is a continuous function that is strictly decreasing from 1 in $p = 0$ to 0 in $p = 1$. Consequently, the set $\{p\colon \mathrm{P}_p(X \leq x) > \tfrac{1}{2}\alpha\}$ is equal to $[0, p_r)$, where p_r is the solution of the equation

$$\mathrm{P}_{p_r}(X \leq x) = \tfrac{1}{2}\alpha.$$

The desired confidence interval is the intersection (p_l, p_r) of the two intervals we found.

If $x = 0$, then $\mathrm{P}_p(X \geq x) = 1$ for every p, and the equation for p_l does not have any solutions. The confidence interval is then $[0, p_r)$. If $x = n$, then $\mathrm{P}_p(X \leq x) = 1$ for every p, and the equation for p_r does not have any solutions. The confidence interval is then $(p_l, 1]$.

The values p_l and p_r can be solved from the equations using tables or the computer, or even using the normal approximation (though this goes against the aim to have an "exact" interval). For example, for $\alpha = 0.05$, $n = 20$, and $x = 13$, the table gives

$$\begin{aligned} \mathrm{P}_{0.84}(X \leq 13) &= 0.03037 \\ \mathrm{P}_{0.85}(X \leq 13) &= 0.02194 \end{aligned} \quad \Rightarrow p_r \approx 0.845.$$

Likewise, we find $p_l \approx 0.405$, so that the exact confidence interval is $(0.405, 0.845)$. This interval is indicated in Figure 5.4.

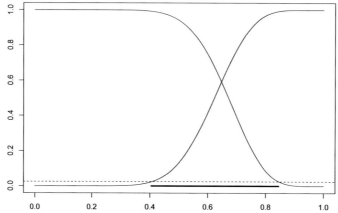

Figure 5.4. Confidence interval for the binomial distribution. The graph shows the functions $p \mapsto P_p(X \geq x)$ (increasing) and $p \mapsto P_p(X \leq x)$ (decreasing) for $n = 20$ and $x = 13$ and a dashed line at height 0.025. The 95% confidence interval contains the values between the intersection points of the curves with the dashed line.

5.6 Likelihood Ratio Regions

The procedure to deduce confidence regions from tests is, in particular, often applied to the likelihood ratio test. This test rejects the null hypothesis $H_0: \theta = \tau$ for large values of the likelihood ratio statistic $p_{\hat{\theta}}(X)/p_\tau(X)$, for $\hat{\theta}$ the maximum likelihood estimator for θ. In many cases, we use the chi-square approximation to find a critical value (compare with Theorem 4.43). The likelihood ratio test rejects the null hypothesis $H_0: \theta = \tau$ concerning a k-dimensional parameter when $2 \log(p_{\hat{\theta}}(X)/p_\tau(X)) \geq \chi^2_{k,1-\alpha}$, for $\chi^2_{k,1-\alpha}$ the $(1 - \alpha)$-quantile of the χ^2_k-distribution (this result is a generalization of the one-dimensional restriction in Example 4.45, which can be extended to a k-dimensional restriction). The procedure in Theorem 5.19 leads to the confidence region

$$\left\{ \theta : \log p_\theta(X) - \log p_{\hat{\theta}}(X) \geq -\tfrac{1}{2}\chi^2_{k,1-\alpha} \right\}.$$

The "inversion of the likelihood ratio test" has the intuitively attractive property that the confidence region contains those values of the parameter θ that maximize the likelihood function.

We can visualize the confidence region using a plot of the log-likelihood function minus its maximum: minus the log-likelihood ratio $\theta \mapsto \log p_\theta(x) - \log p_{\hat{\theta}}(x)$. For a one-dimensional parameter, this is a function with a "normal," two-dimensional graph. If we draw a horizontal line at height $-\tfrac{1}{2}\chi^2_{1,1-\alpha}$, then the confidence regions consists precisely of the values of θ where minus the log-likelihood ratio statistic rises above the horizontal line (see Figure 5.5 for an illustration). For multidimensional parameters, the log-likelihood function is a hypersurface, and the confidence region is the set of values where the hypersurface rises above height $-\tfrac{1}{2}\chi^2_{k,1-\alpha}$, for k the dimension of the parameter.

198

It is clear from the graphical description of the confidence region that the maximum likelihood estimate always lies in the confidence region, and that the form of this region is determined by the form of the likelihood function. In particular, a likelihood ratio confidence region is not necessarily symmetric about the maximum likelihood estimator; see Figure 5.5 as an illustration. In general, the asymmetry, when it occurs, is viewed as desirable, as an expression of a different measure of uncertainty over the parameter in different directions. Note, however, that the likelihood ratio surface may have several local maxima, and that in extreme situations, this can lead to a confidence region that consists of more than one disconnected component. It is unclear whether disconnected confidence regions are desirable.

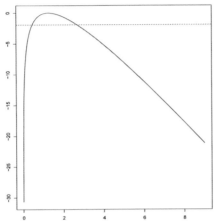

Figure 5.5. Minus the log-likelihood ratio statistic as a function of θ for a sample of size 4 from the Poisson distribution with expectation 1. The dotted line is at height $-\frac{1}{2}\chi^2_{1,0.95}$. The values of θ for which the curve rises above the line belong to the approximate 95% confidence interval.

Example 5.23 Exponential distribution

Let $X = (X_1, \ldots, X_n)$ be a sample from the exponential distribution with unknown parameter $\lambda > 0$. The log-likelihood function is then

$$\lambda \mapsto n \log \lambda - \lambda \sum_{i=1}^{n} X_i,$$

and the maximum likelihood estimator for λ is $\hat{\lambda} = 1/\overline{X}$ (see Example 3.12). The set

$$\left\{ \lambda: n \log \lambda - \lambda \sum_{i=1}^{n} X_i - \left(n \log \hat{\lambda} - \hat{\lambda} \sum_{i=1}^{n} X_i \right) \geq -\tfrac{1}{2}\chi^2_{1,1-\alpha} \right\}$$

$$= \left\{ \lambda: n \log \lambda - \lambda \sum_{i=1}^{n} X_i + n \log \overline{X} + n \geq -\tfrac{1}{2}\chi^2_{1,1-\alpha} \right\}$$

is then the approximate confidence region for λ based on the likelihood ratio test of size α.

199

Often, we are interested in a confidence region for a component θ_1 of a higher-dimensional parameter $\theta = (\theta_1, \ldots, \theta_k)$, instead of a region for the full parameter vector θ. We can easily provide this using the likelihood ratio statistic, by inverting the test for the hypothesis $H_0: \theta_1 = \tau$, instead of the hypothesis for the full parameter used earlier (now with $\tau \in \mathbb{R}$). The likelihood ratio test rejects the null hypothesis $H_0: \theta_1 = \tau$ for large values of the test statistic

$$2 \log \frac{\sup_{\theta \in \Theta} p_\theta(X)}{\sup_{\theta \in \Theta: \theta_1 = \tau} p_\theta(X)}.$$

We can often choose the critical value equal to the $(1 - \alpha)$-quantile of the chi-square distribution with 1 degree of freedom, because the dimension k of the full model and the dimension $k_0 = k - 1$ of the null hypothesis $\Theta_0 = \{\theta: \theta_1 = \tau\}$ differ by 1 (see Example 4.45). The confidence region for θ_1 consists of the values of τ that have not been rejected.

This confidence region can be visualized using the so-called profile likelihood function.

Definition 5.24 Profile likelihood

The profile likelihood function is given by

$$L_1(\tau; X) = \sup_{\theta \in \Theta: \theta_1 = \tau} p_\theta(X).$$

For a fixed value of θ_1, the profile likelihood $L_1(\theta_1; X)$ is equal to the maximum of the "usual" likelihood $p_\theta(X)$ over the remaining parameters $\theta_2, \ldots, \theta_k$. By maximizing the profile likelihood $\theta_1 \mapsto L_1(\theta_1; X)$ with respect to θ_1, we find the maximum of the "usual" likelihood over the full parameter; the maximum is taken on in the maximum likelihood estimator $\hat{\theta}_1$ for θ_1. (This procedure splits finding the overall maximum of the likelihood in two steps but gives the same maximum.) The likelihood ratio statistic for testing $H_0: \theta_1 = \tau$ can therefore be written in the form $L_1(\hat{\theta}_1; X)/L_1(\tau; X)$, and when we use the chi-square approximation, the confidence region for θ_1 is of the form

$$\left\{\theta_1: \log L_1(\theta_1; X) - \log L_1(\hat{\theta}_1; X) \geq -\tfrac{1}{2}\chi^2_{1,1-\alpha}\right\}.$$

Using the profile likelihood, we can visualize the likelihood ratio region for θ_1 in a way analogous to that used for the usual likelihood for the full parameters. We plot minus the logarithm of the profile likelihood ratio function, $\log L_1(\theta_1; x) - \log L_1(\hat{\theta}_1; x)$, and take as confidence region the values of θ_1 where the function rises above a certain level.

This procedure can be extended to general functions g of the parameter θ by defining the profile likelihood for g as the function $\tau \mapsto L_g(\tau; X)$ given by

$$L_g(\tau; X) = \sup_{\theta \in \Theta: g(\theta) = \tau} p_\theta(X).$$

$*$ **Example 5.25** Application: compound Poisson process

In Example 3.22, we modeled the monthly payout by a health insurance company and estimated the unknown parameters μ and θ using the maximum likelihood estimators. Suppose that we want to construct a confidence interval for θ. In Example 4.49, we discussed the likelihood ratio test for testing $H_0 \colon \theta = \theta_0$ against $H_0 \colon \theta \neq \theta_0$. The test statistic does not depend on the parameter μ and, under H_0, asymptotically follows the chi-square distribution with 1 degree of freedom. The approximate confidence interval for θ can now easily be deduced from the above.

$*$ ## 5.7 Bayesian Confidence Regions

The Bayesian approach gives an alternative way to quantify the uncertainty of an estimate. In addition to a point estimator, this approach also gives the posterior distribution. This distribution is an expression of the uncertainty we have about the value of the parameter after carrying out the observation. The parameter value is viewed as a random vector distributed following the posterior distribution.

If we want to express our uncertainty through a margin or region around a point estimate, then a logical choice would be a region with probability $1 - \alpha$ under the posterior distribution. This is not uniquely determined, but in general, we will choose a symmetric region or the smallest possible region with this property.

This way of constructing an uncertainty margin is completely different from the methods discussed previously, and there is no guarantee that such a Bayesian region is also a confidence region in the sense of Definition 5.1. To express this difference, we speak of a *credible region* instead of a confidence region. We can show that in many cases, a credible region based on a large sample is approximately a confidence region in the usual sense.

The reason for this phenomenon is that Bayesian estimators are asymptotically normally distributed, and that the differences with maximum likelihood estimators disappear as the number of observations increases. A credible region is therefore asymptotically the same as the confidence region based on the maximum likelihood estimator, discussed in Section 5.4. The basic theorem that explains this is the *Bernstein-von Mises theorem*, according to which a posterior distribution is asymptotically a normal distribution centered in the maximum likelihood estimator. For simplicity, we again restrict ourselves to the case where the observation $X = (X_1, \ldots, X_n)$ is a sample from partial observations X_i with marginal probability density p_θ, for $\theta \in \Theta \subset \mathbb{R}^k$. Let $\overline{\Theta}_n$ be a random variable that has the prior distribution, so that the posterior distribution is equal to the conditional distribution of $\overline{\Theta}_n$ given $X_1, \ldots, X_n$.

Theorem 5.26 Bernstein–von Mises

Suppose that the map $\vartheta \mapsto \log p_\vartheta(x)$ is continuously differentiable for all x with gradient $\dot{\ell}_\vartheta(x)$ such that $\|\dot{\ell}_\vartheta(x)\| \leq L(x)$ for every ϑ in a neighborhood of θ, where L is a function with $\mathrm{E}_\theta L^2(X_1) < \infty$. Suppose, moreover, that the Fisher information matrix i_ϑ is invertible for all ϑ and depends continuously on ϑ, and that the maximum likelihood estimator $\hat{\theta}_n$ is consistent for θ. Then for every prior probability distribution that is continuous with strictly positive density on Θ, we have

$$\limsup_{n\to\infty} \mathrm{E}_\theta \sup_B \left| \mathrm{P}\left(\overline{\Theta}_n \in B \mid X_1, \ldots, X_n\right) - N_k\left(\hat{\theta}_n, \frac{1}{n}i_\theta^{-1}\right)(B) \right| = 0.$$

Here, we denote by N_k the k-dimensional normal distribution, see Appendix B, and by $N_k(\mu, \Sigma)(B)$ the probability that an $N_k(\mu, \Sigma)$-distributed variable takes on a value in B. In the theorem, the supremum of an absolute difference between two probabilities is taken over all events (sets) $B \subset \mathbb{R}^k$. This difference can be seen as a distance between the posterior distribution and a certain normal distribution that also depends on the observations through the maximum likelihood estimator $\hat{\theta}_n$. The theorem says that the expectation of this distance converges to 0. The variance of the approximate normal distribution is exactly the (limit) variance of the maximum likelihood estimator. By choosing the event B suitably, we can transform the statement into a statement on a credible region.

Let us specify this for the estimation of a real-valued parameter $g(\theta)$ based on a sample from the density p_θ. A natural credible interval is then the interval between two symmetrically chosen quantiles of the posterior distribution of $g(\theta)$. If $F_{g(\overline{\Theta}_n)|X_1,\ldots,X_n}$ is the distribution function of this posterior distribution and

$$Q_{g(\overline{\Theta}_n)|X_1,\ldots,X_n}(\alpha) = \inf\left\{x : F_{g(\overline{\Theta}_n)|X_1,\ldots,X_n}(x) \geq \alpha\right\}$$

is the corresponding quantile function, then the credible interval is

$$(5.3) \qquad \left[Q_{g(\overline{\Theta}_n)|X_1,\ldots,X_n}\left(\frac{\alpha}{2}\right), Q_{g(\overline{\Theta}_n)|X_1,\ldots,X_n}\left(1-\frac{\alpha}{2}\right)\right].$$

We can compare this with the confidence interval based on the maximum likelihood estimator, which equals

$$\left[g(\hat{\theta}_n) - \frac{\xi_{1-\alpha/2}}{\sqrt{n}}\sqrt{g'_{\hat{\theta}_n} i_{\hat{\theta}_n}^{-1}(g'_{\hat{\theta}_n})^T}, \; g(\hat{\theta}_n) + \frac{\xi_{1-\alpha/2}}{\sqrt{n}}\sqrt{g'_{\hat{\theta}_n} i_{\hat{\theta}_n}^{-1}(g'_{\hat{\theta}_n})^T}\right]$$

by Theorem 5.16. The endpoints of the two intervals agree asymptotically.

Theorem 5.27

In the situation of Theorem 5.26, for a differentiable function g and $\alpha \in (0,1)$, we have

$$Q_{g(\overline{\Theta}_n)|X_1,\ldots,X_n}(1-\alpha) - g(\hat{\theta}_n) - \frac{\xi_{1-\alpha}}{\sqrt{n}}\sqrt{g'_{\hat{\theta}_n} i^{-1}_{\hat{\theta}_n}(g'_{\hat{\theta}_n})^T} = o_{P^n_\theta}(1/\sqrt{n}).$$

Consequently, under θ the probability that the credible interval (5.3) contains the parameter θ converges to $1-\alpha$ as $n \to \infty$.

Proof. The last statement follows from the first and the analogous statement concerning the coverage probability of the confidence interval based on the maximum likelihood estimator. See Theorem 5.9 and the discussion following it, or Theorem 5.16 for the case of higher-dimensional parameters. We also use that the difference between the two types of intervals is of order $o(1/\sqrt{n})$, so that we can reduce the difference to 0 by changing the confidence level $1-\alpha$ of the maximum likelihood interval to a value $1-\hat{\alpha}_n$ with $\hat{\alpha}_n \to \alpha$. The confidence level of the maximum likelihood interval still converges to $1-\alpha$ as $n \to \infty$.

For the proof of the first statement of the theorem, we restrict ourselves to the case of a parameter $\theta \in \mathbb{R}$ and the identity function, given by $g(\theta) = \theta$. (The general case requires a second step, based on the delta method.) For $F_n(x) = P(\overline{\Theta}_n \le x | X_1,\ldots,X_n)$, by Theorem 5.26 applied with the set $B = (-\infty, x]$, we find that $\text{E}\sup_x |F_n(x) - \Phi((x-\mu_n)/\sigma_n)| \to 0$ for $\mu_n = \hat{\theta}_n$ and $\sigma_n = 1/\sqrt{ni_\theta}$. For $x = Q_{\overline{\Theta}_n|X_1,\ldots,X_n}(1-\alpha)$, we have $F_n(x-1/n) \le 1 - \alpha \le F_n(x)$ because $Q_{\overline{\Theta}_n|X_1,\ldots,X_n}$ is the quantile function corresponding to F_n. Consequently,

$$\Phi\left(\frac{x - 1/n - \mu_n}{\sigma_n}\right) + \delta_n \le 1 - \alpha \le \Phi\left(\frac{x - \mu_n}{\sigma_n}\right) + \Delta_n,$$

where both rest terms δ_n and Δ_n have an absolute value that is less than $\sup_x |F_n(x) - \Phi((x-\mu_n)/\sigma_n)|$ and therefore converge in probability to 0. By the continuity of the standard normal quantile function Φ^{-1}, we can now invert the inequalities to obtain $(x - 1/n - \mu_n)/\sigma_n \le \Phi^{-1}(1-\alpha-\delta_n)$ and $(x - \mu_n)/\sigma_n \ge \Phi^{-1}(1-\alpha-\Delta_n)$, which imply $x = \mu_n + \sigma_n[\Phi^{-1}(1-\alpha) + o_P(1)]$. ∎

As when we apply Bayesian approximation methods, the weakness of the credible regions lies in the choice of the prior distribution. This choice can significantly influence the form of the posterior distribution. A "wrong" choice of prior distribution can thus lead to "wrong" credible regions. Theorem 5.27 shows that this problem is small when we have sufficiently many observations. In that case, a possible wrong prior choice is corrected by the observations.

Example 5.28 Binomial distribution

Let X be binomially distributed with parameters n (known) and θ (unknown). In Example 3.38, we computed that the posterior distribution with respect to the beta distribution with parameters α and β is equal to the beta distribution with parameters $X + \alpha$ and $n - X + \beta$. A credible interval of confidence level $1 - \alpha_0$ with respect to a beta prior distribution is therefore the interval between the $(\alpha_0/2)$- and $(1 - \alpha_0/2)$-quantiles of the beta distribution with parameters $X + \alpha$ and $n - X + \beta$. Figure 5.6 shows a realization of the posterior density, where the credible interval is indicated by the thick solid line.

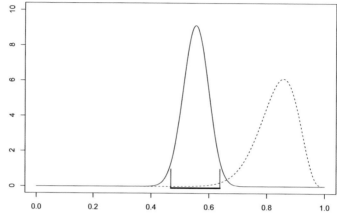

Figure 5.6. Realization of the posterior density (solid) based on an observation from the binomial distribution with parameters 100 en 1/2 with respect to the beta prior distribution with parameters $\alpha = 25$ and $\beta = 5$ (dashed). The 95% credible interval is indicated by the thick solid line.

5.8 Summary

Let X be an observation with distribution P_θ that depends on an unknown parameter $\theta \in \Theta$. A *confidence region* for $\theta \in \Theta$ of confidence level $1 - \alpha$ is a map $X \mapsto G_X$ such that

$$P_\theta(G_X \ni \theta) \geq 1 - \alpha \qquad \text{for all } \theta \in \Theta.$$

The region G_X depends on X and is therefore stochastic. The *confidence level* $1 - \alpha$ is the probability that this region contains the true value of the parameter.

Confidence regions based on (near-)pivots:
- A *pivot* is a random variable or vector $T = T(X, \theta)$ that has a fixed distribution that does not depend on the parameter θ. The set of θ for which the pivot belongs to a fixed region (for example, the interval between two quantiles of its fixed distribution) is a confidence region for θ. A *near-pivot* is a variable or vector $T = T(X, \theta)$ that has approximately a fixed distribution.
- Maximum likelihood estimators can act as near-pivots. Let p_θ be the marginal density of a sample $X_1, \ldots, X_n$. The *Fisher information* is the number

$$i_\theta = \operatorname{var}_\theta \dot{\ell}_\theta(X_1),$$

with $\theta \mapsto \dot{\ell}_\theta(x) = \frac{\partial}{\partial \theta} \log p_\theta(x)$ the *score function* of the model. Under certain conditions, the *maximum likelihood estimator* $\hat{\theta}_n$ for θ based on the sample $X = (X_1, \ldots, X_n)$ satisfies $\sqrt{n}(\hat{\theta}_n - \theta) \rightsquigarrow N(0, i_\theta^{-1})$. Hence $\sqrt{n i_\theta}(\hat{\theta}_n - \theta)$ is a near-pivot. When we use an estimator $\hat{i}_\theta$ for i_θ, this gives the approximate confidence interval for θ:

$$\theta = \hat{\theta} \pm \frac{1}{\sqrt{n \hat{i}_\theta}} \xi_{1-\alpha/2}.$$

Confidence regions based on tests:
- The set of all values τ for which the hypothesis $H_0 \colon g(\theta) = \tau$ is not rejected is a confidence region for $g(\theta)$. If the test has size α, then the confidence region has confidence level $1 - \alpha$. Conversely, rejecting parameter values that do not belong to a confidence region gives a test of corresponding size.
- A *likelihood ratio region* for θ is a confidence region based on the likelihood ratio test for $H_0 \colon \theta = \tau$. This test does not reject for small values of the likelihood ratio statistic $\lambda(X)$, so that the confidence region of (approximate) confidence level $1-\alpha$ becomes

$$\{\theta \colon \lambda(X) \leq \chi^2_{k,1-\alpha}\} = \{\theta \colon \log p_\theta(X) - \log p_{\hat{\theta}}(X) \geq -\tfrac{1}{2}\chi^2_{k,1-\alpha}\},$$

where $\hat{\theta}$ is the maximum likelihood estimator for θ and k is the dimension of θ. This region contains the parameter values where the likelihood function exceeds a certain bound.

Exercises

1. Lab workers are trying to measure a certain quantity θ. Normally distributed measurement errors occur with a standard deviation of 2.3 and expectation 0. The lab workers carry out 25 independent measurements and find an average value of 18.61. Determine a (numerical) confidence interval for θ of confidence level 0.98.

2. If in the previous exercise, the standard deviation is not assumed known, and the sample standard deviation is 2.3, what is the (numerical) confidence interval for θ of confidence level 0.98?

3. Let $X_1, \ldots, X_m$ and $Y_1, \ldots, Y_n$ be independent random samples from, respectively, a normal $N(\mu, \sigma^2)$-distribution and a normal $N(\nu, \sigma^2)$-distribution. Determine a confidence interval for $\mu - \nu$ of confidence level $1 - \alpha$
   (i) if σ^2 is known,
   (ii) if σ^2 is unknown.

4. Let $X_1, \ldots, X_n$ be a sample from the $N(\mu, \sigma^2)$-distribution. Determine a confidence interval for σ^2 based on a suitable pivot.

5. Suppose that in 100 independent Bernoulli experiments with unknown probability of success p, we find 36 cases of success. Determine an (approximate) numerical confidence interval for p of confidence level 0.95.

6. Let $X_1, \ldots, X_m$ and $Y_1, \ldots, Y_n$ be independent random samples from, respectively, a normal $N(\mu, \sigma^2)$-distribution and a normal $N(\nu, \tau^2)$-distribution. Determine a confidence interval for σ^2/τ^2 of confidence level $1 - \alpha$.

7. Let $X_1, \ldots, X_n$ be a sample from the exponential distribution with parameter λ.
   (i) Determine an exact confidence interval for λ based on a suitable pivot.
   (ii) Determine an approximate confidence interval for λ based on the maximum likelihood estimator and the large sample method.

8. Let $X_1, \ldots, X_{10}$ be a sample from the Poisson distribution with unknown expectation θ. We find $x_1 = x_3 = x_6 = x_8 = x_9 = 0$, $x_2 = x_5 = x_{10} = 1$, $x_4 = 2$, $x_7 = 3$.
   (i) Determine an exact (numerical) confidence interval for θ of confidence level 0.9.
   (ii) Determine an approximate (numerical) confidence interval for θ of confidence level 0.9 based on the maximum likelihood estimator by applying the large sample method.

9. Let $X_1, \ldots, X_n$ be a sample from the distribution with probability density

$$p_\theta(x) = 2x\theta e^{-\theta x^2} \qquad \text{for } x \geq 0,$$

and 0 for $x < 0$. The parameter $\theta > 0$ is unknown. We can prove that $X_1^2, \ldots, X_n^2$ are exponentially distributed with parameter θ and that if $\theta = 1$, the random variable $2 \sum_{i=1}^{n} X_i^2$ has a chi-square distribution with $2n$ degrees of freedom.
   (i) Show that $2 \sum_{i=1}^{n} \theta X_i^2$ is a pivot.
   (ii) Determine a $(1-\alpha)100\%$ confidence interval for θ based on the pivot from the previous part, expressed in quantiles of a chi-square distribution.

10. By Example 5.2, the square length of the confidence interval for μ based on a sample from the $N(\mu, \sigma^2)$-distribution when σ^2 is known is equal to $4(\sigma^2/n)\xi_{1-\alpha/2}^2$. Compare this length to the expected square length of the interval from Example 5.4 for the case where σ is unknown.

11. Let $X_1, .., X_n$ be a sample from the geometric distribution with parameter p.
    (i) Determine the Fisher information for p.
    (ii) Determine the observed information.
    (iii) Determine an approximate confidence interval for p of confidence level $1 - \alpha$ based on the maximum likelihood estimator.
    (iv) What is the realization of this interval if $x_1 + ... + x_{40} = 100$ and $\alpha = 0.05$?

12. Let $X_1, .., X_n$ be a sample from the Bernoulli distribution with parameter p.
    (i) Determine the Fisher information for p.
    (ii) Determine the observed information.
    (iii) Determine an approximate confidence interval for p of confidence level $1 - \alpha$ based on the maximum likelihood estimator.
    (iv) What is the realization of this interval if $x_1 + ... + x_{100} = 32$ and $\alpha = 0.05$?

13. Let $X_1, \ldots, X_n$ be a sequence of independent random variables with probability density p_θ given by $p_\theta(x) = \theta^2 x e^{-\theta x}$ for $x \geq 0$, where $\theta > 0$ is an unknown parameter.
    (i) Determine the maximum likelihood estimator for θ.
    (ii) Compute the plug-in estimator for i_θ.
    (iii) Compute the observed (Fisher) information for θ.
    (iv) Give an approximate confidence interval for θ of confidence level $1 - \alpha$ based on the maximum likelihood estimator for θ.

14. Let $X_1, \ldots, X_n$ be a sample from the probability distribution with density $p_\lambda(x) = x\lambda^{-2} e^{-x/\lambda} 1_{x>0}$, where $\lambda > 0$ is an unknown parameter.
    (i) Determine the maximum likelihood estimator for λ.
    (ii) Determine an approximate confidence interval for λ of confidence level $1 - \alpha$ based on the maximum likelihood estimator.
    (iii) Compare this interval with the interval for $\theta = 1/\lambda$ from the previous exercise.

15. We carry out 25 independent Bernoulli experiments, each with unknown success probability p. We find 18 cases of success. Take confidence level 0.95.
    (i) Determine an exact confidence interval for p.
    (ii) Determine an approximate confidence interval for p based on the large sample method. Can 25 be viewed as "large" in this context?

16. Let $X_1, \ldots, X_n$ be a sequence of independent random variables with probability density p_θ given by $p_\theta(x) = \theta^2 x e^{-\theta x}$ for $x \geq 0$, where $\theta > 0$ is an unknown parameter.
    (i) Determine the likelihood ratio statistic λ_n for testing the null hypothesis $H_0: \theta = \theta_0$ against the alternative hypothesis $H_1: \theta \neq \theta_0$.
    (ii) Determine an approximate confidence interval for θ of confidence level $1 - \alpha$ based on the likelihood ratio statistic.

17. Let $X_1, \ldots, X_n$ be a sequence of independent random variables with probability density p_θ given by $p_\theta(x) = \frac{\theta}{2\sqrt{x}} e^{-\theta\sqrt{x}}$ for $x \geq 0$, where $\theta > 0$ is an unknown parameter.
    (i) Determine the maximum likelihood estimator for θ.
    (ii) Determine the likelihood ratio statistic λ_n for testing the null hypothesis $H_0: \theta = \theta_0$ against the alternative hypothesis $H_1: \theta \neq \theta_0$.
    (iii) Determine an approximate confidence interval for θ of confidence level $1 - \alpha$ based on the likelihood ratio statistic.

18. Let $X_1, \ldots, X_n$ be a sample from the $N(\theta, \theta^2)$-distribution, with $\theta > 0$ unknown. Determine an approximate confidence interval for θ based on the likelihood ratio statistic.

19. A manufacturer of scales claims that his scales have an accuracy of 2 per mille. This means that if X denotes the weight of a 1000 mg object measured on an arbitrary scale, the variance of X is equal to $2^2 = 4$ mg^2. We want to study whether the manufacturer is right. To do this, we take a 1000 mg object and determine its mass using 50 randomly chosen scales. The different measurements are denoted by $X_1, \ldots, X_{50}$. We assume that the observations $X_1, \ldots, X_{50}$ are independent and normally distributed with expectation μ and unknown variance σ^2. The observed sample variance is equal to 4.8. We have also observed that $\sum_{i=1}^{50} (x_i - 1000)^2 / 50 = 4.8$.

(i) Construct a confidence interval for σ^2 of confidence level 0.95 under the assumption that we know that $\mu = 1000$ mg.

(ii) Construct a confidence interval for σ^2 of confidence level 0.95 under the assumption that μ is unknown.

(iii) Describe a test for testing whether σ^2 deviates significantly from the variance given by the manufacturer. Do this for the case where μ is known as well as the case where it is unknown. Give the null hypothesis. Carry out this test with the confidence interval from the previous part, a critical region, or a p-value. Take confidence level equal to 0.95.

20. Let X and Y be independent, binomially distributed variables with respective parameters $(200, p_1)$ and $(725, p_2)$.

(i) Construct an approximate confidence interval for $p_1 - p_2$ of confidence level 0.95.

(ii) Test, using the confidence region from the previous part, the null hypothesis $H_0: p_1 = p_2$ at level 0.05 if we have observed $x = 121$ and $y = 391$.

21. Let $X_{(n)}$ be the maximum of a sample of size n from the uniform distribution on $[0, \theta]$. Determine numbers c and d such that the length of the interval $[X_{(n)}/d, X_{(n)}/c]$ is minimal and the interval is also a confidence interval of confidence level $1 - \alpha$.

22. Let X and Y be independent binomially distributed variables with respective parameters (n_1, p_1) and (n_2, p_2). Determine the profile likelihood function for the parameter $g(p_1, p_2) = p_1/p_2$.

THE SALK VACCINE

Polio (or infantile paralysis) is an infectious disease that was virtually eradicated by vaccination in Western countries in the second half of the twentieth century. The first vaccines against polio were developed and tested in the 1950s. Jonas Salk's vaccine was the most promising of these. After research in a lab, the United States Public Health Service decided, in 1954, to study this vaccine by carrying out a large field trial under the American population.

This experiment consisted of vaccinating a large number of children with either the Salk vaccine or a placebo (an inactive substance) and carrying out a statistical comparison of the measure of infection by the poliovirus in the two groups. Using a placebo is standard procedure in these types of clinical trials, and is meant to rule out possible (usually positive) effects on a patient resulting from the suggestions that they are being treated. Neither the children nor the doctors administrating the shots knew whether a placebo or vaccine was administered: it was a double-blind experiment.

The composition of the group of "cases" (the children treated with the vaccine) and of the control group (the children treated with the placebo) lead to the necessary complications. A significant problem was that a large number of parents did not give permission for participation in the trial. Since it was not excluded, and was even expected, that a positive correlation existed between permission to participate and susceptibility for polio, the researchers decided to first form a group of children whose parents had given permission for participation in the trial, and only then decide how to split the group into a case group and control group. The latter came about through complete randomization, that is, each child had probability $1/2$ of being assigned to one of the two groups, independently of the other children.

The results were as follows. For a group of about $750\,000$ children, the parents of $401\,974$ children gave permission for participation in the trial. Of these children, $200\,745$ were administered the vaccine and $201\,229$ the placebo. Of the treated children, 57 still contracted polio, while in the placebo group, 142 children contracted polio.[b]

This data seems to show that the Salk vaccine significantly reduces the occurrence of polio. How hard can we make this conclusion? Recall that 57 of the treated children also contracted polio. Can we say that the Salk vaccine reduces the probability of contracting polio by a factor of 2.5 ($\approx (142/201\,229)/(57/200\,745)$)?

Even for a carefully planned trial, these questions are not at all trivial. The factor 2.5 that shows up in the data at first glance has the highest uncertainty, but the statement "the Salk vaccine works" also needs further explanation. What do we mean by "works"? In principle, we are looking for a causal conclusion: just like a moving billiard ball touching a stationary billiard ball is the reason why the second ball starts moving, we would like the Salk vaccine to be the reason why there are fewer cases of polio. A clinical trial as the one carried out here is seen as the best method

---

[b] The date can be found in the paper *Evaluation of the 1954 Poliomyelitis vaccine field trial*, T. Francis, Journal of American Medical Association 158, 1955.

for drawing such a conclusion, but to a certain level, speaking in terms of causes may be a linguistic question. By the set-up of the experiment, other explanations for the observed difference have been ruled out as much as possible. Note, however, that the experiment gives hardly any information on children whose parents did not give permission for administering the vaccine. It could, for example, be possible that this group corresponds to the group for which the vaccine does not work. On medical grounds, this is very improbable, but we should not unquestioningly view the factor 2.5 as applicable to this group of children. For example, it turns out that richer parents often refused participation, and there is a speculation that children of richer parents are more susceptible to polio because they build up less resistance in their younger year because of better hygiene.

The contribution of statistics is the analysis of the data based on a statistical model. Roughly speaking, the question is: suppose that we repeat the entire experiment, would we still find similar results (including the factor 2.5), or was this by chance? What statistical model should we use? It seems impossible to set up an experiment in which we also include the possibility that a parent (of a randomly chosen child?) refuse permission to participate. It seems simplest to restrict ourselves to the group of 401 974 participating children. Although, theoretically, we may then not extend our conclusions to all children, such a generalization does seem reasonable.

Let p_1 and p_2 be the probabilities that a randomly chosen child from the given 401 974 children contracts polio if they receive, respectively, the vaccine and the placebo. For every child $i = 1, 2, \ldots, n = 401\,974$, we observe the pair (C_i, P_i), where C_i is defined as

$$C_i = \begin{cases} 1 & \text{if child } i \text{ is in the case group,} \\ 2 & \text{if child } i \text{ is in the control group,} \end{cases}$$

and P_i as

$$P_i = \begin{cases} 0 & \text{if child } i \text{ does not contract polio,} \\ 1 & \text{if child } i \text{ contracts polio.} \end{cases}$$

The marginal distribution of C_i is $\mathrm{P}(C_i = 1) = \mathrm{P}(C_i = 2) = 1/2$, because of the set-up of the experiment: every child has the same probability of being assigned to the case group or control group. The conditional distribution of P_i given $C_i = j$ is Bernoulli(p_j) for $j = 1, 2$, by the definitions of p_1 and p_2. This fixes the probability distribution of (C_i, P_i). We further fix the joint probability distribution of $(C_1, P_1), \ldots, (C_n, P_n)$ by postulating that these vectors are independent. This is a bad assumption, because polio is infectious and does not occur independently in different children. We still make the assumption, for want of a better one.

The observations $C_1, \ldots, C_n$ are the result of the randomization and do not give any information on the parameters p_1 and p_2. The relevant information in the observations $P_1, \ldots, P_n$ is (intuitively) contained in

$$X = \sum_{i:C_i=1} P_i, \qquad Y = \sum_{i:C_i=2} P_i.$$

These are the numbers of cases of polio in the case group and control group, respectively. Given the vector $(C_1, \ldots, C_n)$, the variables X and Y are independent and binomially distributed with respective parameters $(M_1 := \#\{i : C_i = 1\}, p_1)$ and $(M_2 := \#\{i : C_i = 2\}, p_2)$. The simple approach is now to carry out the statistical analysis conditionally on the observed values m_1 and m_2 of M_1 and M_2. In that case, we have reduced the problem to the statistical approach in which we observe the independent variables X and Y with binomial distributions with respective parameters (m_1, p_1) and (m_2, p_2).

To test whether the vaccine has a protective effect, we want to test the null hypothesis $H_0 : p_1 \geq p_2$ against the alternative hypothesis $H_1 : p_1 < p_2$. Within the statistical model described above, there exists a standard test, the Fisher test for the (2×2)-table, based on the fact that under the null hypothesis, X given $X + Y$ has a hypergeometric distribution. We will not discuss this here. Because the number of observations is very large, we are content with an approximate test.

The natural estimator for $p_1 - p_2$ is $X/m_1 - Y/m_2$. This has expectation $p_1 - p_2$ and variance

$$\mathrm{var}\left(\frac{X}{m_1} - \frac{Y}{m_2}\right) = \frac{p_1(1 - p_1)}{m_1} + \frac{p_2(1 - p_2)}{m_2}.$$

We can estimate this variance by replacing p_1 and p_2 with X/m_1 and Y/m_2. By the central limit theorem (Theorem A.28), under $p_1 = p_2$, the random variable

$$T = \frac{X/m_1 - Y/m_2}{\sqrt{\frac{X/m_1(1 - X/m_1)}{m_1} + \frac{Y/m_2(1 - Y/m_2)}{m_2}}}$$

approximately has the standard normal distribution; see Section A.7. If we use T as test statistic to test the null hypothesis stated above, then we find a left p-value 9.09×10^{-9}, which is less than any interesting level of the test. The conclusion is that the vaccine indeed has a protective effect.

To say something about the size of the effect $p_1 - p_2$, we estimate this difference and deduce a 95% confidence interval. As near-pivot, we use

$$\frac{X/m_1 - Y/m_2 - (p_1 - p_2)}{\sqrt{\frac{X/m_1(1 - X/m_1)}{m_1} + \frac{Y/m_2(1 - Y/m_2)}{m_2}}},$$

which approximately has the standard normal distribution. For the given data from the Salk experiment, $p_1 - p_2$ is estimated to be -0.000422 and the approximate 95% confidence interval is equal to -0.000422 ± 0.000137.

Since both p_1 and p_2 are small, it seems natural to study the relative size p_1/p_2. A reasonable estimator is $(X/m_1)/(Y/m_2)$. In a manner similar to that used for the difference, we can deduce a confidence interval for p_1/p_2, but this requires more knowledge of asymptotic methods than we wish to introduce here, so we refrain from doing this.

211

6 Optimality Theory

6.1 Introduction

This chapter is dedicated to optimality theory for estimators and tests. There are, generally, many possible choices for estimators and test statistics. If we are looking for the best estimator or test, it would be useful to reduce the set of possible estimators and test statistics. We can do this by reducing the observation beforehand by filtering out irrelevant information on the parameter. We then base the estimator or test statistic on the reduced observation. This is the subject of Section 6.2.

In Section 6.3, we consider how to find the best estimator and how good the best estimator is, measured in the measure of quality discussed in Chapter 3, the mean square error.

Finally, we discuss the quality of tests in Section 6.4. In Chapter 4, we constructed different tests using ad hoc arguments. Intuitively, most of these tests are reasonable, but are they also the best possible tests? In the last section of this chapter, we will show that some of the tests we discussed are uniformly most powerful; this means that the power of these tests under the alternative hypothesis is maximal.

6.2 Sufficient Statistics

If instead of seeing the full observation X, we only see the value of a statistic $V(X)$, we have, in principle, lost information. For example, $X = (X_1, \ldots, X_n)$ gives more information than $V(X) = \sum_{i=1}^{n} X_i$. We call a statistic V sufficient if, given the model, no relevant information on the unknown parameter has been lost.

Example 6.1 Bernoulli distribution

For a quality control inspection, n items are randomly chosen from a large batch and tested. We observe $X = (X_1, \ldots, X_n)$, where

$$X_i = \begin{cases} 0 & \text{if the } i\text{th article is rejected,} \\ 1 & \text{if the } i\text{th article is approved.} \end{cases}$$

The result of the inspection is therefore a sequence of symbols consisting of 0s and 1s. The unknown proportion p of defect items in the large batch clearly affects the number $V = \sum_{i=1}^{n} X_i$ of observed 1s (the number of approved articles in the sample), but intuitively, the order in which we see the 0s and 1s has little to do with the size of p. Intuitively, $V = \sum_{i=1}^{n} X_i$ therefore suffices.

The technical definition of a sufficient statistic when X has a discrete distribution is as follows.

Definition 6.2 Sufficient statistic for discrete probability distributions

Suppose that the statistical model for X consists of discrete probability distributions that depend on the parameter θ. A statistic $V = V(X)$ is called sufficient if the conditional probabilities

$$P(X = x \,|\, V = v)$$

do not depend on θ, for all possible values of x and v.

The property in the definition is truly special. The distribution of X depends on the unknown parameter θ, hence so does the joint distribution of (X, V). For a general statistic V that is not sufficient, the conditional probabilities $P_\theta(X = x \,|\, V = v)$ will also depend on θ.

Intuitively, we can show as follows that it is plausible that a sufficient quantity possesses all relevant information on θ. We could generate an observation x in two steps:
- First generate v from the marginal distribution of V; for this, we need the "true" parameter θ.
- Given v, generate x from the conditional distribution of X given $V = v$; provided that V is sufficient, we do not need the true θ for this.

213

We can view the result of these two steps as drawing a sample from the distribution of X because we always have

$$P_\theta(X = x) = \sum_v P_\theta(X = x | V = v) P_\theta(V = v),$$

where the conditional probabilities $P_\theta(X = x | V = v)$ do not depend on θ when V is sufficient. The result therefore contains just as much information as a direct observation of X in the original experiment. Apparently, all relevant information on θ is contained in V. If desired, we can always "convert" v to x, by following the second step of the procedure described above. We do not need to know the parameter for this.

Example 6.3 Bernoulli distribution, continued

In Example 6.1, we showed that intuitively, the variable $V = \sum_{i=1}^n X_i$ is sufficient. To make this more precise, we must specify the underlying statistical model. We assume that $X_1, \ldots, X_n$ are independent and have the Bernoulli distribution with parameter p. Then for $x_i \in \{0, 1\}$ and $v \in \{0, 1, \ldots, n\}$,

$$
\begin{aligned}
P_p\big(X_1 = x_1, \ldots, X_n = x_n | V = v\big) &= \frac{P_p\big(X_1 = x_1, \ldots, X_n = x_n, V = v\big)}{P_p(V = v)} \\
&= \begin{cases} \dfrac{p^v(1-p)^{n-v}}{\binom{n}{v}p^v(1-p)^{n-v}} & \text{if } \sum_{i=1}^n x_i = v \\ 0 & \text{otherwise} \end{cases} \\
&= \begin{cases} \binom{n}{v}^{-1} & \text{if } \sum_{i=1}^n x_i = v \\ 0 & \text{otherwise.} \end{cases}
\end{aligned}
$$

Because the last expression does not depend on p, the variable V is indeed sufficient. Note that, as a safeguard, we have included p on the left-hand side. Only at the end of the computation, where p does play a role in the intermediate steps, do we see that we can leave out p.

6.2.1 Factorization Theorem

How do we determine sufficient statistics? The definition is not very useful, because we first need to guess which statistic V might be sufficient, and then compute conditional probabilities that are sometimes rather complicated. The following theorem offers a solution.

Theorem 6.4 Factorization theorem

Suppose that the statistical model for X consists of discrete distributions. A statistic $V = V(X)$ is sufficient if and only if there exist functions g_θ and h such that for all x and θ,

$$p_\theta(x) = g_\theta\big(V(x)\big)h(x),$$

where p_θ is the probability density of X.

Proof. Suppose that V is sufficient. Then

$$P_\theta(X = x) = P_\theta\big(X = x, V = V(x)\big)$$
$$= P\big(X = x \mid V = V(x)\big)P_\theta\big(V = V(x)\big).$$

The first term on the right-hand side does not depend on θ, because V is sufficient. This term can therefore be used for $h(x)$. The second term does depend on θ, but only through $V(x)$, and can therefore be used for $g_\theta\big(V(x)\big)$.

Conversely, suppose that the required functions g_θ and h exist. The conditional probability

$$P_\theta\big(X = x_0 \mid V = v\big) = \frac{P_\theta\big(X = x_0, V = v\big)}{P_\theta(V = v)}$$

is equal to 0 if $V(x_0) \neq v$. In the other case, where $V(x_0) = v$, the expression is equal to

$$\frac{P_\theta(X = x_0)}{P_\theta(V = v)} = \frac{P_\theta(X = x_0)}{\sum_{x:V(x)=v} P_\theta(X = x)}$$
$$= \frac{g_\theta\big(V(x_0)\big)h(x_0)}{\sum_{x:V(x)=v} g_\theta\big(V(x)\big)h(x)}$$
$$= \frac{g_\theta(v)h(x_0)}{g_\theta(v)\sum_{x:V(x)=v} h(x)}$$
$$= \frac{h(x_0)}{\sum_{x:V(x)=v} h(x)}.$$

Neither the last expression nor the condition $V(x_0) = v$ depends on θ. Hence V is sufficient. ∎

Example 6.5 Bernoulli distribution

For the situation in Example 6.3, we have

$$P_\theta(X_1 = x_1, \ldots, X_n = x_n) = \theta^{\sum_{i=1}^n x_i}(1 - \theta)^{n - \sum_{i=1}^n x_i}.$$

This is a function of $\sum_{i=1}^n x_i$. We can take $h(x) \equiv 1$ and $g_\theta(s) = \theta^s(1 - \theta)^{n-s}$. By the factorization theorem, the variable $\sum_{i=1}^n X_i$ is sufficient.

It is mathematically difficult to extend the definition of sufficiency given above to continuously distributed random variables X, because the definition of the conditional probability of X given $V(X)$ is not simple mathematically. To avoid this difficulty, we choose the factorization formula as the definition.

Definition 6.6 Sufficient statistic

A statistic $V(X)$ is called *sufficient* for the observation X with probability density p_θ if there exist functions g_θ and h such that for all θ and x,

$$p_\theta(x) = g_\theta\big(V(x)\big)h(x).$$

For discretely distributed observations, we now have two definitions of sufficiency, but these agree by the factorization theorem. This theorem or the last definition says that a statistic V is sufficient if the likelihood function (based on the observation X) depends on θ only through $V(X)$. This also suggests that observing V is "sufficient."

Sufficient statistics are not at all unique. The observation X itself is, for example, always sufficient, but this is not an interesting sufficient statistic. An interesting sufficient statistic is one that is simple and low-dimensional, a statistic that is sufficient but reduces the data as much as possible. We call a sufficient statistic V *minimally sufficient* if V is a function of every other sufficient statistic. In that case, we know the value of V as soon as we know the value of any sufficient statistic; V therefore contains less information. The following lemma shows that this is a meaningful definition.

Lemma 6.7

Let V be a sufficient statistic, and suppose $V = f(V^*)$ for a map f. Then V^* is also sufficient. If f is a 1-1 function, then $V = f(V^*)$ is sufficient if and only if V^* is sufficient.

Proof. The first assertion is immediate from the factorization theorem or, in the continuous case, the definition. We simply note that $g_\theta\big(V(x)\big)$ can be further written as $g_\theta \circ f\big(V^*(x)\big)$. The second assertion follows by applying the first one in both directions. ∎

Example 6.8 Normal distribution

Let $X = (X_1, \ldots, X_n)$ be a sample from the $N(\mu, \sigma^2)$-distribution with unknown parameters μ and σ^2. We take the natural parameter space for the parameter $\theta = (\mu, \sigma^2)$, namely $\Theta = \mathbb{R} \times (0, \infty)$. The joint density of $X_1, \ldots, X_n$ is

$$\prod_{i=1}^{n} \frac{1}{\sqrt{2\pi\sigma^2}} e^{-\frac{1}{2\sigma^2}(x_i-\mu)^2} = \left(\frac{1}{\sqrt{2\pi\sigma^2}}\right)^n e^{-\frac{1}{2\sigma^2}\sum_{i=1}^{n}(x_i-\mu)^2}$$

$$= \left(\frac{1}{\sqrt{2\pi\sigma^2}}\right)^n e^{-\frac{n}{2\sigma^2}\mu^2} e^{-\frac{1}{2\sigma^2}\sum_{i=1}^{n}x_i^2 + \frac{\mu}{\sigma^2}\sum_{i=1}^{n}x_i}.$$

So the density depends on $X_1, \ldots, X_n$ only through $(\sum_{i=1}^{n}x_i, \sum_{i=1}^{n}x_i^2)$. The vector $(\sum_{i=1}^{n}X_i, \sum_{i=1}^{n}X_i^2)$ is therefore sufficient.

The vector $(\overline{X}, S_X^2)$ has a 1-1 relation with this sufficient vector and is therefore also sufficient. For a random sample from the normal distribution, the sample mean and sample variance therefore contain all information on μ and σ^2. Note that $(\overline{X}, S_X^2, X_1)$ is also sufficient, but not minimally sufficient!

Example 6.9 Uniform distribution

Let $X = (X_1, \ldots, X_n)$ be a sample from the $\mathrm{U}[0, \theta]$-distribution with unknown parameter $\theta > 0$. The joint density of $X_1, \ldots, X_n$ is

$$p_\theta(x_1, \ldots, x_n) = \prod_{i=1}^{n} \frac{1}{\theta} 1_{\{0 \leq x_i \leq \theta\}} = \left(\frac{1}{\theta}\right)^n 1_{\{x_{(n)} \leq \theta\}}.$$

Apparently, $X_{(n)}$ is sufficient: the largest observation contains all information on the parameter θ.

For discretely distributed observations, we have given a thought experiment (generating an observation in two steps) to make intuitively plausible that a sufficient statistic indeed contains all information on the parameter. The name "sufficient statistic" suggest that if we want to estimate an unknown parameter or say something about its value using a test, the information in the data that is not in the sufficient statistic is superfluous. As far as estimating is concerned, this is proved in the Rao–Blackwell theorem (Theorem 6.16) in Section 6.3. In this theorem, we prove that for every estimator $T = T(X)$, there exists an estimator $T^* = T^*(V)$ that depends only on the sufficient quantity V and is at least as good as T (measured using the mean square error). For tests, it is a bit more complicated. The quality of a test is determined by the power function. We thus want to prove that for every test based on X, there exists a test based on V with power function at least as powerful. This is only true if we also allow so-called randomized statistics.

* 6.2.2 Randomized Statistics

The proof that a sufficient statistic contains all relevant information when we are dealing with tests requires the definition of randomized test statistics.

Definition 6.10 Randomized Statistic

A *randomized statistic* $T = T(X, U)$ *is a random vector that depends only on* X *and an independently generated* $\mathrm{U}[0, 1]$-*distributed variable* U.

Every "ordinary" statistic is also a randomized statistic. A randomized statistic may depend not only on the observation but also on a random number U that must be generated independently of the actual experiment and parameter. This random number therefore does not contain any information on the parameter. Without this action that seems useless at first, Theorem 6.11 would not be true. The reason is that what "remains" of X after the sufficient statistic V is known, also does not contain any relevant information, and therefore works as a random number generator. In Theorem 6.11, we need U to match this irrelevant source of random numbers. We can show that if the quality of the estimators is measured incorrectly using the mean square error, randomizing estimators is never meaningful; there is always a nonrandomized estimator with a smaller mean square error (namely $\int_0^1 T(X, u)\, du$). For tests, randomizing can be meaningful.

Theorem 6.11

Let $V = V(X)$ be sufficient for the observation X. For every randomized statistic $T = T(X, U)$, there exists a randomized statistic $T^*(V, U)$ based only on V (and the randomization U), such that the probability distributions of T^* and T are the same under every parameter θ.

We leave out the proof of this theorem. We can apply this theorem to both the estimation problem and the test problem, and obtain the following corollary. We show that knowing only V, we can construct estimators and tests that are as good (measured using the mean square error for the estimators and using the power function for the tests) as those obtained using the full observation X.

Corollary 6.12

Let $V = V(X)$ be sufficient for the observation X. For every estimator $T = T(X)$, there exists an estimator $T^* = T^*(V, U)$ based only on V (and the randomization U) with $\mathrm{MSE}(\theta; T) = \mathrm{MSE}(\theta; T^*)$ under every parameter θ.

Corollary 6.13

Let $V = V(X)$ be sufficient for the observation X. For every test statistic $T = T(X)$ there exists a test statistic $T^* = T^*(V, U)$ based only on V (and the randomization U), such that the tests $\{T \geq c\}$ and $\{T^* \geq c\}$ have the same power function: $\mathrm{P}_\theta(T \geq c) = \mathrm{P}_\theta(T^* \geq c)$ for every c and under every parameter θ.

Proofs. Both the mean square error and the power function depend only on the probability distribution of the statistics T or T^*. For example, in the case of a test, we have (if T is continuously distributed)

$$\mathrm{P}_\theta(T \geq c) = \int_c^\infty p_\theta^T(t)\, dt,$$

with p_θ^T the probability density of T. Equality of probability distributions implies equality of densities $p_\theta^T = p_\theta^{T^*}$, and therefore equality of power functions.

By Theorem 6.11, we can choose the probability distributions of T and T^* equal to each other. Consequently, we can also choose the mean square errors and the power functions equal to each other. ∎

Applying the theorem to the estimation problem is not truly necessary because the Rao–Blackwell theorem (Theorem 6.16) already convincingly shows the "sufficiency" of sufficient statistics. Note, however, that the proof of the first corollary holds for every estimation criterion, hence also for other criteria than the mean square error.

6.3 Estimation Theory

The quality of an estimator is quantified by its mean square error (see Section 3.2). An estimator for an unknown parameter $g(\theta)$ is good if its mean square error is small compared to that of other estimators. An estimator T_0 for $g(\theta)$ would be the absolutely best estimator if

$$\mathrm{MSE}(\theta; T_0) \leq \mathrm{MSE}(\theta; T) \qquad \text{for all } T, \theta.$$

However, such an estimator T_0 does not exist. We can see this by realizing that a trivial estimator $T(X) = g(\theta_0)$, for a fixed θ_0, is also an estimator. The mean square error of this estimator for the estimation of $g(\theta)$ is equal to 0 in $\theta = \theta_0$ (but is very bad for $g(\theta)$ far from $g(\theta_0)$). An absolutely best estimator should therefore also have mean square error 0, in every θ, which is impossible as soon as there are two different values $g(\theta)$.

The problem is that the measure $\theta \mapsto \mathrm{MSE}(\theta; T)$ for the quality of an estimator is a function of the (unknown) parameter, which we want to minimize with respect to "all" parameters. That is impossible. We need additional criteria for the choice of an estimator. We give three examples. As the basic criterion for quality, we again take the mean square error, although most of the theory also holds for other measures of quality, such as $\mathrm{E}_\theta |T - g(\theta)|$.

We already discussed the *Bayes criterion* in Section 3.5. For a given prior density π on Θ, we are looking for the estimator T that minimizes

$$\int \mathrm{MSE}(\theta; T)\, \pi(\theta)\, d\theta.$$

This is by definition the Bayes estimator with respect to π, which was found in Theorem 3.36.

The *minimax criterion* takes the maximum of the mean square error,

$$\sup_{\theta \in \Theta} \mathrm{MSE}(\theta; T),$$

as measure. An estimator T is called *minimax* if T minimizes this maximal risk over all estimators. Like the Bayes criterion, the minimax criterion reduces the function $\theta \mapsto \mathrm{MSE}(\theta; T)$ to a number. A "best" estimator can be found by minimizing this number over T. In principle, this is almost always possible.

Example 6.14 Binomial distribution

Suppose that the observation X has the $\mathrm{bin}(n, p)$-distribution. Then the minimax estimator for p is equal to

$$T(X) = \frac{X + \frac{1}{2}\sqrt{n}}{n + \sqrt{n}}.$$

We can deduce this from the fact that T is a Bayes estimator with mean square error $\mathrm{MSE}(p; T)$ that is constant in $p \in [0, 1]$ (see Example 3.38). The proof is by contradiction. If T were not minimax, there would exist an estimator S with a smaller maximum risk, and we would have

$$\mathrm{MSE}(p; S) \leq \sup_{0 \leq q \leq 1} \mathrm{MSE}(q; S) \leq \sup_{0 \leq q \leq 1} \mathrm{MSE}(q; T) = \mathrm{MSE}(p; T)$$

for all $0 \leq p \leq 1$. The first inequality follows from the definition of the supremum, the second inequality expresses the smaller maximum risk of S, and the equality follows from the fact that $\mathrm{MSE}(p; T)$ is constant in p. In summary, we have $\mathrm{MSE}(p; S) \leq \mathrm{MSE}(p; T)$ for $p \in [0, 1]$. It follows that for every prior distribution, the Bayes risk of S is less than or equal to the Bayes risk of T, because the Bayes risk is a weighted version of the mean square error. Since T is the Bayes estimator for p for the $\mathrm{beta}(\frac{1}{2}\sqrt{n}, \frac{1}{2}\sqrt{n})$ prior distribution, T minimizes the Bayes risk for this prior distribution over all estimators. The Bayes risk of S therefore cannot be smaller; hence, the Bayes risks of the two estimators are equal, so that S is also a Bayes estimator for p with respect to the same prior distribution. Theorem 3.36 then implies that $S = T$.

A third criterion, which we will discuss in detail in the next section, is the criterion of *minimum-variance unbiased estimators*. The idea is to look for a best estimator in the class of all unbiased estimators. Since the mean square error of unbiased estimators is equal to the variance, this means that were are looking for an unbiased estimator with minimal variance.

6.3.1 UMVU Estimators

In this section, we look for the so-called UMVU estimators in an estimation problem.

Definition 6.15 Uniformly minimum-variance unbiased (UMVU)

An estimator T is called *uniformly minimum-variance unbiased* or *UMVU for $g(\theta)$* if T is an unbiased estimator for $g(\theta)$ and $\text{var}_\theta T \leq \text{var}_\theta S$ for all θ and all other unbiased estimators S for $g(\theta)$.

How do we determine UMVU estimators? The Rao–Blackwell theorem is a first step in the right direction. This theorem says that for every estimator T for $g(\theta)$, there exists an estimator $T^* = T^*(V)$ that depends only on the sufficient quantity V, has the same bias as T, and has variance less than or equal to that of T. If we are looking for a UMVU estimator, then the Rao–Blackwell theorem says that we can restrict ourselves to unbiased estimators that depend only on a sufficient statistic.

When the distribution of X is discrete, we can construct T^* explicitly: given the estimator T, we define

$$T^*(v) = \text{E}(T \mid V = v) = \sum_x T(x)\text{P}(X = x \mid V = v).$$

Since V is sufficient, we may leave θ out of the subscripts of E_θ and P_θ. Hence T^* is indeed an estimator; it is a function of the observations and not of the unknown parameter θ.

Theorem 6.16 Rao–Blackwell

Let $V = V(X)$ be a sufficient statistic, and let $T = T(X)$ be an arbitrary real-valued estimator for $g(\theta)$. Then there exists an estimator $T^* = T^*(V)$ for $g(\theta)$ that depends only on V, such that $\text{E}_\theta T^* = \text{E}_\theta T$ and $\text{var}_\theta T^* \leq \text{var}_\theta T$ for all θ. In particular, we have $\text{MSE}(\theta; T^*) \leq \text{MSE}(\theta; T)$. This inequality is strict unless $\text{P}_\theta(T^* = T) = 1$.

Proof. We give the proof only in the case where the distribution of X is discrete. Let $T^* = \text{E}(T \mid V)$. In the paragraph before the theorem, we already saw that T^* does not depend on the parameter θ and is therefore an estimator for $g(\theta)$. By the rules for conditional expectations, we have

$$\text{E}_\theta T^* = \sum_v T^*(v)\text{P}_\theta(V = v) = \sum_v \text{E}(T \mid V = v)\text{P}_\theta(V = v) = \text{E}_\theta T.$$

221

This proves the assertion $E_\theta T^* = E_\theta T$. We moreover have

$$
\begin{aligned}
E_\theta TT^* &= \sum_v E(TT^* \,|\, V = v) P_\theta(V = v) \\
&= \sum_v T^*(v) E(T \,|\, V = v) P_\theta(V = v) \\
&= \sum_v T^*(v)^2 P_\theta(V = v) \\
&= E_\theta(T^*)^2.
\end{aligned}
$$

This implies

$$
\begin{aligned}
E_\theta T^2 &= E_\theta(T - T^*)^2 + 2E_\theta(T - T^*)T^* + E_\theta(T^*)^2 \\
&= E_\theta(T - T^*)^2 + 0 + E_\theta(T^*)^2 \\
&\geq E_\theta(T^*)^2.
\end{aligned}
$$

Since T and T^* have the same expectation, it immediately follows that $\mathrm{var}_\theta\, T^* \leq \mathrm{var}_\theta\, T$.

The inequality in the last display is strict unless $E_\theta(T - T^*)^2 = 0$. This is equivalent to having $T = T^*$ with probability 1. ∎

When we restrict ourselves to the class of unbiased estimators, then by the Rao–Blackwell theorem, it suffices to consider only unbiased estimators based on a sufficient quantity. Suppose that for a given sufficient statistic V, there exists only one estimator $T = T(V)$ that is based on V and unbiased. Then T is automatically UMVU. This method, based on finding a special sufficient statistic, works in many cases. The special property of the sufficient statistic is completeness.

Definition 6.17 Complete statistic

A statistic V is called *complete* if $E_\theta f(V) = 0$ (and therefore $E_\theta|f(V)| < \infty$) for all $\theta \in \Theta$ can hold only for functions f such that $P_\theta\big(f(V) = 0\big) = 1$ for all $\theta \in \Theta$.

We can prove that if there exists a minimally sufficient statistic, then a sufficient and complete statistic is also minimally sufficient (see Exercise 6.10). In that case, the complete statistic contains all necessary, and no superfluous, information from the data to estimate the model parameter (see Example 6.19).

Theorem 6.18

Let V be sufficient and complete, and let $T = T(V)$ be an unbiased estimator for $g(\theta)$ that depends only on V. Then T is a UMVU estimator for $g(\theta)$.

Proof. By the Rao–Blackwell theorem, for every unbiased estimator S for $g(\theta)$, there exists an unbiased estimator $S^* = S^*(V)$ that depends only on V and whose variance is less than or equal to that of S. Now, $S^* - T$ is a statistic that depends only on V, with $\mathrm{E}_\theta(S^* - T) = \mathrm{E}_\theta S^* - \mathrm{E}_\theta T = 0$ for all θ because both estimators are unbiased. By the completeness, we have $\mathrm{P}_\theta(S^* - T = 0) = 1$ for all θ. Hence $T = S^*$ with probability 1 and $\mathrm{var}_\theta T \le \mathrm{var}_\theta S$. ∎

Example 6.19 Uniform distribution

Let $X = (X_1, \ldots, X_n)$ be a sample from the $\mathrm{U}[0, \theta]$-distribution. In Example 6.9, we saw that the maximum $X_{(n)}$ is sufficient. If the parameter space is equal to $\Theta = (0, \infty)$, then $X_{(n)}$ is also complete. Therefore, assume that

$$0 = \mathrm{E}_\theta f(X_{(n)}) = \int_0^\theta f(x) \frac{1}{\theta^n} n x^{n-1} \, dx \qquad \text{for all } \theta > 0.$$

This implies $\int_0^\theta f(x) x^{n-1} \, dx = 0$ for all $\theta > 0$. If f is continuous, we can differentiate this equality with respect to θ, which gives $f(\theta)\theta^{n-1} = 0$ for all θ. Hence $f \equiv 0$. For noncontinuous f, the same conclusion holds, but the deduction requires measure theory instead of calculus. Hence $X_{(n)}$ is complete.

Since $(n + 1)/n X_{(n)}$ is an unbiased estimator for θ and depends only on the sufficient and complete quantity $X_{(n)}$, it immediately follows from Theorem 6.18 that this estimator is a UMVU estimator for θ. This is a nice result, which indicates that we cannot find a better unbiased estimator than $(n+1)/n X_{(n)}$. The biased estimator $(n+2)/(n+1)X_{(n)}$, however, has a slightly smaller mean square error (see Example 3.6) and is therefore preferable to the UMVU estimator. The difference between the mean square errors of these two estimators is, however, negligible.

Note that the statistic $W = (X_{(n)}, \overline{X})$ is also sufficient, and that $2\overline{X}$ is an unbiased estimator for θ based on W. We cannot conclude that $2\overline{X}$ is UMVU, because W is not complete. For example, $\mathrm{E}_\theta f(W) = 0$ for all $\theta > 0$ for $f(w) = (n + 1)w_1/n - 2w_2$.

It is not always easy to give a direct proof that a given statistic is complete. The following theorem applies to many of the standard models. It concerns probability densities that belong to an "exponential family" of probability densities.

Definition 6.20 Exponential family

A family of probability densities p_θ that depends on a parameter θ is called a k-dimensional exponential family if there exist functions c, h, Q_j, and V_j such that

$$p_\theta(x) = c(\theta)h(x) \, e^{\sum_{j=1}^k Q_j(\theta)V_j(x)}.$$

It follows from the factorization theorem that the statistic $V = (V_1, \ldots, V_k)$ in a given exponential family is sufficient. This statistic is also complete provided that the parameter space is "sufficiently rich" (see Theorem 6.21).

Theorem 6.21

Suppose that a statistical model is given by a k-dimensional exponential family such that the set

$$\left\{ (Q_1(\theta), \ldots, Q_k(\theta)) : \theta \in \Theta \right\} \subset \mathbb{R}^k$$

has an interior point. Then $V = (V_1, \ldots, V_k)$ is sufficient and complete.

---

* **Proof.** We only sketch the proof, restricting ourselves to the case $k = 1$ and assuming that the model is continuous, so that expectations can be given by integrals with respect to the probability density $x \mapsto p_\theta(x)$.

The equation $E_\theta f(V) = 0$ implies $K(Q(\theta)) = 0$ for K the function defined by $K(z) = \int f(V(x)) h(x) e^{zV(x)} \, dx$. The assumptions that this holds for every $\theta \in \Theta$ and that the set $\{Q(\theta) : \theta \in \Theta\}$ contains an interior point imply that the function K must be well defined and finite for all z in an interval $(a, b) \subset \mathbb{R}$ with $a < b$. Since for $z \in \mathbb{C}$ and $v \in \mathbb{R}$ we have $|e^{zv}| = e^{\operatorname{Re} zv}$, the function K is also well defined and finite for all *complex* numbers z with $\operatorname{Re} z \in (a, b)$. We can, moreover, prove that the derivative $K'(z)$ exists for every z in this region. (The derivative is given by $K'(z) = \int f(V(x)) h(x) V(x) e^{zV(x)} \, dx$, computed by "differentiating under the integral." The integral on the right-hand side is indeed finite, because $|v| \le \varepsilon^{-1}(e^{-\varepsilon v} + e^{\varepsilon v})$ for every $\varepsilon > 0$ and $v \in \mathbb{R}$, so that $|V| e^{zV} \le \varepsilon^{-1}(e^{(z-\varepsilon)V} + e^{(z+\varepsilon)V})$, where for given z with $\operatorname{Re} z \in (a, b)$, we choose the value ε sufficiently small that both $\operatorname{Re} z - \varepsilon$ and $\operatorname{Re} z + \varepsilon$ belong to (a, b).) In other words, the function K is holomorphic in the region $\{z \in \mathbb{C} : \operatorname{Re} z \in (a, b)\}$. By complex analysis, the zero function is the only holomorphic function on a given region that is 0 in a set that has a limit point. Hence, from the assumption $K(Q(\theta)) = 0$ for all $\theta \in \Theta$ now follows $K = 0$.

By taking $z = Q(\theta) + it$ with $Q(\theta) \in (a, b)$ and $t \in \mathbb{R}$, we find that $E_\theta e^{itV} f(V) = 0$ for every $t \in \mathbb{R}$. By the theory of "characteristic functions" of probability distributions, it follows that $f(V) = 0$ with probability 1 and that V is consequently complete. Let us now give a direct proof of the conclusion that $f = 0$. Using the equality $\int e^{itv} e^{-t^2 \sigma^2/2} \, dt = e^{-v^2/(2\sigma^2)} \sqrt{2\pi}/\sigma$, we find, after exchanging the order of $\int$ and E_θ, that for $y \in \mathbb{R}$ and $\sigma > 0$,

$$0 = \int E_\theta e^{itV} f(V) \, e^{-ity - t^2\sigma^2/2} \, dt = \frac{\sqrt{2\pi}}{\sigma} E_\theta f(V) e^{-(V-y)^2/(2\sigma^2)}$$

$$= 2\pi \int \phi_\sigma(v - y) f(v) q_\theta(v) \, dv,$$

for ϕ_σ the density of the $N(0, \sigma^2)$-distribution and q_θ the probability density of V. Setting $f^+ = f 1_{f > 0}$ and $f^- = -f 1_{f < 0}$, so that $f = f^+ - f^-$, we find that $\phi_\sigma *$ $(f^+ q_\theta)(y) = \phi_\sigma * (f^- q_\theta)(y)$ for every $y \in \mathbb{R}$ and $\sigma > 0$, where $*$ is the convolution of two densities. The functions $f^+ q_\theta$ and $f^- q_\theta$ are both nonnegative. If we integrate them up to 0, we have $f^+ = f^-$, and then $f = 0$, so the proof is complete. In the other case, we can multiply f by a constant in order for $f^+ q_\theta$ and $f^- q_\theta$ to be probability densities of random variables X^+ and X^-. The functions $\phi_\sigma * (f^+ q_\theta)$ and $\phi_\sigma * (f^- q_\theta)$ are then the probability densities of random variables $X^+ + \sigma Z$ and $X^- + \sigma Z$, for Z a variable with the standard normal distribution that is independent of X^+ and X^-. Equality of the probability densities of the variables $X^+ + \sigma Z$ and $X^- + \sigma Z$ implies equality in distribution, for every $\sigma > 0$. As $\sigma \to 0$, the quantities converge with probability 1 to X^+ and X^-. It follows that the latter are also equal in distribution. The probability densities $f^+ q_\theta$ and $f^- q_\theta$ therefore agree, from which we conclude that $f^+ = f^-$ and therefore $f = 0$. ∎

The condition of the theorem indirectly requires the parameter space of Θ to be sufficiently rich. It is a logical condition, because completeness means that the system of equations

$$\mathrm{E}_\theta f(V) = 0 \qquad \text{for all } \theta \in \Theta$$

has only one solution in f, namely $f \equiv 0$. If there are "too few" θ, then there are too few equations to determine f uniquely, and V is not complete. The condition of the theorem is flexible: the existence of an arbitrarily small open set in the codomain of Q suffices.

Example 6.22 Binomial distribution

Let X be binomially distributed with parameters n and p. The binomial probability density can be written as

$$\binom{n}{x} p^x (1-p)^{n-x} = (1-p)^n \binom{n}{x} e^{x \log(p/(1-p))}.$$

This statistical model therefore forms a one-dimensional exponential family, with $c(p) = (1-p)^n$, $h(x) = \binom{n}{x}$, $V(x) = x$, and $Q(p) = \log(p/(1-p))$. If we take $[0, 1]$ as the parameter space for p, then the set

$$\{Q(p) : 0 \leq p \leq 1\} = \{\log(p/(1-p)) : 0 \leq p \leq 1\}$$

of Theorem 6.21 is equal to $\mathbb{R}$ and certainly contains an interior point. The statistic $V(X) = X$ is therefore both sufficient and complete. The estimator X/n for p is unbiased and based only on the sufficient and complete random variable, and is therefore a UMVU estimator by Theorem 6.18.

For the same reason, $(X/n)^2$ is a UMVU estimator for the expectation $\mathrm{E}_p(X/n)^2 = p(1-p)/n + p^2$. Can we use this to deduce a UMVU estimator for p^2? ☐

225

Example 6.23 Poisson distribution

The probability density of a sample $X = (X_1, \ldots, X_n)$ from the Poisson(θ)-distribution can be written as

$$\prod_{i=1}^{n} \frac{e^{-\theta}\theta^{x_i}}{x_i!} = e^{-n\theta} \frac{1}{\prod_{i=1}^{n} x_i!} e^{\sum_{i=1}^{n} x_i \log \theta}.$$

We conclude that this model forms a one-dimensional exponential family, with $c(\theta) = e^{-n\theta}$, $h(x) = (\prod_{i=1}^{n} x_i!)^{-1}$, $V(x) = \sum_{i=1}^{n} x_i$, and $Q(\theta) = \log \theta$. The set

$$\{Q(\theta) \colon \theta > 0\} = \{\log \theta \colon \theta > 0\} = (-\infty, \infty)$$

contains an interior point. The sum $V(X) = \sum_{i=1}^{n} X_i$ is therefore sufficient and complete. The estimator $\overline{X}$ for θ is unbiased and based only on the sufficient and complete random variable, and is therefore a UMVU estimator for θ (see Theorem 6.18).

Example 6.24 Normal distribution

The probability density of a sample $X = (X_1, \ldots, X_n)$ from the $N(\mu, \sigma^2)$-distribution can be written as

$$\prod_{i=1}^{n} \frac{1}{\sqrt{2\pi\sigma^2}} e^{-\frac{1}{2\sigma^2}(x_i - \mu)^2} = \left(\frac{1}{\sqrt{2\pi\sigma^2}}\right)^n e^{-\frac{n}{2\sigma^2}\mu^2} e^{\frac{\mu}{\sigma^2} \sum_{i=1}^{n} x_i - \frac{1}{2\sigma^2} \sum_{i=1}^{n} x_i^2}.$$

If we take the natural parameter space $\Theta = \mathbb{R} \times (0, \infty)$ for the parameter $\theta = (\mu, \sigma^2)$, then the set from Theorem 6.21 is equal to

$$\left\{\left(\frac{\mu}{\sigma^2}, \frac{-1}{2\sigma^2}\right) \colon \mu \in \mathbb{R}, \sigma^2 > 0\right\} = \mathbb{R} \times (-\infty, 0)$$

and contains an interior point. We conclude that the statistic $(\sum_{i=1}^{n} X_i, \sum_{i=1}^{n} X_i^2)$ is sufficient and complete. Because the sample variance can be rewritten as $S_X^2 = (n-1)^{-1}(\sum_{i=1}^{n} X_i^2 - n(\overline{X})^2)$, it immediately follows that $\overline{X}$ and S_X^2 are UMVU estimators for μ and σ^2.

Example 6.25 Curved normal family

Let $X = (X_1, \ldots, X_n)$ be a sample from the $N(\theta, \theta^2)$-distribution. The joint density is then given by

$$\prod_{i=1}^{n} \frac{1}{\sqrt{2\pi\theta^2}} e^{-\frac{1}{2}(x_i - \theta)^2/\theta^2} = \left(\frac{1}{\sqrt{2\pi\theta^2}}\right)^n e^{-\frac{1}{2}n} e^{\sum_{i=1}^{n} x_i/\theta - \frac{1}{2}\sum_{i=1}^{n} x_i^2/\theta^2}.$$

This probability density belongs to the two-dimensional exponential family with

$$Q(\theta) = \left(\frac{1}{\theta}, -\frac{1}{2\theta^2}\right),$$

and $V(X) = (\sum_{i=1}^{n} X_i, \sum_{i=1}^{n} X_i^2)$. However, the condition of Theorem 6.21 is not satisfied. For θ varying over $\mathbb{R}$, the function $\theta \mapsto Q(\theta)$ is a "one-dimensional curve" in $\mathbb{R}^2$; as a subset of $\mathbb{R}^2$, it does not contain any interior points.

The previous examples, as well as other examples, give a large number of interesting cases where a UMVU estimator exists and is reasonable. The UMVU criterion is therefore very appealing. We do have a few comments:
- Sometimes there is no unbiased estimator.
- Even if there exist (many) unbiased estimators, a UMVU estimator does not necessarily exist.
- There can exist a biased estimator with an overall smaller mean square error than that of the UMVU estimator.
- The bias is not invariant under nonlinear transformations: if T is UMVU for θ, then $g(T)$ is in general not unbiased for $g(\theta)$, and therefore also not UMVU.

In other words, always (only) looking for UMVU estimators is not wise, and can sometimes mean looking for a nonexistent estimator. The UMVU criterion is therefore not the answer to *all* questions. Unfortunately, there is no criterion in statistics that always "works" and pleases everyone. In practice, it is wise to apply several methods that seem reasonable. If the results do not diverge too much, we can confidently use our favorite criterion. Otherwise, there is a problem that may not be solvable objectively.

6.3.2 Cramér–Rao Lower Bounds

Instead of looking for a best estimator according to a specific criterion, we can also try to give a lower bound for the mean square error of an arbitrary estimator. For a given estimator, we can then compare the mean square error with the lower bound, and it is then clear how much this estimator may still be improved. Such a lower bound can then only depend on the given statistical model.

Such lower bounds naturally lead to the same shortcoming as "best estimators": unless we restrict the class of estimators, the absolute lower bound for the mean square error is 0, and therefore useless.

The Cramér–Rao lower bound is restricted to unbiased estimators. First, consider the case of a real-valued parameter θ. If p_θ is the probability density of the (full) observation X, $\ell_\theta = \log p_\theta$, and $\dot{\ell}_\theta = \partial/\partial\theta \log p_\theta = \dot{p}_\theta/p_\theta$ is the score function, then the *Fisher information* is defined as

$$I_\theta = \operatorname{var}_\theta \dot{\ell}_\theta(X).$$

Unlike the notation in Chapter 5, we have denoted the Fisher information with a capital I_θ. This is to distinguish between the Fisher information in the full observation and that in partial observations.

Theorem 6.26 Cramér–Rao inequality

Suppose that $\theta \mapsto p_\theta(x)$ is differentiable for every x. Under certain regularity conditions, the variance of every unbiased estimator T of $g(\theta) \in \mathbb{R}$ satisfies

$$\operatorname{var}_\theta T \geq \frac{g'(\theta)^2}{I_\theta},$$

with g' the derivative of g.

Proof. We write the formulas under the assumption that X is continuously distributed. (For a discrete probability density, we replace the integrals by sums.) Since $g(\theta) = \mathrm{E}_\theta T$ for all θ, we have

$$g'(\theta) = \frac{\partial}{\partial \theta} \int T(x) p_\theta(x)\, dx = \int T(x) \dot{p}_\theta(x)\, dx$$

$$= \int T(x) \dot{\ell}_\theta(x) p_\theta(x)\, dx = \mathrm{E}_\theta \left(T \dot{\ell}_\theta(X) \right).$$

The fact that we may change the order of differentiation and integration comes from the regularity conditions. (Explicit conditions are given in calculus, or rather in measure theory.) We already saw in Lemma 5.10 that $\mathrm{E}_\theta \dot{\ell}_\theta(X) = 0$. Combining these two equalities gives $g'(\theta) = \mathrm{E}_\theta(T \dot{\ell}_\theta(X)) - \mathrm{E}_\theta T\, \mathrm{E}_\theta \dot{\ell}_\theta(X) = \mathrm{cov}_\theta(T, \dot{\ell}_\theta(X))$. By the Cauchy–Schwarz inequality, we now have

$$\mathrm{cov}_\theta\left(T, \dot{\ell}_\theta(X)\right)^2 \leq \mathrm{var}_\theta T\, \mathrm{var}_\theta\, \dot{\ell}_\theta(X) = \mathrm{var}_\theta T I_\theta.$$

The inequality of the theorem now follows by replacing $\mathrm{cov}_\theta\left(T, \dot{\ell}_\theta(X)\right)^2$ on the left-hand side by $g'(\theta)^2$ and then dividing by I_θ. ∎

The number $g'(\theta)^2 / I_\theta$ is called the *Cramér–Rao lower bound* for estimating $g(\theta)$. For estimating θ, it of course reduces to $1/I_\theta$. We call the lower bound sharp if there exists an unbiased estimator T whose variance is equal to the lower bound. In that case, T is automatically a UMVU estimator for $g(\theta)$ because T is an unbiased estimator for $g(\theta)$ and has minimal variance.

The greater the Fisher information, the smaller the Cramér–Rao lower bound. Theorem 6.26 suggests that in that case, we can give a more accurate estimate of θ. Since the lower bound is not always sharp, this suggestion is not entirely correct. However, at the end of the chapter, we will see that the bound is sharp for (infinitely) large samples.

The theorem can be extended to multidimensional parameters θ. In that case, the Fisher information is not a number but a matrix, the *Fisher information matrix*

$$I_\theta = \left(\mathrm{cov}_\theta \left(\frac{\partial}{\partial \theta_i} \ell_\theta(X), \frac{\partial}{\partial \theta_j} \ell_\theta(X) \right) \right)_{i,j=1,\ldots k}.$$

We still restrict ourselves to real-valued functions g, and denote the gradient of g in θ by $g'(\theta)$ (a row vector). Then for every unbiased estimator T of $g(\theta)$, we have

$$\mathrm{var}_\theta T \geq g'(\theta) I_\theta^{-1} g'(\theta)^T.$$

In particular, the lower bound for the first coordinate $g(\theta) = \theta_1$ is equal to the $(1,1)$-element of I_θ^{-1}, because in that case the gradient is the vector $g'(\theta) = (1, 0, \ldots, 0)$.

When the full observation X is made up of independent subobservations $X_1, \ldots, X_n$, we can use that the information is additive.

> **Lemma 6.27**
>
> Let X and Y be independent random variables. Then the Fisher information in the observation (X, Y) is equal to the sum of the information in X and Y separately.

Proof. We give the proof only for the case where the parameter θ is real valued. The (joint) density of (X, Y) is the product $(x, y) \mapsto p_\theta(x) q_\theta(y)$ of the (marginal) densities of X and Y. The Fisher information in (X, Y) is the variance of the score function

$$\frac{\partial}{\partial \theta} \log p_\theta(x) q_\theta(y) = \frac{\partial}{\partial \theta} \log p_\theta(x) + \frac{\partial}{\partial \theta} \log q_\theta(y).$$

Because of the independence, this variance is the sum of the variances of the two terms on the right-hand side, namely the Fisher information in X and Y. ∎

In particular, the Fisher information in a vector $X = (X_1, \ldots, X_n)$ of independent, identically distributed observations $X_1, \ldots, X_n$ is equal to n times the Fisher information in one X_i, that is, $I_\theta = n i_\theta$, where i_θ denotes the Fisher information in one observation. The Cramér–Rao inequality then becomes: for every unbiased estimator of $g(\theta)$ based on $X_1, \ldots, X_n$, we have

$$\mathrm{var}_\theta \, T_n \geq \frac{g'(\theta) i_\theta^{-1} g'(\theta)^T}{n}.$$

Example 6.28 Normal distribution

The Fisher information for μ in one observation from the $N(\mu, \sigma^2)$-distribution (with σ^2 unknown) is equal to

$$i_\mu = \mathrm{var}_\mu \left(\frac{\partial}{\partial \mu} \left[\log\left(\frac{1}{\sigma \sqrt{2\pi}} e^{-\frac{1}{2}(X_1 - \mu)^2/\sigma^2} \right) \right] \right) = \mathrm{var}_\mu \left(\frac{X_1 - \mu}{\sigma^2} \right) = \frac{1}{\sigma^2}.$$

The Cramér–Rao lower bound for estimating μ based on a sample $X_1, \ldots, X_n$ of size n from the $N(\mu, \sigma^2)$-distribution is therefore

$$\frac{1}{n i_\mu} = \frac{\sigma^2}{n}.$$

This is exactly the variance of the unbiased estimator $\overline{X}$ for μ. In this case, the Cramér–Rao lower bound is therefore sharp. We have again proved that $\overline{X}$ is a UMVU estimator for μ, without using the theory of sufficient and complete statistics from Section 6.2 and Theorem 6.18.

The estimator $\overline{X}^2 - \sigma^2/n$ is unbiased for μ^2 (and an estimator because we assume σ^2 known) and a function of the sufficient, complete variable $\overline{X}$, hence UMVU. Some computation gives

$$\mathrm{var}_\mu \left(\overline{X}^2 - \frac{\sigma^2}{n} \right) = \frac{4\mu^2 \sigma^2}{n} + \frac{2\sigma^4}{n^2}.$$

The Cramér–Rao lower bound for the variance of an unbiased estimator of μ^2 is equal to

$$\frac{\left((\mu^2)'\right)^2}{ni_\mu} = \frac{4\mu^2\sigma^2}{n}.$$

Hence in this case, this lower bound is not attained. However, the extra term $2\sigma^4/n^2$ is small, and becomes negligible with respect to the first term as $n \to \infty$. ☐

Example 6.29 Binomial distribution

The Fisher information for p in a $\mathrm{bin}(n, p)$-distributed observation X is equal to

$$\mathrm{var}_p\left(\frac{\partial}{\partial p}\left[\log\left(\binom{n}{X}p^X(1-p)^{n-X}\right)\right]\right) = \mathrm{var}_p\left(\frac{X-np}{p(1-p)}\right) = \frac{n}{p(1-p)}.$$

The Cramér–Rao lower bound for the variance of an unbiased estimator for p based on X is therefore

$$\frac{p(1-p)}{n}.$$

This is exactly the variance of the unbiased estimator X/n. Hence in this case, the Cramér–Rao lower bound is sharp, and we may conclude that X/n is a UMVU estimator for p. ☐

Example 6.30 Uniform distribution

Let $X = (X_1, \ldots, X_n)$ be a sample from the uniform distribution on the interval $[0, \theta]$. The estimator $(n+1)/nX_{(n)}$ is unbiased and has variance

$$\mathrm{var}_\theta \frac{n+1}{n}X_{(n)} = \frac{\theta^2}{n(n+2)}.$$

For large n (and every given θ), this variance is much smaller than a bound of the form $1/(ni_\theta)$. So in this case the Cramér–Rao lower bound does not hold. The reason is that the density does not depend on the parameter in a differentiable manner. An expression such as $\dot{\ell}_\theta(x)$ is not well defined for all x. ☐

Upon further consideration, it turns out that the Cramér–Rao lower bound is seldom sharp. Nevertheless, we conclude this section with the essential assertion that, in a sense, the Cramér–Rao lower bound is asymptotically sharp and that the bound is attained by the maximum likelihood estimator.

We can see this as follows. We already know from Theorem 5.9 that under θ, the maximum likelihood estimator $\hat{\theta}_n$ based on a sample of size n from a density that depends on the parameter in a differentiable manner satisfies

$$\sqrt{n}(\hat{\theta}_n - \theta) \rightsquigarrow N(0, i_\theta^{-1})$$

as $n \to \infty$. A rough interpretation of this result is that, for large n, the random vector $\sqrt{n}(\hat{\theta}_n - \theta)$ is normally distributed with $E_\theta \sqrt{n}(\hat{\theta}_n - \theta) \approx 0$ and $\text{var}_\theta \sqrt{n}(\hat{\theta}_n - \theta) \approx i_\theta^{-1}$. It immediately follows that

$$E_\theta \hat{\theta}_n \approx \theta, \qquad \text{var}_\theta \, \hat{\theta}_n \approx \frac{i_\theta^{-1}}{n}.$$

In other words, the maximum likelihood estimator is (asymptotically) unbiased for θ with (asymptotic) variance equal to the Cramér–Rao lower bound, hence equal to the minimal variance for unbiased estimators. Conclusion: *maximum likelihood estimators are asymptotically UMVU*. This result is a strong motivation for using maximum likelihood estimators.

Maximum likelihood estimators are not the only types of estimators that are asymptotically UMVU. For example, by the Bernstein–von Mises theorem, Theorem 5.26, the median of the posterior distribution has the same asymptotic distribution provided that the prior density is positive on the entire parameter space Θ. Since, by this theorem, the posterior distribution is asymptotically normal and therefore symmetric, it moreover follows that, under certain conditions, most Bayes estimators are asymptotically normal.

Therefore, based on these asymptotic arguments, we cannot express a preference for maximum likelihood estimators over Bayes estimators or vice versa. On the other hand, these arguments do show that these two classes of estimators are preferable to method of moments estimators, which in general are not asymptotically efficient. The method of moments is interesting because of its simplicity, and also in cases where the theoretical moments can be specified but the full probability density cannot. In the latter case, we cannot implement the maximum likelihood estimators or Bayes estimators.

6.4 Testing Theory

According to the theory discussed in Chapter 4, a good test has size less than or equal to the given level and a power function that is as large as possible. A test is uniformly most powerful (at a given level) if, under the alternative hypothesis, the power function is maximal in all possible parameter values. In this section, we discuss several special, but important, cases in which a uniformly most powerful test exists.

6.4.1 Simple Hypotheses

A "simple" hypothesis is one that consists of only one parameter value. In most cases, for tests of a simple null hypothesis against a simple alternative, there exists an optimal test, that is, a test with a maximal power function in the parameter value, under the alternative hypothesis. This is the statement of the following "fundamental lemma" of the theory of tests.

Suppose that, for a given parameter space $\Theta = \{\theta_0, \theta_1\}$, p_{θ_0} and p_{θ_1} are the two possible probability densities of the observation X. Let $L(\theta_1, \theta_0; X) = p_{\theta_1}(X)/p_{\theta_0}(X)$ be the quotient of these densities, evaluated in the observation.

Theorem 6.31 Neyman–Pearson

Suppose that there exists a number c_{α_0} such that $\mathrm{P}_{\theta_0}\big(L(\theta_1, \theta_0; X) \geq c_{\alpha_0}\big) = \alpha_0$. Then the test with critical region $K = \{x : L(\theta_1, \theta_0; x) \geq c_{\alpha_0}\}$ is most powerful at level α_0 for testing $H_0 : \theta = \theta_0$ against $H_1 : \theta = \theta_1$.

Proof. By the assumption on the number c_{α_0}, the size of the critical region K mentioned in the theorem is exactly α_0. Suppose that K' is another critical region of size at most α_0, that is, $\mathrm{P}_{\theta_0}(X \in K') \leq \alpha_0$. We must prove that $\mathrm{P}_{\theta_1}(X \in K') \leq \mathrm{P}_{\theta_1}(X \in K)$.

We claim that for all x,

$$\big(1_{K'}(x) - 1_K(x)\big)\big(p_{\theta_1}(x) - c_{\alpha_0}p_{\theta_0}(x)\big) \leq 0.$$

Indeed, if $x \in K$, then $1_{K'}(x) - 1_K(x) = 1_{K'}(x) - 1 \leq 0$ and $p_{\theta_1}(x) - c_{\alpha_0}p_{\theta_0}(x) \geq 0$ by the definition of K. If $x \notin K$, then both inequalities hold in the other direction. In both cases, the expression on the left-hand side of the inequality is the product of a nonpositive and a nonnegative term, and is therefore nonpositive.

The integral of this nonpositive function over the sample space (or the sum if the distribution is discrete) is then also nonpositive. We can write this as

$$\int \big(1_{K'}(x) - 1_K(x)\big) p_{\theta_1}(x)\, dx \leq c_{\alpha_0} \int \big(1_{K'}(x) - 1_K(x)\big) p_{\theta_0}(x)\, dx$$

$$= c_{\alpha_0}\big(\mathrm{P}_{\theta_0}(X \in K') - \mathrm{P}_{\theta_0}(X \in K)\big)$$

$$\leq c_{\alpha_0}(\alpha_0 - \alpha_0) = 0.$$

It follows that $\mathrm{P}_{\theta_1}(X \in K') \leq \mathrm{P}_{\theta_1}(X \in K)$, and therefore the test with critical region K is most powerful at level α_0. ∎

The test from Theorem 6.31 is intuitively reasonable because it rejects the null hypothesis $H_0 : \theta = \theta_0$ in favor of the alternative $H_1 : \theta = \theta_1$ when under H_1, the density $p_{\theta_1}(X)$ in the observation is large with respect to the density $p_{\theta_0}(X)$ under the null hypothesis. The motivation for this is the same as in the case of the likelihood ratio test. We view $p_\theta(x)$ as a measure for the probability that the realization x occurs if θ is the true parameter, and a small value of $p_\theta(x)$ means that it is improbable that θ is the true parameter. (When $c_{\alpha_0} \geq 1$, the test from Theorem 6.31 reduces to the likelihood ratio test.) Test of the same form as that in Theorem 6.31 are also called likelihood ratio tests or *Neyman–Pearson tests*.

Example 6.32 Gauss test

Let $X = (X_1, \ldots, X_n)$ be a sample from the normal distribution with unknown expectation μ and known variance σ^2. We want to test the simple null hypothesis $H_0\colon \mu = \mu_0$ against the simple alternative $H_1\colon \mu = \mu_1$. The Neyman–Pearson lemma says that the test with test statistic

$$L(\mu_1, \mu_0; X) = \exp\left(-\frac{1}{2\sigma^2}\sum_{i=1}^{n}(X_i - \mu_1)^2 + \frac{1}{2\sigma^2}\sum_{i=1}^{n}(X_i - \mu_0)^2\right)$$

$$= \exp\left(n\overline{X}(\mu_1 - \mu_0)/\sigma^2 + n(\mu_0^2 - \mu_1^2)/(2\sigma^2)\right)$$

and critical region $K = \{x = (x_1, \ldots, x_n)\colon L(\mu_1, \mu_0; x) \geq c_{\alpha_0}\}$, with c_{α_0} such that $P_{\mu_0}(L(\mu_1, \mu_0; X) \geq c_{\alpha_0}) = \alpha_0$, is the most powerful test at level α_0 for testing the null hypothesis in question. The null hypothesis is rejected for large values of $L(\mu_1, \mu_0; X)$, that is, for large values of $\overline{X}(\mu_1 - \mu_0)$. This means that if $\mu_1 > \mu_0$, the null hypothesis is rejected for large values of $\overline{X}$ or, equivalently, for large values of $T = \sqrt{n}(\overline{X} - \mu_0)/\sigma$. The most powerful test is therefore the test that rejects the null hypothesis for $\sqrt{n}(\overline{X} - \mu_0)/\sigma$ greater than a value d_{α_0} such that $P_{\mu_0}(\sqrt{n}(\overline{X} - \mu_0)/\sigma \geq d_{\alpha_0}) = \alpha_0$. Since under $\mu = \mu_0$, the quantity $\sqrt{n}(\overline{X} - \mu_0)/\sigma$ has the standard normal distribution, we have $d_{\alpha_0} = \xi_{1-\alpha_0}$, and the null hypothesis is rejected for $\sqrt{n}(\overline{X} - \mu_0)/\sigma \geq \xi_{1-\alpha_0}$. This is exactly the Gauss test from Example 4.12. The conclusion is that the Gauss test is the most powerful test for testing the simple null hypothesis $H_0\colon \mu = \mu_0$ against the simple alternative $H_1\colon \mu = \mu_1$ based on a sample from the normal distribution with unknown expectation μ and known variance σ^2.

Under the null hypothesis, the condition of Theorem 6.31 that there exist a number c_{α_0} such that $P_{\theta_0}(L(\theta_1, \theta_0; X) \geq c_{\alpha_0}) = \alpha_0$ is always satisfied when the likelihood ratio statistic $L(\theta_1, \theta_0; X)$ has a continuous distribution function. Namely, the condition is equivalent to the condition that the distribution function of $L(\theta_1, \theta_0; X)$ is equal to $1 - \alpha_0$ in c_{α_0}. The size of the optimal test is exactly α_0.

If the distribution function of $L(\theta_1, \theta_0; X)$ has jumps, then there will not exist a c_{α_0} for every α_0. The statement of Theorem 6.31 can then be incorrect. The idea that an optimal test can be based on the likelihood ratio statistic $L(\theta_1, \theta_0; X)$ does remain true, however. In all cases, we can find a value c_{α_0} such that

$$P_{\theta_0}(L(\theta_1, \theta_0; X) > c_{\alpha_0}) \leq \alpha_0 \leq P_{\theta_0}(L(\theta_1, \theta_0; X) \geq c_{\alpha_0}).$$

If these inequalities are strict, then the test with critical region

$$K = \{x\colon L(\theta_1, \theta_0; x) > c_{\alpha_0}\}$$

has size strictly less than α_0 and the test with critical region $K = \{x\colon L(\theta_1, \theta_0; x) \geq c_{\alpha_0}\}$ has size strictly greater than α_0. The second test is not admissible, but the first test is not necessarily most powerful because we could further enlarge the critical region. We can construct a more powerful test by sometimes also rejecting the null hypothesis when $L(\theta_1, \theta_0; x) = c_{\alpha_0}$.

In some examples, the set $\{x\colon L(\theta_1, \theta_0; x) = c_{\alpha_0}\}$ can be split into two subsets R_1 and R_2, and the test that rejects when $L(\theta_1, \theta_0; X) > c_{\alpha_0}$ and when $L(\theta_1, \theta_0; X) = c_{\alpha_0}$ and $X \in R_1$ is most powerful. In general, we can extend Theorem 6.31 to likelihood ratio statistics with jumps in the distribution function by generalizing the notion of test.

Definition 6.33 Randomized test

A randomized test is a statistic ψ with values in $[0, 1]$. If x is observed, then we reject H_0 with probability $\psi(x)$. The power function of the randomized test ψ is by definition equal to $\pi(\theta; \psi) = \mathrm{E}_\theta \psi(X)$, and the size is equal to $\sup_{\theta \in \Theta_0} \pi(\theta; \psi)$.

A test with critical region K is a special case of a randomized test, through the identification $\psi(x) = 1_K(x)$. If we allow randomized tests, there always exists a most powerful test. The proof of the following theorem is analogous to that of Theorem 6.31.

Theorem 6.34 Neyman–Pearson

There exist numbers c_{α_0} and $\delta \in [0, 1]$ such that

$$\mathrm{P}_{\theta_0}\big(L(\theta_1, \theta_0; X) > c_{\alpha_0}\big) + \delta \mathrm{P}_{\theta_0}\big(L(\theta_1, \theta_0; X) = c_{\alpha_0}\big) = \alpha_0.$$

For every choice of these numbers, the randomized test

$$\psi = 1_{\{x\colon L(\theta_1, \theta_0; x) > c_{\alpha_0}\}} + \delta 1_{\{x\colon L(\theta_1, \theta_0; x) = c_{\alpha_0}\}}$$

is the most powerful test at level α_0 for testing $H_0\colon \theta = \theta_0$ against $H_1\colon \theta = \theta_1$.

As the theorem shows, the optimal test only uses the randomization to sometimes reject or not observations in the "boundary region" $\{x\colon p_{\theta_1}(x)/p_{\theta_0}(x) = c_{\alpha_0}\}$. If the likelihood ratio $L(\theta_1, \theta_0; X)$ is strictly greater than c_{α_0}, then we always reject, and if the ratio is strictly smaller, then we never reject. In the intermediate case, we reject with probability δ. The randomization with a constant probability δ as in the theorem is one way to "split" the boundary region, and often the optimal test is not unique with respect to this aspect.

The randomized test mostly has a theoretical interest. In practice, it is rarely used.

Example 6.35 Binomial distribution

Let X be binomially distributed with parameter n and unknown probability $p \in [0, 1]$. The likelihood ratio for testing the simple null hypothesis $H_0\colon p = p_0$ against $H_1\colon p = p_1$ is given by

$$L(p_1, p_0; X) = \frac{\binom{n}{X} p_1^X (1 - p_1)^{n-X}}{\binom{n}{X} p_0^X (1 - p_0)^{n-X}} = \left(\frac{p_1}{p_0}\right)^X \left(\frac{1 - p_1}{1 - p_0}\right)^{n-X}.$$

In this example, we assume $p_1 > p_0$, so that $L(p_1, p_0; x)$ is increasing in x. A large value of X therefore implies a large value of $L(p_1, p_0; X)$ (and vice versa). The question is now for which values of X we must reject the null hypothesis.

Suppose $n = 100$, $p_0 = 1/2$, and $\alpha_0 = 0.05$; then $P_{0.5}(X \geq 59) = 0.044$ and $P_{0.5}(X \geq 58) = 0.067$ (see Example 4.11). It follows that the test given by the critical region $\{59, 60, \ldots, 100\}$ has level 0.05, while the test that rejects for $X \geq 58$ is not admissible at this level. The size of the given test, 0.044, is strictly less than the level $\alpha_0 = 0.05$. This means that the Neyman–Pearson lemma (Theorem 6.31) cannot be applied and the test may not be optimal. The randomized test defined by

$$\psi(x) = 1_{\{x \geq 59\}} + 0.26 \, 1_{\{x = 58\}},$$

on the other hand, does have size exactly equal to 0.05:

$$E_{0.5}\psi(X) = P_{0.5}(X \geq 59) + 0.26 P_{0.5}(X = 58) = 0.05.$$

By Theorem 6.34, the randomized test ψ is now most powerful for testing $H_0: p = 1/2$ against $H_1: p = p_1$ for $p_1 > 1/2$.

Note that in this example, we only assume $p_1 > p_0$; we do not make any assumptions about the exact value of p_1.

Example 6.36 Uniform distribution

The likelihood ratio for testing the null hypothesis $H_0: \theta = \theta_0$ against the alternative hypothesis $H_1: \theta = \theta_1$ for $\theta_1 > \theta_0$ based on a sample $X = (X_1, \ldots, X_n)$ from the uniform distribution on $[0, \theta]$ is given by

$$L(\theta_1, \theta_0; X) = \frac{(1/\theta_1)^n 1_{\{X_{(n)} \leq \theta_1\}}}{(1/\theta_0)^n 1_{\{X_{(n)} \leq \theta_0\}}} = \begin{cases} \left(\frac{\theta_0}{\theta_1}\right)^n & \text{if } X_{(n)} \leq \theta_0, \\ \infty & \text{if } \theta_0 < X_{(n)} \leq \theta_1. \end{cases}$$

Under the null hypothesis, we are always in the first case, and the likelihood ratio has a degenerate probability distribution; all probability mass lies in the point $(\theta_0/\theta_1)^n$. The value c_{α_0} from Theorem 6.34 is therefore equal to the constant value $(\theta_0/\theta_1)^n$ of the likelihood ratio, and the sets of values of the observation for which the likelihood ratio is strictly greater than or equal to c_{α_0} are, respectively, the sets $\{(x_1, \ldots, x_n): x_{(n)} > \theta_0\}$ and $\{(x_1, \ldots, x_n): x_{(n)} \leq \theta_0\}$.

By Theorem 6.34, the randomized test $\psi(X_1, \ldots, X_n) = 1_{\{X_{(n)} > \theta_0\}} + \delta 1_{\{X_{(n)} \leq \theta_0\}}$ is optimal, where the randomization value δ must be such that the size is equal to α_0. Since the size is equal to $P_{\theta_0}(X_{(n)} > \theta_0) + \delta P_{\theta_0}(X_{(n)} \leq \theta_0) = \delta P_{\theta_0}(X_{(n)} \leq \theta_0)$ and $P_{\theta_0}(X_{(n)} \leq \theta_0) = 1$, we have $\delta = \alpha_0$. This test corresponds to rejecting when $X_{(n)}$ takes on a value that is impossible under the null hypothesis (that is, when $X_{(n)} > \theta_0$) and always randomizing with probability α_0 when $X_{(n)}$ takes on a value that is possible under H_0. The first is quite natural, while the randomization does not seem reasonable intuitively.

The optimal test is not unique. In particular, we can avoid randomization by taking $K = \{(x_1, \ldots, x_n) : x_{(n)} > d_{\alpha_0}\}$ with $d_{\alpha_0} = \theta_0 \sqrt[n]{1 - \alpha_0}$ as the critical region, so that the size of this test is equal to α_0. This test and the previously described randomized test both have power function $1 - (1 - \alpha_0)(\theta_0/\theta_1)^n$ in θ_1. We have then replaced the randomization with probability α_0 when $X_{(n)} \in [0, \theta_0]$ with always rejecting when $X_{(n)} \in [\theta_0 \sqrt[n]{1 - \alpha_0}, \theta_0]$. (Note that both tests always reject when $L(\theta_1, \theta_0; X) = \infty$ and reject with probability α_0 when $L(\theta_1, \theta_0; X) = (\theta_0/\theta_1)^n$. In terms of the likelihood ratio, the optimal test is therefore unique.)

6.4.2 Monotone Likelihood Ratio

In the previous section, we saw that for testing simple hypotheses, there always exists an optimal test. For general hypotheses, this unfortunately does not hold. A test is optimal for a composite alternative hypothesis if the test is uniformly most powerful in the sense of the following definition.

Definition 6.37 Uniformly most powerful test

A test with power function $\theta \mapsto \pi(\theta; K)$ is called *uniformly most powerful* or *UMP* at size α_0 for testing $H_0 : \theta \in \Theta_0$ against $H_1 : \theta \in \Theta_1$ if $\sup_{\theta \in \Theta_0} \pi(\theta; K) \leq \alpha_0$ and the power function $\theta \mapsto \pi(\theta; K')$ of every other test with $\sup_{\theta \in \Theta_0} \pi(\theta; K') \leq \alpha_0$ satisfies $\pi(\theta; K) \geq \pi(\theta; K')$ for all $\theta \in \Theta_1$.

The qualification of "uniform" in "uniformly most powerful" refers to the fact that the power function of an optimal test must be maximal for all parameter values under the alternative hypothesis: a uniformly most powerful test for $H_0 : \theta \in \Theta_0$ against $H_1 : \theta \in \Theta_1$ must be most powerful for testing $H_0 : \theta \in \Theta_0$ against $H_1 : \theta = \theta_1$ for *all* $\theta_1 \in \Theta_1$. The greater the parameter space under the alternative hypothesis, the stronger this condition.

Nevertheless, in a number of important examples, there do exist uniformly most powerful tests. First, consider testing a simple null hypothesis $H_0 : \theta = \theta_0$ against a composite alternative hypothesis $H_1 : \theta \in \Theta_1$. A test of level α_0 for this composite problem is also a test of level α_0 for every simple testing problem of $H_0 : \theta = \theta_0$ against $H_1 : \theta = \theta_1$, for every $\theta_1 \in \Theta_1$. By Theorems 6.31 and 6.34, the most powerful test for such a simple problem is the Neyman–Pearson test based on the likelihood ratio $p_{\theta_1}(X)/p_{\theta_0}(X)$. It follows from the proofs that the Neyman–Pearson test is also the unique most powerful test when the likelihood ratio statistic has a continuous distribution function. We conclude that when the Neyman–Pearson tests for different alternatives $\theta_1 \in \Theta_1$ differ, there cannot exist a uniformly most powerful test.

Conversely, we can also apply this reasoning in the positive direction and conclude that if the Neyman–Pearson test for $H_0 : \theta = \theta_0$ against $H_1 : \theta = \theta_1$ is the same for every $\theta_1 \in \Theta_1$, then this test is automatically uniformly most powerful. At first glance, the Neyman–Pearson test with test statistic $p_{\theta_1}(X)/p_{\theta_0}(X)$ seems to always depend on θ_1. This is not the case because the critical value of this test will also depend on θ_1, and these two dependencies can cancel each other out.

Example 6.38 Gauss test, continued from Example 6.32

Let $X = (X_1, \ldots, X_n)$ be a sample from the normal distribution with unknown expectation μ and known variance σ^2. We are looking for the uniformly most powerful test for testing the simple null hypothesis $H_0 : \mu = \mu_0$ against the composite alternative $H_1 : \mu > \mu_0$.

We already saw in Example 6.32 that for a simple null hypothesis and a simple alternative hypothesis, the Gauss test is most powerful. In this example, we show that this also holds for the composite alternative hypothesis given above.

The most powerful test for testing $H_0 : \mu = \mu_0$ against $H_1 : \mu = \mu_1$ for $\mu_1 > \mu_0$ rejects the null hypothesis for $\sqrt{n}(\overline{X} - \mu_0)/\sigma > \xi_{1-\alpha_0}$. This criterion does not depend on the value μ_1, and this test is therefore most powerful for every value of $\mu_1 \in (\mu_0, \infty)$. We conclude that the Gauss test is uniformly most powerful for testing $H_0 : \mu = \mu_0$ against $H_1 : \mu > \mu_0$.

Example 6.39 Binomial distribution, continued from Example 6.35

Let X be binomially distributed with parameter n and unknown probability $p \in [0, 1]$. In Example 6.35, we deduced that for $n = 100$ and $\alpha_0 = 0.05$, the most powerful test for testing the simple hypotheses $H_0 : p = 1/2$ against $H_1 : p = p_1$ with $p_1 > 1/2$ is the randomized test that rejects H_0 when $X \geq 59$ and rejects it with probability 0.26 when $X = 58$. We already noted in that example that the test does not depend on the value of p_1 provided $p_1 > p_0 = 1/2$. We can immediately conclude that the randomized test described here is uniformly most powerful for testing $H_0 : p = 1/2$ against $H_1 : p > 1/2$.

Example 6.40 Uniform distribution, continued from Example 6.36

In Example 6.36, we saw that the test that rejects H_0 when $X_{(n)} \geq d_{\alpha_0}$ for $d_{\alpha_0} = \theta_0 \sqrt[n]{1 - \alpha_0}$ is uniformly most powerful for testing $H_0 : \theta = \theta_0$ against $H_1 : \theta = \theta_1$ for every $\theta_1 > \theta_0$. This test does not depend on θ_1. We conclude that the test is uniformly most powerful for testing $H_0 : \theta = \theta_0$ against $H_1 : \theta > \theta_0$.

Through a similar reasoning, we can sometimes also deduce a uniformly most powerful test for a composite null hypothesis from uniformly most powerful tests for simple null hypotheses. The relevant criterion is the size. A test that is uniformly most powerful for testing $H_0 : \theta = \theta_0$ against $H_1 : \theta \in \Theta_1$ for a given $\theta_0 \in \Theta_0$ is also uniformly most powerful for testing $H_0 : \theta \in \Theta_0$ against $H_1 : \theta \in \Theta_1$ provided that the test is of level α_0 for this problem, so that the test is admissible. The latter is not necessarily the case, because the size for the null hypothesis $H_0 : \theta \in \Theta_0$ (a supremum over Θ_0) is greater than that for a simple hypothesis $H_0 : \theta = \theta_0$. It is, however, sufficient if we can justify the reasoning for one parameter value $\theta_0 \in \Theta_0$, namely the value in which the supremum is taken on.

Example 6.41 Gauss test, continued from Example 6.38

We already saw in Example 6.38 that the Gauss test is a uniformly most powerful test for a simple null hypothesis and composite alternative hypothesis. In this example, we will see that even when the null hypothesis is composite, the Gauss test remains uniformly most powerful. Consider the hypothesis $H_0\colon \mu \leq \mu_0$ against the alternative hypothesis $H_1\colon \mu > \mu_0$. We only need to show that the size of the Gauss test is equal to α_0:

$$\sup_{\mu \leq \mu_0} \mathrm{P}_\mu\left(\sqrt{n}\frac{\overline{X} - \mu_0}{\sigma} \geq \xi_{1-\alpha_0}\right) = \mathrm{P}_{\mu_0}\left(\sqrt{n}\frac{\overline{X} - \mu_0}{\sigma} \geq \xi_{1-\alpha_0}\right) = \alpha_0,$$

as we already saw in Example 4.12.

Example 6.42 Binomial distribution, continued from Example 6.39

Let X be binomially distributed with parameters n and unknown probability $p \in [0, 1]$. In Example 6.39, we gave a uniformly most powerful randomized test for testing the simple null hypothesis $H_0\colon p = 1/2$ against the composite alternative $H_1\colon p > 1/2$. In this example we show that this randomized test is also uniformly most powerful for testing the composite null hypothesis $H_0\colon p \leq 1/2$ against the alternative $H_1\colon p > 1/2$. We must show that the randomized test also has level $\alpha_0 = 0.05$ for this null hypothesis; in other words, we must show that $\sup_{p \leq 1/2} \mathrm{E}_p \psi(X) \leq 0.05$ for $\psi(x) = 1_{\{x \geq 59\}} + 0.26\, 1_{\{x = 58\}}$. The size of the test is given by

$$\sup_{p \leq 1/2} \mathrm{E}_p \psi(X) = \sup_{p \leq 1/2} \left(\mathrm{P}_p(X \geq 59) + 0.26\mathrm{P}_p(X = 58)\right).$$

The supremum is taken on at $p = 1/2$ (see Example 4.11) and $\mathrm{E}_{0.5}\psi(X) = 0.05$ (see Example 6.35).

Example 6.43 Uniform distribution, continued from Example 6.36

The size of the test from Example 6.36 for testing the null hypothesis $H_0\colon \theta \leq \theta_0$, which rejects when $X_{(n)} \geq d_{\alpha_0} = \theta_0 \sqrt[n]{1 - \alpha_0}$, is given by

$$\sup_{\theta \leq \theta_0} \mathrm{P}_\theta(X_{(n)} \geq d_{\alpha_0}) = \mathrm{P}_{\theta_0}(X_{(n)} \geq d_{\alpha_0}) = \alpha_0$$

by the construction of d_{α_0}. We conclude that the test is uniformly most powerful for testing $H_0\colon \theta \leq \theta_0$ against $H_1\colon \theta > \theta_0$.

We can use the previous arguments to show that uniformly most powerful tests exist for testing one-sided hypotheses for one-dimensional exponential families. By Definition 6.20, a family of probability densities p_θ belongs to a one-dimensional exponential family if there exist functions c, h, and Q such that the density in the family is of the following form:

$$p_\theta(x) = c(\theta)h(x)e^{Q(\theta)V(x)},$$

for a one-dimensional sufficient statistic $V(X)$. In the following theorem, we assume that the density of the observation X is of this form.

Theorem 6.44 Exponential family

Suppose that the density of X belongs to a one-dimensional exponential family with sufficient statistic $V = V(X)$ and that there exists a number d_{α_0} such that $P_{\theta_0}(V(X) > d_{\alpha_0}) = \alpha_0$. Then the test with critical region $K = \{x : V(x) > d_{\alpha_0}\}$ is uniformly most powerful at level α_0 for testing $H_0 : Q(\theta) \leq Q(\theta_0)$ against $H_1 : Q(\theta) > Q(\theta_0)$.

Proof. The Neyman–Pearson test for testing the simple null hypothesis $H_0 : \theta = \theta_0$ against the alternative hypothesis $H_1 : \theta = \theta_1$ is based on the likelihood ratio

$$L(\theta_1, \theta_0; x) = \frac{c(\theta_1)}{c(\theta_0)} e^{(Q(\theta_1) - Q(\theta_0))V(x)}.$$

By Theorem 6.34, the most powerful test for testing $H_0 : \theta = \theta_0$ against $H_1 : \theta = \theta_1$ is the randomized test

$$\psi(x) = 1_{\{L(\theta_1, \theta_0; x) > c_{\alpha_0}\}} + \delta 1_{\{L(\theta_1, \theta_0; x) = c_{\alpha_0}\}}$$

for constants c_{α_0} and δ such that the size of the test is equal to α_0, that is, such that $P_{\theta_0}(L(\theta_1, \theta_0; X) > c_{\alpha_0}) + \delta P_{\theta_0}(L(\theta_1, \theta_0; X) = c_{\alpha_0}) = \alpha_0$. For $Q(\theta_1) > Q(\theta_0)$, the likelihood ratio $L(\theta_1, \theta_0; x)$ is a strictly increasing function of $V(x)$, so that the randomized test ψ is equivalent to the test

$$\psi'(x) = 1_{\{V(x) > d_{\alpha_0}\}} + \delta 1_{\{V(x) = d_{\alpha_0}\}},$$

for d_{α_0} such that the size of the test is α_0. It follows from the assumption that there exists a number d_{α_0} such that $P_{\theta_0}(V(X) > d_{\alpha_0}) = \alpha_0$, so that we can choose $\delta = 0$. Since this test does not depend on the alternative θ_1, provided $Q(\theta_1) > Q(\theta_0)$, the test is automatically uniformly most powerful for testing $H_0 : \theta = \theta_0$ against $H_1 : Q(\theta) > Q(\theta_0)$.

Every test of level α_0 for testing $H_0 : Q(\theta) \leq Q(\theta_0)$ against $H_1 : Q(\theta) > Q(\theta_0)$ is also a test of level α_0 for testing $H_0 : Q(\theta) = Q(\theta_0)$ against $H_1 : Q(\theta) > Q(\theta_0)$. Its power function is therefore not greater than that of the best test for this problem, the test ψ' defined above. It now suffices to prove that the latter has size α_0 for the null hypothesis $H_0 : Q(\theta) \leq Q(\theta_0)$.

The density $p_\theta(x) = c(\theta)h(x) \exp(Q(\theta)V(x))$ is exponential in $V(x)$. The form of this distribution depends on $Q(\theta)$. A larger value of $Q(\theta)$ puts relatively more probability mass on large values of $V(x)$ and less on small values of $V(x)$. This implies that for every d, the probability $P_\theta(V \geq d)$ increases as $Q(\theta)$ increases. In other words, for every parameter value θ with $Q(\theta) \leq Q(\theta_0)$, we have $P_\theta(V(X) \geq d) \leq P_{\theta_0}(V(X) \geq d)$ for every d. This holds, in particular, for $d = d_{\alpha_0}$, which means that the test ψ' has size α_0 for the null hypothesis $H_0 : Q(\theta) \leq Q(\theta_0)$. This concludes the proof of the theorem. ∎

Example 6.45 Gauss test

Let $X = (X_1, \ldots, X_n)$ be a sample from the normal distribution with unknown expectation μ and known variance σ^2. If we assume σ^2 known, then the statistical model is a one-dimensional exponential family with sufficient quantity $\overline{X}$. By Theorem 6.44, there exists a uniformly most powerful test for $H_0: \mu \leq \mu_0$ against $H_1: \mu > \mu_0$, and this rejects H_0 for large values of $\overline{X}$. We again recover the Gauss test from Example 4.12.

The proof of Theorem 6.44 in fact only uses the structure of the exponential family because this implies that there exists a one-dimensional sufficient quantity V whose distributions under the different values of the parameter are ordered stochastically. The reasoning for the existence of a uniformly most powerful test for the uniform distribution in Example 6.43 has the same structure. We can unite the two cases under the notion of "monotone likelihood ratio family."

Definition 6.46 Monotone likelihood ratio

A statistical model $\{p_\theta : \theta \in \Theta \subset \mathbb{R}\}$ is called a family with monotone likelihood ratio if there exist a real-valued statistic V and a monotonically increasing function g_{θ_0, θ_1} for all $\theta_0 \leq \theta_1$ such that

$$\frac{p_{\theta_1}(x)}{p_{\theta_0}(x)} = g_{\theta_0, \theta_1}\left(V(x)\right).$$

Within a family with monotone likelihood ratio, the statistic V is sufficient by the factorization theorem. Moreover, the monotonicity of the function g_{θ_0, θ_1} and the Neyman–Pearson lemma imply that the most powerful test for $H_0: \theta = \theta_0$ against $H_1: \theta = \theta_1$ for a given $\theta_0 < \theta_1$ can be based on V, where the test rejects H_0 for large values of V. More precisely, there exists a most powerful randomized test of the form

$$(6.1) \qquad \psi = 1_{\{x:V(x)>d_{\alpha_0}\}} + \delta 1_{\{x:V(x)=d_{\alpha_0}\}}$$

for some d_{α_0} and δ. This leads to the following theorem, whose proof is analogous to that of Theorem 6.44.

Theorem 6.47 Monotone likelihood ratio

For testing the hypothesis $H_0: \theta \leq \theta_0$ against $H_1: \theta > \theta_0$ based on an observation from a family with monotone likelihood ratio, there exists a uniformly most powerful randomized test for every given level α_0. This test can be taken of the form (6.1), for V the sufficient quantity of the family.

Example 6.48 Binomial distribution, continued

Let X be a bin(n, p)-distributed random variable. Then the likelihood ratio for $H_0: p \leq p_0$ against $H_1: p > p_0$ is equal to

$$L(p_1, p_0; X) = \left(\frac{p_1}{p_0}\right)^X \left(\frac{1 - p_1}{1 - p_0}\right)^{n-X}$$

(see Example 6.35), which is increasing in X when $p_1 > p_0$. By Theorem 6.47, there now exists a uniformly most powerful randomized test of the form (6.1). We already found this randomized test in Example 6.42.

6.4.3 Optimality of the t-Test

For statistical models without a one-dimensional sufficient quantity, there does not, in general, exist a uniformly most powerful test. This concerns, in particular, all models with a two-dimensional parameter. The problem is already apparent with tests of the expectation parameter for the normal distribution when the variance σ^2 is unknown. In this section, we take $X = (X_1, \ldots, X_n)$ to be a sample from the normal distribution with expectation μ and variance σ^2. We already saw in the previous section that when σ^2 is assumed known, the Gauss test is the uniformly most powerful test for $H_0: \mu \leq \mu_0$ against $H_1: \mu > \mu_0$. This test depends on σ and therefore cannot be used when σ is unknown. An intuitively reasonable solution is to replace the unknown parameter σ by the sample standard deviation S_X. This leads to the t-test, which rejects the null hypothesis when $\sqrt{n}(\overline{X} - \mu_0)/S_X \geq t_{n-1, 1-\alpha_0}$. In this section, we will prove that the t-test is uniformly most powerful among the unbiased tests for $H_0: \mu \leq \mu_0$ against $H_1: \mu > \mu_0$.

Definition 6.49 Unbiased test

A test is unbiased for testing $H_0: \theta \in \Theta_0$ against $H_1: \theta \in \Theta_1$ at a given level α_0 if the power function π of the test satisfies $\pi(\theta_0) \leq \alpha_0 \leq \pi(\theta_1)$ for all $\theta_0 \in \Theta_0$ and $\theta_1 \in \Theta_1$.

The randomized test $\psi \equiv \alpha_0$, which rejects H_0 with probability α_0 regardless of the value of the observation, is unbiased. Since a uniformly most powerful test, if it exists, must also dominate this test, it follows that a uniformly most powerful test is automatically unbiased. However, one can prove that at level $\alpha_0 < 1/2$, there do not exist any uniformly most powerful tests. (Surprisingly, the converse is true when $\alpha_0 > 1/2$, but such large levels are not interesting from a practical point of view.) This means that the t-test, although most powerful among the unbiased tests, is not the most powerful test among *all* tests. There exist biased tests that have a greater power function than the t-test in certain alternative values $\mu > \mu_0$.

Theorem 6.50

The t-test is uniformly most powerful among the unbiased tests for testing $H_0: \mu \leq \mu_0$ against $H_1: \mu > \mu_0$.

Proof. Without loss of generality, we assume $\mu_0 = 0$. Indeed, when $\mu_0 \neq 0$, we can base the test on the observations $X_1 - \mu_0, \ldots, X_n - \mu_0$ with expectation $\nu = \mu - \mu_0$. The new, but equivalent, hypotheses become $H_0: \nu \leq 0$ against $H_1: \nu > 0$.

Suppose that ψ is an unbiased (randomized) test that is admissible at level α_0. The absence of bias implies that the power function $(\mu, \sigma^2) \mapsto E_{\mu,\sigma^2} \psi(X)$ is at least α_0 on the set $\{(\mu, \sigma^2): \mu > 0, \sigma^2 > 0\}$, while the admissibility implies that this function is at most α_0 on the set of parameters $\{(\mu, \sigma^2): \mu \leq 0, \sigma^2 > 0\}$. By the continuity of the normal distribution in its parameters, the power function is automatically continuous in (μ, σ^2), and on the boundary between these two parameter regions, that is, on the set of parameters $\{(0, \sigma^2): \sigma^2 > 0\}$, the power function is exactly equal to α_0. We conclude that for every $\sigma^2 > 0$,

$$(6.2) \qquad \alpha_0 = E_{0,\sigma^2} \psi(X) = E_{0,\sigma^2} E_0 \big(\psi(X) \big| \, \overline{X^2} \big).$$

The family of probability distributions of $X = (X_1, \ldots, X_n)$, where the X_i are independent and $N(0, \sigma^2)$-distributed with $\sigma^2 > 0$ is a one-dimensional exponential family, with sufficient and complete variable $\overline{X^2}$. The sufficiency justifies that we have indexed the conditional expectation $E_0 \big(\psi(X) \big| \, \overline{X^2} \big)$ by the parameter $\mu = 0$ only, because the conditional distribution of X given $\overline{X^2}$ is independent of σ^2.

It follows from (6.2) that $E_{0,\sigma^2} E_0 \big(\psi(X) - \alpha_0 \big| \, \overline{X^2} \big) = 0$ for all $\sigma^2 > 0$. The completeness of $\overline{X^2}$ now implies $P \big(E_0(\psi(X) - \alpha_0 | \, \overline{X^2}) = 0 \big) = 1$, that is, for almost all y,

$$(6.3) \qquad E_0 \big(\psi(X) \big| \, \overline{X^2} = y \big) = \alpha_0.$$

In other words, the test ψ is necessarily a test of size α_0 for testing $H_0: \mu = 0$ against $H_1: \mu > 0$ based on an observation X from the conditional distribution of X given $\overline{X^2} = y$, for every y.

Now, consider a fixed y and a fixed parameter (μ, σ^2) with $\mu > 0$ from the alternative hypothesis, and consider the problem of finding a test that satisfies (6.3) and maximizes the conditional power function $E_{\mu,\sigma^2} \big(\psi(X) \big| \, \overline{X^2} = y \big)$ with respect to (μ, σ^2). At first glance, this test will depend on the chosen y and (μ, σ^2), but we will show that this is not the case. Since

$$(6.4) \qquad E_{\mu,\sigma^2} \psi(X) = \int E_{\mu,\sigma^2} \big(\psi(X) \big| \, \overline{X^2} = y \big) \, dP_{\mu,\sigma^2}^{\overline{X^2}}(y)$$

with $P_{\mu,\sigma^2}^{\overline{X^2}}$ the distribution function of $\overline{X^2}$ under (μ, σ^2), the resulting test ψ automatically also maximizes (6.4) over the class of tests that satisfy (6.3). Since all unbiased tests satisfy (6.3), the resulting test is uniformly most powerful among the unbiased tests.

To conclude the proof of the theorem, it therefore suffices to show that for every y, the test that maximizes $E_{\mu,\sigma^2}\left(\psi(X)|\,\overline{X^2} = y\right)$ among all tests that satisfy (6.3) does not depend on y and is exactly the t-test. We will use Theorem 6.44.

Since the pair $(\overline{X}, \overline{X^2})$ is sufficient, we may assume without loss of generality that the test ψ depends only on this pair. For a fixed value of $\overline{X^2} = y$, the test ψ is then a function of only $\overline{X}$. We now show that the conditional distribution of $\overline{X}$ given $\overline{X^2} = y$ takes on the form of an exponential family of probability distributions.

By Theorem 4.29, the variables $\overline{X}$ and S_X^2 are independent and continuously distributed. We can find the probability density of the pair $(\overline{X}, \overline{X^2}) = (\overline{X}, (n-1)S_X^2/n + \overline{X}^2)$ using the transformation theorem from probability theory, as follows:

$$p_{\mu,\sigma^2}^{(\overline{X},\overline{X^2})}(x,y) = p_{\mu,\sigma^2}^{(\overline{X},(n-1)S_X^2/n)}(x, y - x^2) = \phi\left(\frac{x-\mu}{\sigma/\sqrt{n}}\right)p_{\sigma}^{(n-1)S_X^2/n}(y - x^2).$$

As a function of x, for fixed y, this expression is proportional to the conditional density of $\overline{X}$ given $\overline{X^2} = y$. Only the first of the two factors of the product on the right-hand side contains the parameter μ, and this factor can be written as $\exp(n\mu x/\sigma^2)\exp(-n\mu^2/2\sigma^2)$ times a function that does not depend on μ. We conclude that for fixed y and σ^2, the family of conditional distributions of $\overline{X}$ given $\overline{X^2} = y$ with parameter μ is a one-dimensional exponential family, with sufficient quantity $\overline{X}$ and natural parameter $Q(\mu) = n\mu/\sigma^2$.

By Theorem 6.44, the uniformly most powerful test rejects H_0 for testing $H_0\colon \mu \leq 0$ against $H_1\colon \mu > 0$ in this exponential family for values of $\overline{X}$ greater than a certain critical value $c_{\alpha_0}(y, \sigma^2)$ that can depend on y and σ^2 in the current set-up. Since the function $x \mapsto x/\sqrt{y - x^2}$ is monotonically increasing in x (on the interval $[-\sqrt{y}, \sqrt{y}]$), this test is equivalent to rejecting for large values of $\sqrt{n}\overline{X}/S_X = \sqrt{n}\overline{X}/(y - \overline{X}^2)^{1/2}$ if $\overline{X^2} = y$. The critical value should be chosen such that the size of the test is equal to α_0. In the current situation, this means that H_0 is rejected when $\sqrt{n}\overline{X}/S_X \geq d_{\alpha_0}(y, \sigma^2)$ for the critical value $d_{\alpha_0}(y, \sigma^2)$ that is determined by the equation

$$P_{0,\sigma^2}\left(\sqrt{n}\overline{X}/S_X \geq d_{\alpha_0}(y,\sigma^2)|\,\overline{X^2} = y\right) = \alpha_0.$$

Finally, we prove that $d_{\alpha_0}(y, \sigma^2) = t_{n-1,1-\alpha_0}$.

Since under $\mu = 0$, the quantity $\sqrt{n}\overline{X}/S_X$ has a t_{n-1}-distribution, it suffices to prove that $\overline{X}/S_X$ and $\overline{X^2}$ are independent. This is a consequence of Basu's theorem. ∎

Theorem 6.51 Basu's theorem

If $V = V(X)$ is sufficient and complete and $T = T(X)$ is a statistic whose distribution does not depend on the parameter, then V and T are stochastically independent.

Proof. For every event B, the probability $P(T \in B) = E_\theta P(T \in B | V)$ is independent of the parameter. By the sufficiency of V, the conditional probability $P(T \in B | V)$ is also independent of the parameter. The completeness of V then implies that $P(T \in B | V) = P(T \in B)$ with probability 1. It follows that T and V are independent. ∎

6.5 Summary

Let X be an observation with distribution P_θ and density p_θ that depend on an unknown parameter $\theta \in \Theta$.

- A statistic $V(X)$ is called *sufficient* for X if $V(X)$ contains all pertinent information from X about θ. A statistic $V(X)$ is sufficient if there exist functions g_θ and h such that p_θ factors as

$$p_\theta(x) = g_\theta(V(x))h(x) \qquad \text{for all } \theta, x.$$

- A statistic $V = V(X)$ is called *complete* if $E_\theta f(V) = 0$ for all $\theta \in \Theta$ can only hold for functions f with $P_\theta(f(V) = 0) = 1$ for all $\theta \in \Theta$.

Two optimality characteristics for estimators:

- An estimator $T(X)$ is called a *uniformly minimum-variance unbiased* (UMVU) estimator for $g(\theta)$ if it is unbiased and $\mathrm{var}_\theta\, T \le \mathrm{var}_\theta\, S$ for all θ and all other unbiased estimators for $g(\theta)$. If a UMVU estimator exists, it is the best estimator within the class of unbiased estimators. If an unbiased estimator $T = T(V)$ for $g(\theta)$ depends only on a sufficient and complete statistic $V(X)$, then T is automatically a UMVU estimator for $g(\theta)$.
- The *Cramér–Rao lower bound*: Under certain conditions, every unbiased estimator $T = T(X)$ for $g(\theta) \in \mathbb{R}$ satisfies

$$\mathrm{var}_\theta\, T \ge g'(\theta)^2 / I_\theta$$

with g' the derivative of g and I_θ the Fisher information of the full observation X. If the variance of an unbiased estimator $T = T(X)$ for $g(\theta)$ is equal to this lower bound, then T is automatically UMVU. Under certain conditions, maximum likelihood estimators are asymptotically unbiased with variance equal to the Cramér–Rao lower bound, and therefore asymptotically UMVU.

We say that a test is *uniformly most powerful* for testing $H_0: \theta \in \Theta_0$ against $H_1: \theta \in \Theta_1$ if in all $\theta \in \Theta_1$, the value of the power function is maximal among the values of all power functions of all tests of the same level.

Optimal tests for simple and composite hypotheses:

- *Neyman–Pearson lemma* for $H_0: \theta = \theta_0$ against $H_1: \theta = \theta_1$: if there exists a number c_{α_0} with $P_{\theta_0}(p_{\theta_1}(X)/p_{\theta_0}(X) \ge c_{\alpha_0}) = \alpha_0$, then the test with critical region $K = \{x: p_{\theta_1}(X)/p_{\theta_0}(X) \ge c_{\alpha_0}\}$ is the most powerful test of level α_0.
- For testing the hypothesis $H_0: \theta \le \theta_0$ against $H_1: \theta > \theta_0$ based on an observation from a family with *monotone likelihood ratio*, there exists a uniformly most powerful randomized test based on a sufficient statistic V for the family.

Exercises

1. Let $X_1, \ldots, X_n$ be a sample from the exponential distribution with unknown parameter $\lambda > 0$. Determine a sufficient random variable.

2. Let $X_1, \ldots, X_n$ be a sample from the Poisson distribution with unknown parameter $\theta > 0$. Determine a sufficient random variable.

3. Let $X_1, \ldots, X_n$ be a sample from a distribution with probability density $p_\theta(x) = \theta x^{\theta-1} 1_{(0,1)}(x)$, where $\theta > 0$ is an unknown parameter. Determine a sufficient random variable.

4. Let $X_1, \ldots, X_n$ be a sample from the uniform distribution on $[\theta_1, \theta_2]$, where $\theta = (\theta_1, \theta_2)$ is an unknown parameter. Show that $(X_{(1)}, X_{(n)})$ is a sufficient random vector.

5. Let $X_1, \ldots, X_n$ be a sample from the $N(\theta, \theta^2)$-distribution, where $\theta > 0$ is an unknown parameter. Determine a sufficient two-dimensional random vector.

6. Let $X_1, \ldots, X_n$ be a sample from a distribution with probability density $p_{\lambda,\mu}(x) = \lambda e^{-\lambda(x-\mu)} 1_{\{x > \mu\}}$, where $\lambda > 0$ and $\mu \in \mathbb{R}$ are unknown parameters. Determine a sufficient random vector.

7. Show that if V is sufficient, then the maximum likelihood estimator (based on X) depends only on V.

8. Show that if V is sufficient, then the Bayes estimator (based on X with respect to a given prior distribution) depends only on V.

9. Show that if V is sufficient, then the likelihood ratio statistic (based on X) depends only on V.

10. Let $X = (X_1, \ldots, X_n)$ be a sample from a distribution with density p_θ, with θ unknown. Let U be a sufficient and complete statistic, and suppose that there exists a minimally sufficient statistic T. Show that U is also minimally sufficient. [Hint: Give a proof by contradiction. Suppose that U is not minimally sufficient; then there exists a function ψ such that $P_\theta(\psi(U) \neq E[\psi(U)|T]) > 0$.]

11. Do the geometric distributions with parameter p form an exponential family?

12. Determine whether the multinomial probability distributions with parameters n and p form an exponential family for fixed n.

13. Let $X_1, \ldots, X_n$ be a sample from the exponential distribution with parameter λ. Determine a UMVU estimator for $1/\lambda$.

14. Determine a UMVU estimator for p^2 based on a bin(n, p)-observation X ($n \geq 2, 0 \leq p \leq 1$).

15. Determine a UMVU estimator for μ^2 based on a sample $X_1, \ldots, X_n$ from the $N(\mu, \sigma^2)$-distribution.

16. Let $X_1, \ldots, X_n$ be a sample from the Poisson distribution with parameter θ. Determine a UMVU estimator for θ^2.

17. Let $X_1, \ldots, X_n$ be a sample from the probability distribution with density

$$p_\theta(x) = \theta x^{-2} 1_{\{x > \theta\}},$$

where $\theta > 0$ is an unknown parameter.
   (i) Determine a sufficient and complete statistic.
   (ii) Determine a UMVU estimator for θ.

18. Let $X_1, \ldots, X_n$ be a sample from the uniform distribution on $[0, \theta]$. Determine a UMVU estimator for θ^2.

19. Let $X_1, \ldots, X_n$ be a sample from the shifted exponential distribution with probability density

$$p_{\mu,\lambda}(x) = \frac{1}{\lambda} \exp\left(-\frac{x - \mu}{\lambda}\right) 1_{\{x \geq \mu\}},$$

with $\lambda > 0$ and $\mu \in (-\infty, \infty)$ unknown. The function $x \mapsto 1_{\{x \geq \mu\}}$ is equal to 1 for $x \geq \mu$ and 0 for $x < \mu$.
   (i) Determine a sufficient two-dimensional statistic for (μ, λ).
   (ii) Determine a UMVU estimator for λ under the assumption $\mu = 1$.

20. Let $X_1, \ldots, X_n$ be a sample from the beta distribution with density

$$p_{\alpha,\beta}(x) = B(\alpha, \beta)^{-1} x^{\alpha-1} (1 - x)^{\beta-1} 1_{\{0 < x < 1\}}.$$

   (i) Do the probability distributions of $X = (X_1, \ldots, X_n)$ form an exponential family?
   (ii) Determine a sufficient and complete statistic.
   (iii) Determine a UMVU estimator for $\mathrm{E}_{\alpha,\beta} \log X_1$.

21. Let $X_1, \ldots, X_m$ and $Y_1, \ldots, Y_n$ be independent samples from the Bernoulli distributions with parameters p_1 and p_2, respectively, where p_1 and p_2 are unknown parameters in $[0, 1]$. Determine a UMVU estimator for $p_1 - p_2$.

22. Let $X_1, \ldots, X_n$ be a sample from the $N(\mu, \sigma^2)$-distribution, with σ^2 known and $\mu \in \mathbb{R}$ unknown.
   (i) Show that $\overline{X}$ is sufficient and complete.
   (ii) Show that $(\overline{X}, S_X^2)$ is not complete.

23. Let $X_1, \ldots, X_n$ be a sample from the U$[-\theta, \theta]$-distribution.
   (i) Show that $(X_{(1)}, X_{(n)})$ is sufficient.
   (ii) Show that $(X_{(1)}, X_{(n)})$ is not complete.
   (iii) Determine whether $(X_{(1)}, X_{(n)})$ is minimally sufficient.

24. Let $X_1, \ldots, X_n$ be a sample from the probability distribution with $\mathrm{P}_\theta(X_i = x) = 2^{-x/\theta}$ for $x = \theta, \theta + 1, \theta + 2, \ldots$, where $\theta > 0$ is an unknown parameter.
   (i) Determine a sufficient statistic.
   (ii) Determine whether the probability distributions of $(X_1, \ldots, X_n)$ form an exponential family.

25. Let $X_1, \ldots, X_m$ be a sample from the $N(\mu, \sigma^2)$-distribution, and let $Y_1, \ldots, Y_n$ be a sample from the $N(\mu, \tau^2)$-distribution, where μ, σ^2, and τ^2 are unknown. Suppose that the two samples are independent.

(i) Show that for every $\alpha \in \mathbb{R}$, the estimator $\alpha \overline{X} + (1 - \alpha)\overline{Y}$ is an unbiased estimator for μ.

(ii) The vector $(\overline{X}, \overline{Y}, S_X^2, S_Y^2)$ is sufficient and $\frac{1}{2}\overline{X} + \frac{1}{2}\overline{Y}$ is an unbiased estimator for μ. Does it follow from Theorems 6.18 and 6.21 that this estimator is UMVU?

(iii) Determine the α minimizing the variance of the given estimator.

(iv) Is the situation different if we assume beforehand that σ^2 and τ^2 are equal?

26. Determine the Cramér–Rao lower bound for the variance of unbiased estimators of θ based on a sample from the Poisson(θ)-distribution. Is the bound sharp?

27. Let $Y_1, \ldots, Y_n$ be independent and suppose that Y_i has an $N(x_i\theta, 1)$-distribution, for known constants $x_1, \ldots, x_n$.

(i) Determine the Fisher information for θ in Y_i.

(ii) Determine the Fisher information for θ in $(Y_1, \ldots, Y_n)$.

(iii) Determine the Cramér–Rao lower bound for estimating θ.

(iv) Is this lower bound sharp?

28. Let $X_1, \ldots, X_n$ be a sample from the $N(\theta, \theta)$-distribution, with $\theta > 0$ unknown. Determine the Cramér–Rao lower bound for estimating $g(\theta) = \sqrt{\theta}$.

29. Let $X_1, \ldots, X_n$ be a sample from the exponential distribution with unknown parameter $\lambda > 0$.

(i) Determine the Cramér–Rao lower bound for estimating $g(\lambda) = 1/\lambda$.

(ii) Show that, in this case, the lower bound is sharp.

30. Let $X_1, \ldots, X_n$ be a sample from the gamma distribution with parameters k and λ, where k is known and $\lambda > 0$ is unknown.

(i) Determine the Cramér–Rao lower bound for estimating $g(\lambda) = 1/\lambda$.

(ii) Show that, in this case, the lower bound is sharp.

31. Let $X_1, \ldots, X_n$ be a sample from a probability density that belongs to an exponential family. Show that probability distributions of $X = (X_1, \ldots, X_n)$ also belong to an exponential family.

32. Let $X_1, \ldots, X_n$ be a sample from a probability distribution with density

$$p_\theta(x) = \theta \exp\left(x - \theta(e^x - 1)\right),$$

for $x > 0$ and 0 elsewhere, with θ an unknown parameter.

(i) Determine a most powerful test for $H_0: \theta = 1$ against $H_1: \theta = 2$ at level $\alpha_0 = 0.05$.

(ii) Determine a most powerful test for $H_0: \theta = 1$ against $H_1: \theta = 3$ at level $\alpha_0 = 0.05$

33. Let $X_1, \ldots, X_n$ be a sample from a probability distribution with density

$$p_\theta(x) = 2\theta^2 x^{-3} 1_{x > \theta},$$

where $\theta > 0$ is an unknown parameter.

(i) Determine a most powerful test for $H_0: \theta = 1$ against $H_1: \theta = 1/2$ at level $\alpha_0 = 0.05$.

(ii) Determine a most powerful test for $H_0: \theta = 1$ against $H_1: \theta = 2$ at level $\alpha_0 = 0.05$.

34. Let $X_1, \ldots, X_n$ be a sample from a discrete probability distribution with probability density $p_\theta(x) = 1/\theta$ when $x \in \{1, 2, \ldots, \theta\}$, where $\theta \in \mathbb{N}$.

(i) Determine a most powerful test for $H_0: \theta = 2$ against $H_1: \theta = 3$ at level $\alpha_0 = 0.05$.

(ii) Determine a uniformly most powerful test for $H_0: \theta = 2$ against $H_1: \theta > 2$ at level $\alpha_0 = 0.05$.

35. Let $X_1, \ldots, X_m$ be $N(\mu, 1)$-distributed, and let $Y_1, \ldots, Y_n$ be $N(\nu, 1)$-distributed. Suppose that the random variables are all independent. Let $\mu_1 > \nu_1$, and let $\xi_0 = m\mu_1/(m+n) + n\nu_1/(m+n)$.
  (i) Determine a most powerful test for the null hypothesis $H_0: \mu = \nu = \xi_0$ against $H_1: \mu = \mu_1, \nu = \nu_1$ at level α_0.
  (ii) Determine a uniformly most powerful test for the null hypothesis $H_0: \mu \leq \nu$ against $H_1: \mu > \nu$ at level α_0.

36. Let X be hypergeometrically distributed with parameters m, r, and N. We want to test $H_0: r \leq r_0$ against $H_1: r > r_0$. Set $p_r(x) = P(X = x)$.
  (i) Show that for $r_1 > r_0$, the quotient p_{r_1}/p_{r_0} is an increasing function of x.
  (ii) Determine a uniformly most powerful test for testing H_0 against H_1.

37. Let $X_1, \ldots, X_n$ be the incomes of n randomly chosen persons from a certain population. Suppose that X_i has a Pareto distribution, that is, X_i has probability density

$$p_\theta(x) = c^\theta \theta x^{-(1+\theta)} 1_{\{x > c\}},$$

where $\theta > 1$ and $c > 0$. We assume that c is known and θ is unknown.
  (i) Express the expectation μ of X_i in θ (and c).
  (ii) Determine a uniformly most powerful test for $H_0: \mu \leq \mu_0$ against $H_1: \mu > \mu_0$ at level α_0.

Figure 6.1 shows the water flow (m^3/s) in the river the Meuse near the town of Borgharen (Netherlands, in the province of Limburg) during 15 consecutive days in December 1965. In the 20th century, the water flow exceeded 1250 m^3/s a total of 70 times, and each time, the pattern of the water flow over time was known (see Figure 6.1). The form of the extreme peaks is important for the consequences of the high water flow. Prolonged high water flow means, for example, an extended exposure of the dikes to high water, resulting in saturation and a higher probability of a breach or flooding. We will, however, restrict ourselves to analyzing the maxima of the waves. The maximum of the wave in Figure 6.1 is 1892 m (recall that 1 m $\approx$ 3.28 feet).

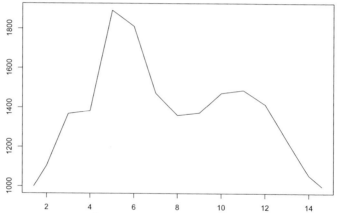

Figure 6.1. Water flow (in m^3/s) in Borgharen on 15 consecutive days in December 1965.

The 70 observed maxima are shown in chronological order in Figure 1.4, and Figure 6.2 gives a histogram of the maximal water flow.[♯] The histogram shows several extremely high values. As follows from Example 1.6, we are greatly interested in the probability of more extreme maxima occurring.

To have a framework for the analysis, we will take the working hypothesis that the 70 observed maximal water flows can be viewed as realizations of independent, identically distributed random variables. This working hypothesis is, of course, debatable. However, since the high water flows occur at separate times, often in different years, the independence of the maxima is not unreasonable. A certain trend over time with, for example, a slowly changing distribution for the maxima, can, however, not be excluded. Think of a climate effect, but more importantly of the effect of the increasing canalization and construction along the Meuse that have influenced the course of the river. We can study a time effect in the data up to a certain point. We will return to this later.

[♯] The data can be found on the book's webpage at http://www.aup.nl under maxflows and flows1965.

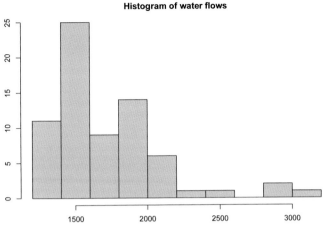

Figure 6.2. Maximal water flow over $1250 \ \mathrm{m}^3/\mathrm{s}$ in Borgharen in the twentieth century.

To determine a suitable probability distribution for the maximal water flow, we can use a theoretical result from probability theory as a starting point. This somewhat surprising theorem gives an approximation for the distribution of a maximum $\max(Y_1, \ldots, Y_m)$ of a large number of independent, identically distributed random variables Y_i. Since each of the 70 maximal water flows is the maximum of a pattern of high water flows as in Figure 6.1, it is not entirely unreasonable to view the maximal water flows as maxima of less extreme water flows.

Theorem 6.52

Suppose that for certain numbers a_m and b_m and independent, identically distributed random variables $Y_1, Y_2, \ldots$, we have that for some distribution function G,

$$\lim_{m \to \infty} \mathrm{P}\big(a_m\big(\max(Y_1, \ldots, Y_m)\big) - b_m\big) \leq x\big) = G(x), \qquad x \in \mathbb{R}.$$

Then G belongs to the location-scale family of one of the following three types of distributions:
  (i)   Gumbel: $G(x) = e^{-e^{-x}}$
  (ii) Fréchet: $G(x) = e^{-(1/x^\alpha)} 1_{\{x > 0\}}$, for $\alpha > 0$
  (iii) Negative Weibull: $G(x) = e^{-(-x)^\alpha} 1_{\{x < 0\}} + 1_{\{x \geq 0\}}$, for $\alpha > 0$

This theorem gives a mathematical limit result and is certainly not conclusive evidence that the maximal water flows must be generated from one of the given distributions. We can, however, use the theorem as motivation to study the fit of the three types of distributions for the flows.

The three families of extreme value distributions can formally be viewed as one family with a parameter $\xi \in \mathbb{R}$. The distribution function of this family is

$$G_{\mu, \sigma, \xi}(x) = e^{-(1 + \xi(x - \mu)/\sigma)^{-1/\xi}} 1_{1 + \xi(x - \mu)/\sigma > 0}.$$

The limit as $\xi \to 0$ corresponds to the Gumbel distribution, $\xi > 0$ corresponds to the Fréchet-distributions, and $\xi < 0$ corresponds to the negative Weibull distributions. The parameter α in the latter two cases corresponds to $1/\xi$.

In addition to the unknown location and scale parameters, the parameter α for the second and third families is also unknown. To study the fit of one of these families using a QQ-plot, we would therefore need to make a QQ-plot for every value of α. Figure 6.3 shows several of these QQ-plots. It is clear from these figures that the negative Weibull distributions (type (iii)) do not fit well. A Gumbel distribution (type (i)) or a Fréchet-distribution with large values of α (in the range from 4 tot 10) does seem to fit the data reasonably well. The lower row of Figure 6.3 gives the QQ-plots against standard (nonextreme values) distributions. We can conclude that an exponential or other gamma distribution with a small shape parameter need not be excluded beforehand. Here, we choose a Fréchet-distribution.

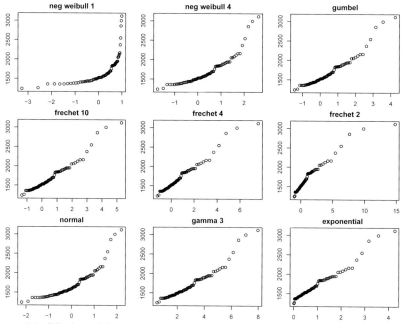

Figure 6.3. QQ-plots of the maximal water flows against a selection of distributions.

A next step in the analysis is to estimate the unknown parameters. The Fréchet family has three unknown parameters, namely the shape parameter α, location, and scale. We can estimate these three parameters using the maximum likelihood method, under the assumption that the maxima are independent. The likelihood function for the Fréchet family with location parameter μ and scale parameter σ is

$$(\mu, \sigma, \alpha) \mapsto \prod_{i=1}^{n} \frac{\sigma^{\alpha}\alpha}{(X_i - \mu)^{\alpha+1}} e^{-((X_i - \mu)/\sigma)^{-\alpha}} 1_{\{X_{(1)} > \mu\}}.$$

Determining the point of maximum of this function requires a numerical optimiza-
tion method, such as the Newton–Raphson (or Fisher-scoring) method. The results
are given in Table 6.1.

```
Estimates
     loc       scale      shape
 1530.4691   214.7064    0.2539


Standard Errors
     loc    scale    shape
 29.4027  24.2129   0.1067


Covariance
                 [,1]         [,2]          [,3]
 loc     864.5207545 436.881289 -0.99632985
 scale   436.8812890 586.265220 -0.17045895
 shape    -0.9963299  -0.170459  0.01139458
```

Table 6.1. R-output containing the parameter estimates for fitting an extreme value distribution to the maximal water flows. The parameters loc, scale, and shape are, respectively, equal to μ, σ, and $1/\alpha$. In addition to the estimated parameter values, under Estimates, we also give the standard errors for the estimates, under Standard Errors, and the estimated covariance matrix of the estimators, under Covariance.

We interpret the standard error 0.11 for the estimate $\hat{\xi} = 0.25$ as defining an
approximate confidence interval $\xi = 0.25 \pm 2 * 0.11$. The Gumbel distribution with
$\xi = 0$ and the negative Weibull distributions with $\xi < 0$ do not seem to qualify.
Given the estimate $\hat{\xi} = 1/\hat{\alpha} = 0.25$, we can verify our interpretation of the QQ-
plots in Figure 6.3. Figure 6.4 gives QQ-plots of several samples of size 70 from the
Fréchet-distribution with shape parameter $\xi = 0.25$ together with the QQ-plot of the
data. This last QQ-plot is the same as "frechet 4" in Figure 6.3. Since the form of the
QQ-plot of the data does not deviate from that of the other QQ-plots, the assumption
that we have a Fréchet-distribution is certainly compatible with the QQ-plots. We
could further support the assumption of a Fréchet-distribution using a goodness-of-fit
test.
Suppose that we are interested in the threshold h such that the probability of a
maximal water flow X greater than or equal to h is equal to p. If X has a Fréchet-
distribution, then this leads to the equation $1 - \exp\left(-((h - \mu)/\sigma)^{-\alpha}\right) = p$, that
is,

$$h = \mu + \frac{\sigma}{(-\log(1-p))^{\xi}}.$$

The maximum likelihood estimator for h is obtained by replacing the unknown μ, σ,
and ξ by their maximum likelihood estimators. For $p = 0.0001$, for example, this
gives $\hat{h} = 3757$, a value that (naturally) lies far above the measured maxima. The

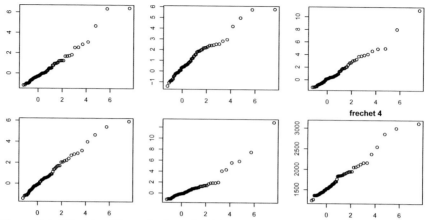

Figure 6.4. QQ-plots of 5 samples from the Fréchet-distribution with shape parameter $\xi = 0.25$ and the maximal water flows (lower right) against the quantiles of the Fréchet-distribution with shape parameter $\xi = 0.25$.

assumption that the distribution is Fréchet plays an essential role in the extrapolation of the data to much more extreme values. The standard error of the estimator for h can be approximated using the numbers in Table 6.1, using the delta method. (An alternative is to compute the profile likelihood for h.) The difference $\hat{h} - h$ is then approximated linearly as a function of the differences $\hat{\mu} - \mu$, $\hat{\sigma} - \sigma$, and $\hat{\xi} - \xi$, that is,

$$\hat{h} - h \approx \hat{\mu} - \mu + \frac{1}{(-\log(1-p))^{\xi}}(\hat{\sigma} - \sigma) - \frac{\sigma \log(-\log(1-p))}{(-\log(1-p))^{\xi}}(\hat{\xi} - \xi).$$

The constants by which we multiply the three differences are the partial derivatives of h viewed as a function $h = h(\mu, \sigma, \xi)$ of the three parameters. We now compute an approximation for the standard error of $\hat{h}$ by expressing the variance of the left-hand side in the covariances of the differences on the right-hand side, which are (approximately) given in the output $varcov *of Table 6.1. We replace the remaining unknown values of ξ and σ in the multiplicative constants by their estimates $\hat{\xi}$ and $\hat{\sigma}$. This gives the standard error 2180, and therefore a confidence interval of the form $h = 3757 \pm 1.96 * 2180$ for h. The interval is extremely long, which indicates that it is very difficult to extrapolate reliably that far into the future.*

The interpretation of this interval is that if the assumption that we have a Fréchet-distribution is correct and we were to repeat the entire experiment of measuring water flows under the same circumstances 100 times and compute the confidence interval the way described above every time, then approximately 95 out of the 100 intervals would contain the desired threshold h. In this case, "repeating" is purely a thought experiment. The restriction that "the assumption of a Fréchet-distribution is correct" is important, because the confidence region does not give any control over a possible systematic error in our analysis.

There are, unfortunately, arguments in favor of such a systematic error. By the "correctness" of the Fréchet-distribution, we mean that the maximal water flows can be viewed as a sample from an extreme value distribution. Based on our earlier analysis, the extreme value assumption is not unreasonable, provided that we can indeed view the data as a (random) sample from a distribution. In particular, there should not exist any time dependence in the data, a dependence that is certainly conceivable for this type of data.

We are thinking, in particular, of a trend over time. Stochastic dependence between consecutive years seems less probable. On the one hand, the 70 maxima are relatively uniformly distributed over the century. On the other hand, the plot of the sample autocorrelation function also does not suggest any dependence (see Figure 6.5).

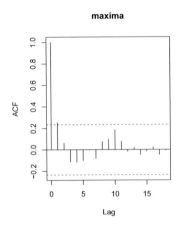

Figure 6.5. Sample autocorrelation of the maximal water flow.

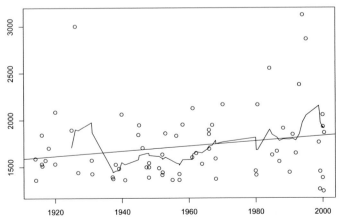

Figure 6.6. Maximal water flow plotted as a function of time, together with the best-fitting linear function and a moving average (averaged over periods of 10 years).

255

A trend in the data is suggested in Figure 6.6, where the maximal flows are plotted over time, with both the best-fitting (least squares) linear function and the moving average (averaged over periods of 10 years). For each year, the moving average gives the average over the previous 10 years. The maxima seem to have increased somewhat in the course of the century. We can study this hypothesis using, for example, a trend test, whose test statistic is equal to the number of indices i such that the maximum at time i is greater than the maximum at time $i-1$. In case of a strictly increasing trend, this number would be equal to 69, while for a random sample, the number would be equal to the number of increases in a random permutation of the indices $1, \ldots, 70$. The observed number of increases 31 does not confirm the impression that there is a trend. This number is even very small (a right p-value of approximately 95%).

Another possible test, with a greater power function, is a two-sample test with the first half of the maxima as first sample and the second half as second sample. Since the maxima are clearly not normally distributed, we use the Wilcoxon test. The two-sided test gives a p-value of 4%, which confirms the belief that the maxima may change over time. The difference between the medians of the two samples is 158. Boxplots of the two half samples confirm the image of the shift (Figure 6.7), although there is a considerable overlap between the two samples.

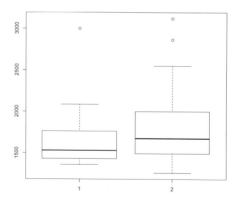

Figure 6.7. Boxplots of the first and last 35 maximal water flows.

If we must take into account a change over time, then it is also possible that the form of the distribution has changed over time. An empirical QQ-plot, where the order statistics of the first half of the observations are set out against the order statistics of the second half, gives an indication for a possible difference in distribution between the two halves. At first glance, Figure 6.8 does seem to show a difference. However, this figure is misleading, as shown by the simulated data (Figure 6.9). The QQ-plot in Figure 6.8 does not deviate essentially from the QQ-plots in Figure 6.9, for which the two samples (of size 35) are both simulated from the same Fréchet-distribution.

We can also apply the idea that the first and second halves of the maxima could differ in distribution within the context of the Fréchet model, by estimating

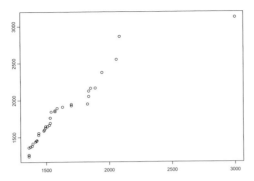

Figure 6.8. Empirical QQ-plot of the first 35 maximum water flows against the last 35 water flows.

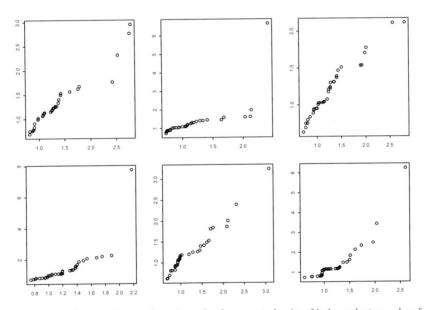

Figure 6.9. Empirical QQ-plots of 6 independently generated pairs of independent samples of size 35 from the Fréchet-distribution with parameter 0.25.

the parameters of this model separately for the two halves. The results are given in Tables 6.2 and 6.3. The estimates are quite different, but not all differences are statistically significant. It is important to note that the estimates are based on only 35 observations, and therefore have relatively large margins of error. The confidence intervals for μ, for example, are $\mu_1 = 1472 \pm 1.96 * 26$ and $\mu_2 = 1607 \pm 1.96 * 57$ and overlap.

To see how possible differences influence the estimate of the threshold h, we can estimate this threshold separately based on the first and second half of the data, using the same method as that applied to the full data. This gives two confidence intervals,

$14\,686 \pm 1.96 * 26\,539$ *and* $2505 \pm 1.96 * 1232$. *The interval based on the first half of the data has a width of approximately* $100\,000$ *and is therefore extremely inaccurate. So we should not put too much weight on the differences between the two point estimates for* h *themselves and with the estimate based on the full data.*

```
Estimates
       loc        scale       shape
   1472.4676    125.5073     0.5056


Standard Errors
      loc     scale     shape
   26.1394   24.7969   0.2195
```

Table 6.2. R-output containing the parameter estimates and standard errors for fitting an extreme value distribution to the first 35 maximal flows.

```
Estimates
       loc        scale       shape
   1606.7206    292.2234     0.1219


Standard Errors
      loc     scale     shape
   57.2150   44.5610   0.1542
```

Table 6.3. R-output containing the parameter estimates and standard errors for fitting an extreme value distribution to the last 35 maximal flows.

7 Regression Models

7.1 Introduction

In contemporary usage, the word *regression* has a negative connotation, even though in statistics it is the standard name for explaining a variable Y using a variable X. A *dependent variable* Y is "regressed on" a *predictor variable* X, also called *independent variable* in this context. Here are some of the many applications:

- predicting the yield of a biochemical process as a function of temperature, amount of catalyst, etc.
- predicting the price of property as a function of size, location, available infrastructure, etc.
- predicting the remaining life span as a function of age, gender, risk of a medical procedure, health indicators, etc.
- predicting the response to a mailing as a function of postal code, education, income, etc.
- predicting the national product from macro-economic variables such as the labor force, interest rates, national deficit, inflation, etc.
- predicting the length of university studies toward a master's degree as a function of the average grade in the last year of high school, choice of major, etc.
- predicting the final adult height of a child based on the heights of the parents and the gender of the child
- predicting the value of shares over 10 days based on the value today, yesterday, etc.

Since the independent variable X can influence the dependent variable Y in many ways and the probability distribution of Y will not be the same for every application, there exist different types of regression models; Sections 7.2, 7.3, and 7.4 treat the most common regression models, where the variable Y is a real-valued random variable. Sections 7.5 and 7.6, on the other hand, describe specific models for application in classification (Y is a 0-1 variable) and life span (Y is a life span).

All applications mentioned above have in common that there does not exist a perfect correlation between the variables X and Y, although we do expect there to be a correlation. For example, we expect that the size and age of a property, its location, and possibly other factors, will influence the price of the property, but, in general, it will not be possible to predict the price exactly from a number of such indicators. This could be due to a lack of information (some relevant variables are still unknown) or to random factors. In both cases, it is not unreasonable to view (x, y) as a realization of a random vector (X, Y). We can then study the correlation between x and y using a probability distribution of the vector (X, Y). We will be interested mostly in the conditional distribution of Y given $X = x$, and to a lesser degree in the marginal distribution of X.

In some cases, we can control the value of the predictor variable X. For example, in determining the optimal production conditions, the different settings x are varied systematically, after which the yield y is analyzed. In such a case, it is not reasonable to view the predictor variable as a realization of a random variable; instead, we view only the dependent variable Y as a random variable. When X is random, on the other hand, we often model the conditional distribution of Y given $X = x$. Therefore, it does not make much practical difference for the different regression models whether we assume X to be random or not. In the next sections, we will indicate each time whether X is assumed to be random. We view the available data $(x_1, y_1), \ldots, (x_n, y_n)$ as either realizations of the random vector (X, Y) or as realizations of the random variable Y in combination with the measured nonrandom observation $(x_1, \ldots, x_n)$.

In the application at the end of this chapter, we will discuss in detail the concept of causality. Because causality is an important subject in the context of regression, let us already discuss two short examples to illustrate this concept. The first example deals with the myth that babies are delivered by storks. In some regions, a positive correlation has been observed in the fluctuations of the stork population and the birth rate; in periods when the stork population shrank, the birth rate dropped, and at times when the stork population grew, the birth rate also rose. This is a remarkable result since we are convinced, and have been for some time already, that babies are not delivered by storks. Despite the correlation, it cannot be expected that if in these regions, the size of the stork population is increased artificially, for example by setting out additional storks, the birth rate will rise. The second example concerns the positive correlation between income and spending: people who earn more in general also spend more. In this case, it is often true that if a person earns more, their spending also increases. What is the difference between the two examples? In both cases, we can say that the predictor variable (number of storks and income) has a predicting value for the dependent variable (birth rate and spending). However, in the stork example, we cannot say that there is a causal correlation, while in the income example there is one: artificially increasing the number of storks will not affect the birth rate, while such an effect is likely to be true in the income example. It is not entirely clear why the birth rate and size of the stork population have a positive correlation in some regions. Possibly both depend on the industrial development: more industry means, on the one hand, more wealth, which traditionally causes the birth rate to drop, and, on the other hand, air pollution and unrest in the region, causing storks to leave.

7.2 Linear Regression

The linear model, the basis for linear regression and analysis of variance, is the workhorse of "classical" statistics, on the one hand, because it is widely applicable (with a bit of sense) and, on the other hand, because the necessary computations are based on simple matrix algebra. Although modern numerical methods have facilitated the application of more flexible models, the linear model is still of great value.

The theory for the linear model is based on the multivariate normal distribution, discussed in Appendix B. In this section, we discuss linear regression models, and in the next, we deal with analysis of variance. The latter is, in fact, a special case of linear regression.

The standard *linear regression model* assumes that, given $X = x = (x_1, \ldots, x_p)$, the variable Y is normally distributed, with conditional expectation and variance

$$ \mathrm{E}(Y|X = x) = \sum_{i=1}^{p} \beta_i x_i, \qquad \mathrm{var}(Y|X = x) = \sigma^2, $$

where the latter is independent of x. When X is not random, Y has this expectation and variance (unconditionally) for predictor variables $(x_1, \ldots, x_p)$. In the remainder of this section, we assume that the predictor variable X is not random, so that we can leave out the conditionality. The model has $p + 1$ real-valued parameters, which we can combine into the parameter vector $\theta = (\beta_1, \ldots, \beta_p, \sigma^2)$.

In the linear regression model, in addition to the observation Y, we also have the predictor variable x, which we use to model the expected value of the dependent variable Y. If we define the "measurement error" as $e = Y - \sum_{i=1}^{p} \beta_i x_i$, then we can write

$$ Y = \sum_{i=1}^{p} \beta_i x_i + e. $$

In the standard regression model, the measurement errors are independent and normally distributed with expectation 0 and variance σ^2. The variable Y then also has a normal distribution, but now with expectation $\sum_{i=1}^{p} \beta_i x_i$ and variance σ^2. We can view the linear regression model as an extension of the measurement error model of Example 1.3.

The linear regression model makes a number of specific assumptions:
- The expectation of Y depends on x, but the variance does not.
- The expectation of Y is a *linear* function of x.
- The measurement error is normally distributed.

In a surprisingly large number of applications, these conditions hold, but this is, of course, not always the case. We should note that the variables are often "preprocessed" into a form that is in line with the linear regression model. For example, the regression can be carried out after transforming the dependent variable Y (for example by taking the logarithm $\log Y$), and the predictor variables, in particular, can be transformed in many ways. When, for example, in the case of a one-dimensional variable x, we expect a polynomial relation between x and Y, we can carry out the regression with predictor variable $(1, x, x^2, \ldots, x^k)$ instead of x. When in the case of a two-dimensional variable $x = (x_1, x_2)$, we expect a quadratic joint correlation relation between the variables (x_1, x_2) and Y, we can use the vector $(1, x_1, x_2, x_1^2, x_2^2, x_1 x_2)$ in the model, etc. We see that the linearity of the linear regression model relates to the linearity in the regression parameters, and not so much in the predictor variables.

In many cases an *intercept* is added to the model. In practice, this corresponds to setting the first predictor variable x_1 equal to 1 and including the measured predictor variables in $x_2, \ldots, x_p$. A linear regression model with intercept has the following form:

$$Y = \beta_1 + \beta_2 x_2 + \ldots + \beta_p x_p + e.$$

The regression parameter β_1 is called the intercept; it is the expectation of Y when the regression parameters $\beta_2, \ldots, \beta_p$ are equal to 0.

It may be worth softening the assumptions of the regression model. Instead of the normality of the measurement errors, we could assume only that the measurement errors have expectation 0, and we could also model the variance of the measurement errors as a function of x. We will not discuss these models.

7.2.1 Simple Linear Regression

Suppose that the variable Y depends on a one-dimensional real-valued predictor variable x and that we have obtained n observations $(x_1, y_1), \ldots, (x_n, y_n)$. A scatter plot of the observations $(x_1, y_1), \ldots, (x_n, y_n)$ can provide insight into the correlation between x and Y. If this correlation seems to be linear, we can model the observations using a so-called simple linear regression model. The *simple linear regression model* with intercept is then described by

(7.1) $$Y_i = \alpha + \beta x_i + e_i, \qquad i = 1, \ldots, n,$$

where the "measurement errors" $e_1, \ldots, e_n$ are independent, $N(0, \sigma^2)$-distributed, unobservable random variables. Under this assumption, the variables $Y_1, \ldots, Y_n$ are also independent and normally distributed, where the variable Y_i has expectation $\alpha + \beta x_i$ and variance σ^2. We see that the observations are not identically distributed. If there were no measurement errors, then there would be an exact linear correlation between Y and x. We take the parameter space for the parameter $\theta = (\alpha, \beta, \sigma^2)$ as large as possible: $(\alpha, \beta) \in \mathbb{R}^2$ and $\sigma^2 > 0$.

7.2.1.1 Estimation

By estimating the parameters α and β, we can determine the (linear) correlation between Y and x.

Theorem 7.1

The maximum likelihood estimators for α, β, and σ^2 in the simple linear regression model (7.1) are equal to

$$\hat{\alpha} = \overline{Y} - \hat{\beta}\overline{x}, \qquad \hat{\beta} = \frac{s_Y}{s_x} r_{x,Y}, \qquad \hat{\sigma}^2 = \frac{1}{n}\sum_{i=1}^{n}(Y_i - \hat{\alpha} - \hat{\beta}x_i)^2,$$

where s_x, s_Y, and $r_{x,Y}$ are the sample standard deviation and the sample correlation (see Definitions 2.2 and 2.15).

Proof. The log-likelihood function for the model in (7.1) is

$$(\alpha, \beta, \sigma^2) \mapsto \log \prod_{i=1}^{n} \frac{1}{\sqrt{2\pi\sigma^2}} e^{-\frac{1}{2}(Y_i - \alpha - \beta x_i)^2/\sigma^2}$$

$$= -\tfrac{1}{2}n\log 2\pi - \tfrac{1}{2}n\log\sigma^2 - \frac{1}{2\sigma^2}\sum_{i=1}^{n}(Y_i - \alpha - \beta x_i)^2.$$

As in Example 3.14, maximizing the log-likelihood function is done in two steps. Maximizing this function with respect to (α, β) is equivalent to minimizing the quadratic form $\sum_{i=1}^{n}(Y_i - \alpha - \beta x_i)^2$ with respect to (α, β). Setting the partial derivatives of the sum with respect to α and β equal to 0 gives the system of equations

$$(7.2) \qquad \sum_{i=1}^{n}(Y_i - \hat{\alpha} - \hat{\beta}x_i) = 0, \qquad \sum_{i=1}^{n}(Y_i - \hat{\alpha} - \hat{\beta}x_i)x_i = 0.$$

After some computation, the estimator for α and β can be solved from this system:

$$\hat{\alpha} = \overline{Y} - \hat{\beta}\overline{x},$$

$$\hat{\beta} = \frac{\sum_{i=1}^{n} x_i(Y_i - \overline{Y})}{\sum_{i=1}^{n} x_i(x_i - \overline{x})} = \frac{\sum_{i=1}^{n}(x_i - \overline{x})(Y_i - \overline{Y})}{\sum_{i=1}^{n}(x_i - \overline{x})^2} = \frac{s_Y}{s_x} r_{x,Y}.$$

We can verify that this solution minimizes the quadratic form for every value of $\sigma^2 > 0$ by, for example, computing the Hessian matrix in $(\hat{\alpha}, \hat{\beta})$. Substituting the resulting values of $\hat{\alpha}$ and $\hat{\beta}$ in the log-likelihood function and maximizing this function with respect to σ^2 gives

$$\hat{\sigma}^2 = \frac{1}{n}\sum_{i=1}^{n}(Y_i - \hat{\alpha} - \hat{\beta}x_i)^2$$

as maximum likelihood estimator for σ^2. ∎

The estimators we found for α and β are so-called *least-squares estimators*, because we minimize the sum of squares $\sum_{i=1}^{n}(Y_i - \alpha - \beta x_i)^2$. Geometrically, this corresponds to minimizing the sum of the squares of the vertical distances from the measurements (x_i, Y_i) to the target regression line $y = \alpha + \beta x$, see Figure 7.1, whence the name. The least-squares estimators $\hat{\alpha}$ and $\hat{\beta}$ are unbiased estimators for α and β (this also holds when we do not assume the measurement errors to be normally distributed). Moreover, there exist simple expressions for the mean square errors (see Exercise 7.4 and Section 7.2.2).

The least-squares estimators $\hat{\alpha}$ and $\hat{\beta}$ are found by minimizing $\sum_{i=1}^{n}(Y_i - \alpha - \beta x_i)^2$, and therefore satisfy the likelihood equations in (7.2). More generally, we can find estimators for α and β by solving the equations

$$\sum_{i=1}^{n}\psi(Y_i - \hat{\alpha} - \hat{\beta}x_i)w(x_i) = 0, \qquad \sum_{i=1}^{n}\psi(Y_i - \hat{\alpha} - \hat{\beta}x_i)x_i w(x_i) = 0$$

for $(\hat{\alpha}, \hat{\beta})$, for suitable functions ψ and w. In general, this leads to different estimators. The role of the function ψ and the weights w is often to reduce the influence of possible extreme values of the residues $Y_i - \hat{\alpha} - \hat{\beta}x_i$ or the variables x_i, or to increase the efficiency of the estimators. This is called *robust regression*.

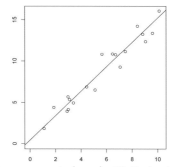

Figure 7.1. A collection of measurements (x_i, y_i) with the least-squares line.

The maximum likelihood estimator for σ^2 can be written as

$$\hat{\sigma}^2 = \frac{1}{n}\sum_{i=1}^{n}(Y_i - \hat{\alpha} - \hat{\beta}x_i)^2 = \frac{1}{n}\sum_{i=1}^{n}(Y_i - \overline{Y})^2 - \frac{1}{n}\hat{\beta}^2\sum_{i=1}^{n}(x_i - \overline{x})^2.$$

If we knew beforehand that $\beta = 0$, then Y_i would not depend on x_i, and the maximum likelihood estimator for σ^2 would be given by the first term on the right-hand side. In the current model, we do not know whether β is equal to 0, and the maximum likelihood estimator for σ^2 is smaller (unless $\hat{\beta} = 0$). Intuitively, this is reasonable: part of the variation in Y now follows from the variation in x, and therefore the sample variance of the Y_i is an overestimate of σ^2.

Definition 7.2 Residuals and sums of squares

The numbers $Y_i - \hat{\alpha} - \hat{\beta} x_i$ for $i = 1, \ldots, n$ are called the *residuals* of the regression of Y on x. The expressions

$$SS_{tot} = \sum_{i=1}^{n} (Y_i - \overline{Y})^2 \qquad \text{and} \qquad SS_{res} = \sum_{i=1}^{n} (Y_i - \hat{\alpha} - \hat{\beta} x_i)^2$$

are called, respectively, the *total sum of squares* and the *residual sum of squares* or, more completely, the "residual sum of squares after linear regression on x."

Definition 7.3 Coefficient of determination

The *coefficient of determination* is

$$1 - \frac{SS_{res}}{SS_{tot}}.$$

The total sum of squares SS_{tot} is the minimum of $\sum_{i=1}^{n} (Y_i - \alpha)^2$ over α, while the residual sum of squares SS_{res} is the minimum of $\sum_{i=1}^{n} (Y_i - \alpha - \beta x_i)^2$ over (α, β). The second minimum is, of course, smaller. If SS_{res} is approximately as large as SS_{tot}, then $SS_{tot} - SS_{res} = \hat{\beta}^2 \sum_{i=1}^{n} (x_i - \overline{x})^2$ is close to 0 (that is, $\hat{\beta}$ is approximately equal to 0 for the normalized x_i) and x_i has little predictive value for Y_i. The coefficient of determination gives the proportion of the variance explained by regression on x (called the *explained variance*) and can be written as

$$1 - \frac{SS_{res}}{SS_{tot}} = \hat{\beta}^2 \frac{\sum_{i=1}^{n} (x_i - \overline{x})^2}{\sum_{i=1}^{n} (Y_i - \overline{Y})^2} = r_{x,Y}^2.$$

When the coefficient of determination is almost equal to 1, this means that the points (x_i, y_i) lie approximately on a straight line. When the coefficient of determination is, for example, 0.2, either the points are widely distributed around a straight line or the linear regression model is not useful because the correlation between x and Y is strongly nonlinear. It is therefore not easy to interpret a coefficient of determination. Note that the scale of the coefficient is quadratic, which is difficult to justify. The coefficient of determination can, however, be viewed as a useful summary of the data and is a standard part of the report on a regression analysis.

The *fitted regression line* is $y = \hat{\alpha} + \hat{\beta} x$. This line can be used to predict the y-value for a certain x. After substituting the formulas for $\hat{\alpha}$ and $\hat{\beta}$, we can rewrite the line in the pleasant form

$$\frac{y - \overline{Y}}{s_Y} = r_{x,Y} \frac{x - \overline{x}}{s_x}.$$

265

Since $|r_{x,Y}| \leq 1$ by the Cauchy–Schwarz inequality, with strict inequality unless the measurement errors are exactly 0, this means that, measured in standard deviations, the predicted y is closer to $\overline{Y}$ than the corresponding x is to $\overline{x}$. We call this *regression to the mean*, in particular in the case where the standard deviations of the two variables are approximately equal. If x is the intelligence of a father and y is the intelligence of a son, then this was once used to deduce that humanity is becoming increasingly mediocre.

An explanation for "regression to the mean" is as follows. We can view an x-value of a randomly selected individual from the population (such as the intelligence of the father) as made up of a random and a systematic component. If the x-value is extremely high, then it is reasonable to assume that the random component has contributed to this relative size. When predicting the derived y-value, it is wise to take this into account: we predict that the y-value will lie at a relatively less extreme position in the population of y-values than that of the x-value in the population of x-values. This interpretation of "regression to the mean" within the setting of prediction is supported by Figure 7.2. This figure shows two regression lines. The dashed line seems to follow the point cloud the best, but the solid line is the least squares line. The slope of the dashed line is s_y/s_x, whereas the least squares line has the smaller slope $r_{x,y}s_y/s_x$. We can see that the least squares line is a better predictor by looking at the region between the two vertical lines. The least squares line divides the points in this strip (as with any other vertical strip) into approximately equal numbers, whereas the dotted line lies too high.

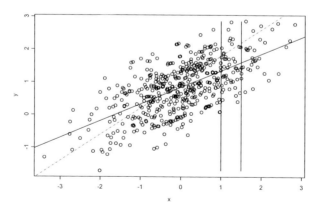

Figure 7.2. The least squares line and "regression to the mean."

7.2.1.2 Tests and Confidence Intervals

Since β is, in general, the most interesting parameter in a simple linear regression model, in this section we will deduce tests and confidence intervals for β. Tests and confidence intervals for the predictor parameter α can be deduced analogously. In this section, we describe the frequently used t-test and the likelihood ratio test.

To determine whether the predictor variable x has a linear influence on Y, we test the null hypothesis $H_0: \beta = 0$. The usual test statistic for the more general null hypothesis $H_0: \beta = \beta_0$ is

$$T = \frac{\hat{\beta} - \beta_0}{\sqrt{\widehat{\mathrm{var}}\hat{\beta}}},$$

where $\hat{\beta}$ is the maximum likelihood estimator for β and the variance of $\hat{\beta}$ is estimated by

$$\widehat{\mathrm{var}}\hat{\beta} = \frac{\frac{1}{n-2}\sum_{i=1}^{n}(Y_i - \hat{\alpha} - \hat{\beta}x_i)^2}{\sum_{i=1}^{n}(x_i - \overline{x})^2}.$$

This last estimator is obtained by writing $\hat{\beta}$ as

$$\hat{\beta} = \frac{\sum_{i=1}^{n}(x_i - \overline{x})(Y_i - \overline{Y})}{\sum_{i=1}^{n}(x_i - \overline{x})^2} = \frac{\sum_{i=1}^{n}(x_i - \overline{x})Y_i}{\sum_{i=1}^{n}(x_i - \overline{x})^2},$$

using that the variance of Y_i is equal to σ^2, and then estimating σ^2 using $\sum_{i=1}^{n}(Y_i - \hat{\alpha} - \hat{\beta}x_i)^2/(n-2)$. (Note that this estimator differs by a factor of $n/(n-2)$ from the maximum likelihood estimator for σ^2.) In Section 7.2.2.3, we prove in a more general model that the test statistic T under $H_0: \beta = \beta_0$ has a t_{n-2}-distribution: the number of degrees of freedom is equal to the number of observations minus the number of estimated regression coefficients. The null hypothesis is then rejected when $|T| \geq t_{n-2,1-\alpha_0/2}$, where α_0 is the level of the test. We can determine a confidence interval for β of size α_0 based on this t-test:

$$\beta = \hat{\beta} \pm t_{n-2,1-\alpha_0/2}\sqrt{\widehat{\mathrm{var}}\hat{\beta}}.$$

The hypothesis $H_0: \beta = 0$ can also be tested using the likelihood ratio test from Section 4.7. The numerator of the likelihood ratio statistic is the value of the likelihood function in the maximum likelihood estimator $(\hat{\alpha}, \hat{\beta}, \hat{\sigma}^2)$. In particular, we have $\hat{\sigma}^2 = SS_{res}/n$. In the denominator, the likelihood is maximized over the restricted parameter space with $\beta = 0$. Under the null hypothesis $H_0: \beta = 0$, the observations $Y_1, \ldots, Y_n$ are independent and $N(\alpha, \sigma^2)$-distributed, and the likelihood function is maximal in $(\hat{\alpha}_0, \hat{\beta}_0, \hat{\sigma}_0^2) = (\overline{Y}, 0, SS_{tot}/n)$; see Example 3.14. The likelihood ratio statistic is then equal to

$$2\log \lambda_n(Y_1, \ldots, Y_n) = -n\log\hat{\sigma}^2 - \frac{SS_{res}}{\hat{\sigma}^2} + n\log\hat{\sigma}_0^2 + \frac{SS_{tot}}{\hat{\sigma}_0^2}$$

$$= -n\log\frac{SS_{res}/n}{SS_{tot}/n}$$

$$= -n\log(1 - r_{xY}^2),$$

Since the observations are not identically distributed, Theorem 4.43 is not directly applicable. However, we can extend the theorem to this case, giving a χ_1^2-limit distribution for $2 \log \lambda_n$, since $k - k_0 = 3 - 2 = 1$, with k and k_0 as in Theorem 4.43. We therefore reject the null hypothesis that $\beta = 0$ when $2 \log \lambda_n$ is greater than or equal to $\chi_{1,1-\alpha_0}^2$, where α_0 is the size of the test. Hence the test also rejects the null hypothesis for large values of $|r_{xy}|$. The transformation of this quantity through $r \mapsto -\log(1 - r^2)$ can be viewed as a way to transform the distribution of the test statistic to approximately a standard distribution, namely a chi-square distribution.

7.2.2 Multiple Linear Regression

In the *multiple linear regression model*, the independent variable is multidimensional instead of one-dimensional as in the simple linear regression model. The multiple linear regression model for n dependent variables $Y_1, \ldots, Y_n$ with corresponding p-dimensional predictor variables $(x_{1,1}, \ldots, x_{1,p}), \ldots, (x_{n,1}, \ldots, x_{n,p})$ is described by

$$Y_i = \sum_{j=1}^{p} \beta_j x_{i,j} + e_i, \qquad i = 1, \ldots, n,$$

where $e_1, \ldots, e_n$ are independent normally distributed random variables with expectation 0 and finite variance σ^2. The predictor variables are once again assumed to be nonrandom, so that we may view the values $x_{i,1}, \ldots, x_{i,p}$ as known constants. It is useful to present this model in matrix notation. The observation is a vector $Y = (Y_1, \ldots, Y_n)^T$ in $\mathbb{R}^n$, and the regression coefficients form a vector $\beta = (\beta_1, \ldots, \beta_p)^T$ in $\mathbb{R}^p$. If we define the $(n \times p)$-matrix X as the matrix with (i, j)-element $x_{i,j}$, then we can write the model as

(7.3) $$Y = X\beta + e,$$

where $e = (e_1, \ldots, e_n)^T$ in $\mathbb{R}^n$ is the error vector. The matrix X is called the *design matrix*. Note that we use the notation X for a nonrandom matrix. In models with an intercept, the elements in the first column of the design matrix are taken equal to 1. The unknown model parameters are the regression coefficients $\beta = (\beta_1, \ldots, \beta_p)$ and the variance σ^2.

7.2.2.1 Dummy Variables

The chosen predictor variables can be both real-valued and categorical. A *categorical predictor variable*, also called a *nominal variable*, is a variable whose values indicate a classification instead of a relevant numerical size. For example, the values 0 and 1 can be used for male and female or can indicate a particular region. A standard technique for incorporating both real-valued and categorical variables in a linear regression model is to use *dummy variables*. A dummy variable is an indicator variable and can only take on the values 0 and 1. When the categorical predictor variable has k possible classes, we add k predictor dummy variables $x_1, \ldots, x_k$ to the regression model (without intercept). When the categorical variable belongs to class i, the dummy variable x_i is given value 1, and the other dummy variables value 0. In the linear regression model, k regression parameters $\beta_1, \ldots, \beta_k$ correspond to these k predictor dummy variables. For an observation corresponding to a predictor variable in the ith class, only the parameter β_i is used in the regression model; see Table 7.1.

x	x_1 x_2 $\cdots$ x_k	$\sum_{i=1}^{k} \beta_i x_i$
"1"	1 0 $\cdots$ 0	β_1
"2"	0 1 $\cdots$ 0	β_2
$\vdots$		$\vdots$
"k"	0 0 $\cdots$ 1	β_k

Table 7.1. Definition of dummy variables $x_1, \ldots, x_k$ for regression on a categorical variable x with k classes labeled "1", $\ldots$, "k".

This way, the parameter β_i is in fact the intercept for the class i. We do not add an intercept to the model. When we do want to add an intercept, the number of dummy variables must be less than the number of classes. In that case, for example the dummy variable for the first class is left out. The parameter β_1 is then the usual intercept and the parameter β_i ($i = 2, \ldots, k$) gives the effect of class i on the dependent variable Y with respect to class 1. In the case of two categorical predictor variables (for example, when both the region and the gender are taken up in the model), for the second variable, the model includes one dummy variable less than the number of corresponding classes. This is necessary to ensure that the design matrix has full rank; see Section 7.2.2.2. Although it is common to allow only the values 0 and 1 for a dummy variable, in some situations, it makes sense to use other values. For instance, in Example 1.5 the model includes a dummy variable that takes on the values -1 and 1. Both the choice of values for the dummy variable and the choice of using an intercept or not depend on the desired interpretation of the parameters β_i. When there are only categorical variables, the model for analysis of variance from Section 7.3 applies.

7.2.2.2 Estimation

The following theorem gives the maximum likelihood estimators for the parameters in the multiple linear regression model.

Theorem 7.4

If the design matrix X in regression model (7.3) has full rank, the maximum likelihood estimators for β and σ^2 are given by

$$\hat{\beta} = (X^T X)^{-1} X^T Y, \qquad \hat{\sigma}^2 = \frac{\|Y - X\hat{\beta}\|^2}{n}.$$

Proof. The log-likelihood function for the multiple linear regression model is given by

$$(\beta, \sigma^2) \mapsto \log \prod_{i=1}^{n} \frac{1}{\sqrt{2\pi\sigma^2}} e^{-\frac{1}{2}(Y_i - \sum_{j=1}^{p} \beta_j x_{i,j})^2/\sigma^2}$$

$$= -\frac{1}{2}n \log 2\pi - \frac{1}{2}n \log \sigma^2 - \frac{1}{2\sigma^2} \|Y - X\beta\|^2,$$

where $\| \cdot \|$ denotes the Euclidean norm. Completely analogously to what we do in the case of simple linear regression, we first deduce the estimator for β for arbitrary σ^2 and then the estimator for σ^2.

The maximum likelihood estimator for β is the least-squares estimator $\hat{\beta}$ that minimizes the function

$$\beta \mapsto \|Y - X\beta\|^2, \qquad \beta \in \mathbb{R}^p.$$

For every β, the vector $X\beta$ belongs to the range (the column space) of the matrix X, viewed as a map $X: \mathbb{R}^p \to \mathbb{R}^n$. The least-squares estimator $\hat{\beta}$ that minimizes $\|Y - X\beta\|^2$ is therefore the vector such that $X\hat{\beta}$ is the element of the range of X that lies as close as possible to the vector Y. In linear algebra, $X\hat{\beta}$ is called the *projection* of Y on the range of X. The projection with respect to the Euclidean norm satisfies the orthogonality relation

$$\langle Y - X\hat{\beta}, X\gamma \rangle = \gamma^T X^T (Y - X\hat{\beta}) = 0 \qquad \forall \gamma \in \mathbb{R}^p.$$

In other words, the "residual" $Y - X\hat{\beta}$ is orthogonal to every arbitrary element of the column space of X, which can, in general, be written as $X\gamma$ for a $\gamma \in \mathbb{R}^p$. Requiring this to be 0 for arbitrary $\gamma \in \mathbb{R}^p$ means that $X^T(Y - X\hat{\beta}) = 0$. This is the so-called *normal equation*. Because of the assumption that X has full rank, X^TX is invertible, and consequently $\hat{\beta} = (X^TX)^{-1}X^TY$. Then $X\hat{\beta}$ is equal to $X(X^TX)^{-1}X^TY$, which is indeed the projection of Y on the column space of X, because $X(X^TX)^{-1}X^T$ is the projection matrix that projects onto this space.

In the second step, we substitute $\hat{\beta}$ for β in the log-likelihood function. We can easily verify that this gives $\hat{\sigma}^2 = \|Y - X\hat{\beta}\|^2/n$ as maximum likelihood estimator for σ^2. ∎

The maximum likelihood estimator $\hat{\beta}$ is unbiased:

$$E\hat{\beta} = (X^TX)^{-1}X^T EY = (X^TX)^{-1}X^T X\beta = \beta.$$

The mean square error of $\hat{\beta}$ is

$$\text{Cov }\hat{\beta} = (X^TX)^{-1}X^T \text{ Cov } Y \, X(X^TX)^{-1} = \sigma^2(X^TX)^{-1}$$

(see Appendix B), since the errors $e_1, \ldots, e_n$ are uncorrelated and therefore so are $Y_1, \ldots, Y_n$. The matrix $X^T X$ is known as the *hat matrix*. The inverse of the hat matrix therefore gives an indication of the accuracy of the least-squares estimators. In particular, after multiplication by σ^2, the diagonal elements are equal to the mean square error of the least-squares estimators $\hat{\beta}_j$ for the regression coefficients β_j. The assumption that X has full rank is required for the existence of the inverse of $X^T X$. By, if necessary, leaving out or combining columns, we can always choose the design matrix such that it has full rank. Linear dependence of the columns of X would lead to the regression coefficients not being uniquely defined; the dependent columns are then *collinear*. This is why in a model with dummy variables, one should be careful with introducing an intercept (see Section 7.2.2.1); the combination of an intercept and a dummy variable for each class leads to a design matrix that does not have full rank, and must therefore be avoided.

Definition 7.5 Residuals and sums of squares

The residuals of the regression of Y on X are the coordinates of the vector $Y - X\hat{\beta}$. The expressions

$$SS_{tot} = \|Y - \overline{Y}1\|^2 \qquad \text{and} \qquad SS_{res} = \|Y - X\hat{\beta}\|^2$$

with $\overline{Y}1 = (\overline{Y}, \ldots, \overline{Y})$ are called the total sum of squares *and the* residual sum of squares.

The coefficient of determination, which is equal to $1 - SS_{res}/SS_{tot}$ (see Definition 7.3) takes on values between 0 and 1, as it does in the case of simple linear regression. This can be seen as follows. The vector $\overline{Y}1 = (\overline{Y}, \ldots, \overline{Y})$ is the best prediction of Y in a model consisting of only an estimated intercept. It is the projection of Y onto the one-dimensional linear space spanned by the vector $1 := (1, 1, \ldots, 1)$. Since we have chosen a model with intercept, this space is contained in the column space of X. Consequently, the residue vector $Y - X\hat{\beta}$ is orthogonal to the vector 1, that is, $\langle Y - X\hat{\beta}, 1 \rangle = 0$. It also follows that $\langle Y - X\hat{\beta}, \overline{Y}1 \rangle = 0$. Moreover, for every γ in $\mathbb{R}^p$, we have $\langle Y - X\hat{\beta}, X\gamma \rangle = 0$, hence in particular $\langle Y - X\hat{\beta}, X\hat{\beta} \rangle = 0$. Consequently, $Y - X\hat{\beta}$ is orthogonal to both the vector $\overline{Y}1$ and the vector $X\hat{\beta}$. So the vectors $Y - X\hat{\beta}$ and $X\hat{\beta} - \overline{Y}1$ are orthogonal to each other. By the Pythagorean theorem, the square of the length of the sum of these two vectors, $Y - \overline{Y}1 = (Y - X\hat{\beta}) + (X\hat{\beta} - \overline{Y}1)$, is equal to

$$\|Y - \overline{Y}1\|^2 = \|Y - X\hat{\beta}\|^2 + \|X\hat{\beta} - \overline{Y}1\|^2.$$

The left-hand side of this equation is equal to SS_{tot}, and the first term on the right-hand side is equal to SS_{res}. We see that $0 \leq SS_{res} \leq SS_{tot}$ and that the coefficient of determination lies between 0 and 1. We can show that, analogously to $1 - SS_{res}/SS_{tot} = r_{x,Y}^2$ in the simple linear regression model, we have $1 - SS_{res}/SS_{tot} = r_{X\hat{\beta},Y}^2$ for the multiple linear regression model.

When we do not include an intercept in the model, we should also not include an intercept when computing the total sum of squares. In that case we use $SS_{tot} = \|Y\|^2$.

7.2.2.3 Tests

As for the simple linear regression model, for a multiple linear regression model, there are two important types of tests to determine the influence of one or more predictor variables on the dependent variable Y. In this section, we discuss the likelihood ratio test and the F-test.

For the likelihood ratio test, it is best to view the multiple linear regression model as a special case of the general linear model. When the design matrix X has full rank, the column space of X is a p-dimensional linear subspace $V \subset \mathbb{R}^n$. The multiple linear regression model can therefore be viewed as a model for the n-dimensional normally distributed observation Y with expectation vector μ in the linear subspace V. This is the general form of a linear model. We take the covariance matrix Σ of Y equal to $\Sigma = \sigma^2 I$. The model is then parameterized by $\theta = (\mu, \sigma^2) \in V \times (0, \infty)$.

The log-likelihood function is given by

$$(\mu, \sigma^2) \mapsto \log \prod_{i=1}^{n} \frac{1}{\sqrt{2\pi\sigma^2}} e^{-\frac{1}{2}(y_i - \mu_i)^2/\sigma^2}$$

$$= -\frac{n}{2}\log 2\pi - \frac{n}{2}\log \sigma^2 - \frac{\|Y - \mu\|^2}{2\sigma^2}.$$

Maximizing the log-likelihood with respect to $\mu \in V$ is equivalent to minimizing the function $\mu \mapsto \|Y - \mu\|^2$ with respect to $\mu \in V$. Analogously to what we saw in Section 7.2.2.2, this minimum is reached in $\hat{\mu} = P_V Y$, with $P_V Y$ the orthogonal projection of Y onto the space V. The square distance $\|Y - \hat{\mu}\|^2 = \|(I - P_V)Y\|^2$ from Y to its projection is then exactly the residual sum of squares. Maximizing the likelihood with respect to σ^2 then gives the maximum likelihood estimator $\hat{\sigma}^2 = \|(I - P_V)Y\|^2/n$.

The null hypothesis that one or more variables do not influence the dependent variable Y can now be viewed as a special case of the null hypothesis $H_0: \mu \in V_0$, with V_0 a p_0-dimensional linear subspace of V. The log-likelihood ratio statistic is based on the maximum likelihood estimators under the null hypothesis and for the full model. Computations analogous to those described above show that the maximum likelihood estimators for μ and σ^2 under the null hypothesis are given by $\hat{\mu}_0 = P_{V_0}Y$ and $\hat{\sigma}_0^2 = \|(I - P_{V_0})Y\|^2/n$. Twice the log-likelihood ratio statistic then becomes

$$2\log \lambda_n(Y) = -n\log \hat{\sigma}^2 - \frac{\|(I - P_V)Y\|^2}{\hat{\sigma}^2} + n\log \hat{\sigma}_0^2 + \frac{\|(I - P_{V_0})Y\|^2}{\hat{\sigma}_0^2}$$

$$= n\log \frac{\|(I - P_{V_0})Y\|^2}{\|(I - P_V)Y\|^2}$$

(verify). We see that the numerator of the likelihood ratio statistic is equal to the residual sum of squares under the null hypothesis, while the denominator is the residual sum of squares under the full model. When the numerator is much larger than the denominator, this is an indication that the null hypothesis is incorrect. We therefore reject the null hypothesis for large values of the statistic. More precisely, we reject the null hypothesis at level α_0 when $2 \log \lambda_n(Y) \geq \chi^2_{p-p_0;1-\alpha_0}$.

In the F-test, we use a different, but related ratio of sums of squares:

$$F = \frac{\|(P_V - P_{V_0})Y\|^2/(p-p_0)}{\|(I-P_V)Y\|^2/(n-p)}.$$

The vector $(P_V - P_{V_0})Y$ is an element of V, since $V_0 \subset V$ and V is a linear space. Since $(I - P_V)Y$ is orthogonal to V, by the Pythagorean theorem we therefore have

$$\|(I-P_{V_0})Y\|^2 = \|(I-P_V)Y + (P_V - P_{V_0})Y\|^2 = \|(I-P_V)Y\|^2 + \|(P_V - P_{V_0})Y\|^2.$$

By substituting this equality in the formula for $2 \log \lambda_n(Y)$, we see that we can write the latter as $2 \log \lambda_n(Y) = n \log(1 + (p-p_0)F/(n-p))$. The log-likelihood ratio is therefore an increasing function of F, and the likelihood ratio test can equivalently be formulated as rejecting the null hypothesis for large values of F. By Cochran's theorem, Theorem B.8, under the null hypothesis F has an F-distribution with $p - p_0$ and $n - p$ degrees of freedom. We consequently reject the null hypothesis at level α_0 when the F-test statistic is greater than or equal to the $(1-\alpha_0)$-quantile of the $F_{p-p_0,n-p}$-distribution, which we denote by $F_{p-p_0,n-p;1-\alpha_0}$.

A common null hypothesis for multiple linear regression models is $H_0 \colon \beta_j = 0$ for some $j \in \{1, \ldots, p\}$. If $\beta_j = 0$, then the jth predictor variable does not influence the dependent variable Y. Under the null hypothesis, the regression model can therefore be simplified by leaving out this predictor variable. Specifically, this means that we remove the jth column from the design matrix, as well as the jth coordinate of β. We denote the new design matrix by X_{-j} and the shortened vector of regression parameters by β_{-j}. The maximum likelihood estimator for β_{-j} is derived analogously to the estimator for β and is equal to $\hat{\beta}_{-j} = (X_{-j}^T X_{-j})^{-1} X_{-j}^T Y$. The projection matrices P_V and P_{V_0} defined for the general linear model are then equal to $X(X^T X)^{-1} X^T$ and $X_{-j}(X_{-j}^T X_{-j})^{-1} X_{-j}^T$, and the test statistic for the likelihood ratio test becomes

$$2 \log \lambda_n(Y) = n \log \frac{\|(I - X_{-j}(X_{-j}^T X_{-j})^{-1} X_{-j}^T)Y\|^2}{\|(I - X(X^T X)^{-1} X^T)Y\|^2}.$$

The null hypothesis that $\beta_j = 0$ is rejected when $2 \log \lambda_n(Y) \geq \chi^2_{1,1-\alpha_0}$. The F-test statistic under $H_0 \colon \beta_j = 0$ can be found by substituting the expressions for the projection matrices; under the null hypothesis, it has an F-distribution with 1 and $n - p$ degrees of freedom. A test equivalent to this F-test is based on the test statistic

(7.4)
$$T = \frac{\hat{\beta}_j}{\hat{\sigma}^2 \left((X^T X)^{-1} \right)_{j,j}},$$

273

for which we can show that $T^2 = F$. Under the null hypothesis, the variable T has a t-distribution with $n - p$ degrees of freedom. We have $F \geq F_{1,n-p;1-\alpha_0}$ if and only if $|T| \geq t_{n-p;1-\alpha_0/2}$ since $F_{1,n-p;1-\alpha_0} = (t_{n-p;1-\alpha_0/2})^2$. In particular, for the simple linear regression model with $p = 2$, the variable F is the square of the test statistic T in Section 7.2.1.2 corresponding to $H_0: \beta = \beta_0$ with $\beta_0 = 0$.

Example 7.6 Height

Example 1.5 describes a multiple linear regression model for estimating the final adult height of a child based on the heights of the (biological) parents and the gender of the child. In that example, the regression parameters for the predictor variables "height of the father" and "height of the mother" are taken equal to $1/2$, so that the estimated model is easy to interpret. In this section, we will leave out this assumption and estimate a multiple linear regression model based on our own data.

For Y the final height of a child, x_2 the height of the father, x_3 the height of the mother, and x_4 a variable for the child's gender, the multiple linear regression model looks as follows:

$$Y = \beta_1 + \beta_2 x_2 + \beta_3 x_3 + \beta_4 x_4 + e,$$

with e a normally distributed random variable with expectation 0 en variance σ^2. We have taken the first predictor variable, x_1, equal to 1, so that the model has an intercept. The predictor variable x_4 indicates whether the child is a boy or a girl. Because the model has one intercept, one dummy variable suffices. We want β_4 to be equal to half the average height difference between men and women, and therefore choose x_4 equal to -1 for a girl and equal to 1 for a boy. Since, on average, boys are taller than girls, β_4 will be positive.

Our data consist of final heights of 111 adolescents (44 male and 67 female), their gender, and the heights of their parents.[†] For each of the regression parameters, we test whether the value deviates significantly from 0, using the t-test described in the previous section. We take the size of the tests equal to 0.05. All test are rejected, and the final model with estimated regression parameters is given by

$$Y = 2.52 + 0.46x_2 + 0.55x_3 + 6.27x_4 + e,$$

where e is assumed to be normally distributed with expectation 0 and (estimated) variance 25.78. The coefficient of determination of the model is 0.69. To study whether the normality assumption is plausible, we can draw scatter plots and possibly carry out additional tests. Despite the fact that the estimated regression parameters in the model above do not match the estimates in the Fourth (Dutch) National Growth Study (see (1.1)), the expected final heights are not far apart. For example, in the model above, the final heights of the children of a 180 cm man and a 172 cm woman are equal to 186.2 cm (son) and 173.7 cm (daughter), while the regression model in (1.1) gives heights 187 cm and 174 cm.

[†] The data can be found on the book's webpage at `http://www.aup.nl` under `heightdata`.

In the model above, the mother's height is of greater influence on the final height of the child than the father's. Under the assumption that the influences are equal, the estimated parameters are $\hat{\beta}_1 = 3.47, \hat{\beta}_2 = \hat{\beta}_3 = 0.50$, and $\hat{\beta}_4 = 6.30$. This model matches the model in (1.1) better. The estimated values for β_1 and β_4 in the model are, however, somewhat lower, so that the predicted final heights will also be somewhat lower. The model estimated in the Fourth National Growth Study is based on much more data. The estimated models in this example are therefore less reliable. ▭

7.3 Analysis of Variance

Analysis of variance (or *ANOVA*) is a technique used to study the influence of discrete experimental variables, called factors, on a given continuous dependent variable. We will focus on analysis of variance with two factors, although the technique is certainly not restricted to this case.

The classical analysis of variance was developed for the analysis of experiments in agriculture, to study which type of fertilizer, which irrigation method, combined with which plant genera, would give the highest yield. Each of the variables fertilizer, irrigation, and genera is referred to as a *factor*, and the "yield" is the dependent variable. It is typical that the factors are categorical variables and take on only a few different values, which are usually not ordered. The values of the factors are fixed before the experiment and are viewed as known constants. The observation is a vector whose coordinates are the measured yield for different combinations of the factors. Table 7.2 gives an example of data classified according to two factors with, respectively, two and three categories.

	A					B					C				
L	101	78	68	80	82	42	41	23	19	37	100	83	101	80	106
	80	85	74	77	87	52	32	53	67	47	73	106	102	109	105
N	94	71	93	103	87	34	52	42	44	58	117	81	83	91	127
	92	86	81	72	87	36	69	82	49	32	99	91	105	91	118

Table 7.2. Data for the analysis of variance with two factors. The first factor has $I = 2$ levels, "L" and "N"; the second factor has $J = 3$ levels, "A", "B", and "C". There are $K = 10$ observations for each combination of the factors. The data give the traveled distance per day in kilometers (1 km = 1.6 mi) for rental cars in three different classes, rented in Leiden or Noordwijk (Netherlands).

Two factors with, respectively, I and J different levels can be combined in IJ different ways. There can be several observations for each combination (i, j), as in Table 7.2. We parameterize the model with expectation values μ_{ij} for the different combinations (i, j) of the two factors. A basic assumption in analysis of variance is that the observation for combination (i, j) is normally distributed with expectation μ_{ij} and variance σ^2. Moreover, all observations are assumed independent, so that the observation vector is a multidimensional normally distributed vector. The goal is to analyze the dependence of μ_{ij} on the two factors.

In an analysis of variance, the expectations μ_{ij} are commonly expressed in so-called main and interaction effects,

$$(7.5) \qquad \mu_{ij} = \mu + \alpha_i + \beta_j + \gamma_{ij}.$$

The parameter μ is the grand mean over all combinations of factors. The *main effects* α_i and β_j give the deviations with respect to the grand mean if the factors are set to i and j, respectively. The parameters γ_{ij} are the parts of the expectations μ_{ij} that cannot be explained by the factors separately, but can be explained by their combinations; they are called *interaction effects*. Without additional conditions on the parameters μ, α_i, β_j, and γ_{ij}, the model cannot be identified, because we have $1 + I + J + IJ$ parameters for the IJ expectations. The usual conditions for the parameters are

$$(7.6) \qquad \sum_i \alpha_i = 0, \quad \sum_j \beta_j = 0,$$
$$\sum_{i=1}^{I} \gamma_{ij} = 0 \quad \text{for } j = 1, \ldots, J, \quad \sum_{j=1}^{J} \gamma_{ij} = 0 \quad \text{for } i = 1, \ldots, I.$$

We can easily verify that the parameters that satisfy (7.6) also satisfy

$$(7.7) \qquad \begin{aligned} \mu &= \mu_{..}, \\ \alpha_i &= \mu_{i.} - \mu, \\ \beta_j &= \mu_{.j} - \mu, \\ \gamma_{ij} &= \mu_{ij} - \mu - \alpha_i - \beta_j = \mu_{ij} - \mu_{i.} - \mu_{.j} + \mu_{..}. \end{aligned}$$

In these formulas, a dot $\cdot$ means that an average was taken over the corresponding index, for example $\mu_{i.} = J^{-1} \sum_{j=1}^{J} \mu_{ij}$. Conversely, we can verify that the parameters $\mu, \alpha_i, \beta_j, \gamma_{ij}$ defined in (7.7) are the only parameters that satisfy all conditions in (7.5) and (7.6).

We can therefore describe the model both in terms of the parameters μ_{ij} and in terms of the parameters $\mu, \alpha_i, \beta_j, \gamma_{ij}$. An advantage of reparameterizing is that we can easily formulate the important hypotheses in terms of main effects and interaction effects. The hypothesis that there is no interaction is $H_0 : \gamma_{ij} = 0$ for $i = 1, \ldots, I, j = 1, \ldots, J$, while the hypotheses $H_0 : \alpha_i = 0$ for $i = 1, \ldots, I$ and $H_0 : \beta_j = 0$ for $j = 1, \ldots, J$ imply that, respectively, the first and second factors of the experiment do not play a role in the value of the observation. Of course, more specific hypotheses concerning the effects can also be interesting.

276

7.3.1 Estimation

When there are K_{ij} observations for each combination (i, j) of the factors, we have an observation vector $Y = (Y_{ijk})$ of length $n = \sum_{i,j} K_{ij}$. The model is then given by

$$(7.8) \qquad Y_{ijk} = \mu + \alpha_i + \beta_j + \gamma_{ij} + e_{ijk},$$

where the errors e_{ijk} are independent and normally distributed with expectation 0 and variance σ^2 for $i = 1, \ldots, I$, $j = 1, \ldots, J$, and $k = 1, \ldots, K_{ij}$. The log-likelihood function is given by

$$(\mu, \alpha, \beta, \gamma, \sigma^2) \mapsto -\frac{n}{2} \log(2\pi\sigma^2) - \frac{1}{2\sigma^2} \sum_{i,j,k} (Y_{ijk} - \mu - \alpha_i - \beta_j - \gamma_{ij})^2,$$

where the vector $(\mu, \alpha, \beta, \gamma)$ is the parameter vector that contains all effects. As with linear regression, the maximum likelihood estimators for the expectation parameters (in this case, the effects) are equal to the least-squares estimators. The likelihood equations for the effects are

$$\sum_{i=1}^{I} \sum_{j=1}^{J} \sum_{k=1}^{K_{ij}} (Y_{ijk} - \mu - \alpha_i - \beta_j - \gamma_{ij}) = 0,$$

$$\sum_{j=1}^{J} \sum_{k=1}^{K_{ij}} (Y_{ijk} - \mu - \alpha_i - \beta_j - \gamma_{ij}) = 0 \quad \text{for } i = 1, \ldots, I,$$

$$\sum_{i=1}^{I} \sum_{k=1}^{K_{ij}} (Y_{ijk} - \mu - \alpha_i - \beta_j - \gamma_{ij}) = 0 \quad \text{for } j = 1, \ldots, J,$$

$$\sum_{k=1}^{K_{ij}} (Y_{ijk} - \mu - \alpha_i - \beta_j - \gamma_{ij}) = 0 \quad \text{for } i = 1, \ldots, I, j = 1, \ldots, J.$$

Using the relations in (7.6), we find the following estimators:

$$\hat{\mu} = Y_{...},$$
$$\hat{\alpha}_i = Y_{i..} - Y_{...},$$
$$\hat{\beta}_j = Y_{.j.} - Y_{...},$$
$$\hat{\gamma}_{ij} = Y_{ij.} - Y_{i..} - Y_{.j.} + Y_{...},$$

where a dot $\cdot$ again means averaging over the corresponding index. The estimator for the variance can be found by substituting these estimators in the likelihood function and maximizing the result with respect to σ^2:

$$\hat{\sigma}^2 = \frac{1}{n} \sum_{i,j,k} (Y_{ijk} - \hat{\mu} - \hat{\alpha}_i - \hat{\beta}_j - \hat{\gamma}_{ij})^2.$$

Finally, we note that it follows from the above that the maximum likelihood estimator for μ_{ij} is equal to $\hat{\mu}_{ij} = \hat{\mu} + \hat{\alpha}_i + \hat{\beta}_j + \hat{\gamma}_{ij} = Y_{ij\cdot}$. This result is, of course, not surprising; if we were to leave out the reparameterization and estimate the parameter μ_{ij} itself, we would have found exactly this estimator.

7.3.2 Tests

The interesting null hypotheses to test in analysis of variance are $H_0: \alpha_i = 0$ for $i = 1, \ldots, I$ and $H_0: \beta_j = 0$ for $j = 1, \ldots, J$, which we can use to study the main effects, and $H_0: \gamma_{ij} = 0$ for $i = 1, \ldots, I, j = 1, \ldots, J$, which we can use to study the significance of the interaction effects. The analysis is particularly appealing if we have the same number $K_{ij} = K > 1$ of replicates for every combination (i, j), a so-called "balanced design with replication." The observations vector $Y = (Y_{ijk})$ is then an n-dimensional multivariate normally distributed random vector, with $n = IJK$. In the remainder of this section, we assume that we have a balanced design with K replicates.

Not all hypotheses can be tested in a meaningful way; whether they can be tested depends on the available data. If, for example, we have only one observation for each combination (i, j) of factors, then we have only one observation for each "free" parameter μ_{ij} (not counting an additional variance parameter). It is intuitively clear that in such a case, we cannot draw meaningful conclusions on the interaction parameters γ_{ij}. We then need to either collect more data or make some prior assumptions. A popular assumption is, for example, that the interaction factors γ_{ij} are equal to 0. The resulting *additive model* $\mu_{ij} = \mu + \alpha_i + \beta_j$ has only $I + J - 1$ parameters and can be fitted to the data in a meaningful way, provided that the prior assumption that there is no interaction is correct. Unfortunately, the latter can rarely be shown (without additional data).

It turns out to be useful and insightful to view analysis of variance as a special case of the general linear model (see Section 7.2.2.3). For this interpretation, we first study the design matrix. Consider, for convenience, the case $I = 2$ and $J = 3$, as in Table 7.2. The expectation vector (μ_{ij}) for a single replicate can be written as

$$(7.9) \quad \begin{pmatrix} \mu_{11} \\ \mu_{12} \\ \mu_{13} \\ \mu_{21} \\ \mu_{22} \\ \mu_{23} \end{pmatrix} = \begin{pmatrix} 1 & 1 & 1 & 0 & 1 & 0 \\ 1 & 1 & 0 & 1 & 0 & 1 \\ 1 & 1 & -1 & -1 & -1 & -1 \\ 1 & -1 & 1 & 0 & -1 & 0 \\ 1 & -1 & 0 & 1 & 0 & -1 \\ 1 & -1 & -1 & -1 & 1 & 1 \end{pmatrix} \begin{pmatrix} \mu \\ \alpha_1 \\ \beta_1 \\ \beta_2 \\ \gamma_{11} \\ \gamma_{12} \end{pmatrix}.$$

There are $IJ = 6$ parameters $(\mu, \alpha_1, \beta_1, \beta_2, \gamma_{11}, \gamma_{12})$ on the right-hand side. In this parameterization, the parameters $(\alpha_2, \beta_3, \gamma_{13}, \gamma_{21}, \gamma_{22}, \gamma_{23})$ are expressed in the other parameters using the relations (7.6). For example,

$$\mu_{23} = \mu + \alpha_2 + \beta_3 + \gamma_{23}$$
$$= \mu - \alpha_1 - \beta_1 - \beta_2 - \gamma_{13}$$
$$= \mu - \alpha_1 - \beta_1 - \beta_2 + \gamma_{11} + \gamma_{12}.$$

The 6 parameters on the right-hand side, and therefore the 6 columns of the design matrix, break up into 4 groups corresponding to the grand mean (column 1), the main effect α (column 2), the main effect β (columns 3 and 4), and the interaction effects (columns 5 and 6). Inspecting the design matrix shows that the four linear spaces spanned by these groups of columns are orthogonal to one another. This property of the design matrix also holds for other values of I and J. The number of columns per group is equal to 1 (grand mean), $I - 1$ (main effects α), $J - 1$ (main effects β), and $(I - 1)(J - 1)$ (interaction effects). The total number of columns in the design matrix, and hence the number of parameters, is therefore equal to IJ. This number of parameters is equal to the number of parameters before the reparameterization (7.5), that is, to the number of expected values μ_{ij}. In the case of a balanced design with K replicates, the expectation vector for the full observation vector Y of length IJK can be obtained by stacking K expectation vectors of length IJ for one replicate of all combinations on top of one another. The design matrix corresponding to this full observation matrix is obtained by repeating the design matrix in (7.9) K times and placing the matrices on top of one another. The distribution of the columns in four orthogonal groups (grand mean, main effect α, main effect β, and interaction effects) is obviously preserved in this combined design matrix.

The distribution in orthogonal column groups of the design matrix described above is important for the application of the F-test from Section 7.2.2.3 to the relevant null hypotheses in analysis of variance. The maximum likelihood estimator for the expectation vector EY in a balanced design is given by $P_V Y$ if V is the linear subspace of $\mathbb{R}^n$ spanned by the columns of the combined design matrix. Since V is spanned by four orthogonal column groups, the projection P_V is equal to the sum of the four orthogonal projections, each corresponding to a column group,

$$(7.10) \qquad P_V Y = P_\mu Y + P_\alpha Y + P_\beta Y + P_\gamma Y,$$

where P_μ, P_α, P_β, and P_γ are the orthogonal projections from $\mathbb{R}^n$ onto the four subspaces. The dimensions of the spaces onto which P_μ, P_α, P_β, and P_γ project are equal to the numbers of columns in the corresponding column groups of the design matrix, that is, 1, $I - 1$, $J - 1$, and $(I - 1)(J - 1)$, respectively. The null hypotheses on the interaction and main effects are linear in the parameters and can therefore be viewed as assertions that the expectation vector EY belongs to a certain linear subspace of the n-dimensional space. For example, the null hypothesis that there is no interaction, $H_0: \gamma_{ij} = 0$ for $i = 1, \ldots, I, j = 1, \ldots, J$, corresponds to the linear subspace V_0 spanned by the three column groups of the design matrix corresponding to P_μ, P_α, and P_β. In that case, the term $P_{V_0} Y$ in the F-test statistic is equal to $P_{V_0} Y = P_\mu Y + P_\alpha Y + P_\beta Y$. To compute the sums of squares in the numerator and denominator of the F-test statistic, we take a closer look at the projections in (7.10).

The four projections on the right-hand side of (7.10) are orthogonal to one another and orthogonal to $(I - P_V)Y$. It therefore follows from the Pythagorean theorem that

$$\|Y\|^2 - \|P_\mu Y\|^2 = \|P_\alpha Y\|^2 + \|P_\beta Y\|^2 + \|P_\gamma Y\|^2 + \left\|(I - P_\mu - P_\alpha - P_\beta - P_\gamma)Y\right\|^2.$$

After some computation, we see that the different terms in this decomposition can be written as

$$\|Y\|^2 - \|P_\mu Y\|^2 = \sum_{i,j,k} (Y_{ijk} - Y_{...})^2,$$

(7.11)

$$\|P_\alpha Y\|^2 = \sum_{i,j,k} (Y_{i..} - Y_{...})^2 = JK \sum_i \hat{\alpha}_i^2,$$

$$\|P_\beta Y\|^2 = \sum_{i,j,k} (Y_{.j.} - Y_{...})^2 = IK \sum_j \hat{\beta}_j^2,$$

$$\|P_\gamma Y\|^2 = \sum_{i,j,k} (Y_{ij.} - Y_{i..} - Y_{.j.} + Y_{...})^2 = K \sum_{i,j} \hat{\gamma}_{ij}^2.$$

Definition 7.7 Sums of squares

The *total sum of squares* is equal to

$$SS_{tot} = \sum_{i,j,k} (Y_{ijk} - Y_{...})^2.$$

The *sums of squares of the first and second factors* are equal to

$$SS_\alpha = \sum_{i,j,k} (Y_{i..} - Y_{...})^2,$$

$$SS_\beta = \sum_{i,j,k} (Y_{.j.} - Y_{...})^2.$$

The *sum of squares of the interaction* is equal to

$$SS_\gamma = \sum_{i,j,k} (Y_{ij.} - Y_{i..} - Y_{.j.} + Y_{...})^2.$$

The *residual sum of squares* is equal to

$$SS_{res} = \sum_{i,j,k} (Y_{ijk} - Y_{ij.})^2.$$

It follows from Definition 7.7 that

$$SS_{tot} = SS_\alpha + SS_\beta + SS_\gamma + SS_{res}.$$

The F-test statistics turn out to be quotients of these sums of squares, also called "variances." This explains the name "analysis of variance." The following theorem follows from the equations in (7.11), Definition 7.7, and Cochran's theorem (Theorem B.8).

Theorem 7.8

Suppose that the errors e_{ijk} in (7.8) are independent and normally distributed with expectation 0 and variance σ^2. Then

(i) *under $H_0: \alpha_i = 0$ for $i = 1, \ldots, I$, the variable*

$$F_\alpha = \frac{SS_\alpha/(I-1)}{SS_{res}/(IJ(K-1))}$$

has an F-distribution with $I - 1$ and $IJ(K-1)$ degrees of freedom;

(ii) *under $H_0: \beta_j = 0$ for $j = 1, \ldots, J$ the variable*

$$F_\beta = \frac{SS_\beta/(J-1)}{SS_{res}/(IJ(K-1))}$$

has an F-distribution with $J - 1$ and $IJ(K-1)$ degrees of freedom;

(iii) *under $H_0: \gamma_{ij} = 0$ for $i = 1, \ldots, I, j = 1, \ldots, J$, the variable*

$$F_\gamma = \frac{SS_\gamma/((I-1)(J-1))}{SS_{res}/(IJ(K-1))}$$

has an F-distribution with $(I-1)(J-1)$ and $IJ(K-1)$ degrees of freedom.

Together with (7.11), this theorem shows that the null hypothesis that the first factor does not have any influence is rejected for large values of $\sum \hat{\alpha}_i^2$, which is intuitively clear. An analogous statement holds for the second main effect and the interaction effects. The results of these three tests are commonly given in an *analysis of variance table*.

In the case of an unbalanced design with replication, the test statistics can be computed analogously using the general theory. However, the linear spaces corresponding to the effects are no longer necessarily orthogonal to one another. The sums of squares corresponding to the hypotheses then do not add up to the total sum of squares, and Theorem 7.8 does not hold.

Example 7.9

We apply the theory given above to the data in Table 7.2. The traveled distance is assumed to be approximately normally distributed. We discuss the results of both the model with interaction and the additive model. For both models, we have $I = 2$ and $J = 3$, and we have a balanced design with $K = 10$.

For the model with interaction, the estimates for the grand mean and the main and interaction effects are given in Table 7.3. We can easily verify that the relations in (7.6) hold for the estimated effects.

```
    Grand mean
      75.95

    City
        Leiden Noordwijk
          -2.95        2.95

    Class
            A      B      C
        7.95 -30.40   22.45

    City:Class
                 Class
    City         A      B      C
        Leiden      0.25 -1.30   1.05
        Noordwijk -0.25   1.30  -1.05
```

Table 7.3. R-output containing the maximum likelihood estimates for the grand mean and the main and interaction effects for the data from Table 7.2.

The results of the three F-tests from Theorem 7.8 are shown in the analysis of variance table in Table 7.4. The sums of squares SS_α, SS_β, SS_γ, and SS_{res} are the successive entries in the column "Sum Sq". The column "Df" has the corresponding degrees of freedom. The column "Mean Sq" gives the quotient of the sum of squares and the number of degrees of freedom. The values of F_α, F_β, and F_γ are the successive entries in the column "F value", and the last column gives the p-values for these test statistics.

```
                Df   Sum Sq   Mean Sq   F value   Pr(>F)
    city         1    522.2     522.2    2.9586   0.09115
    class        2  29827.3   14913.7   84.5028   < 2e-16
    city:class   2     57.1      28.5    0.1618   0.85105
    residuals   54   9530.3     176.5
```

Table 7.4. R-output containing the analysis of variance table for testing the main and interaction effects for the data from Table 7.2 in an analysis of variance model with two factors with interaction.

The result of the tests is that the null hypothesis of no interaction is not rejected. The main effect of the factor "city" is not significant at level 0.05, whereas the main effect of the factor "class" is clearly significant. The conclusion of the tests is that the class influences the traveled distance, while there is no clear influence of the factor "city".

When we make the prior assumption that there are no interaction effects, we can use an additive model. The estimates for the main effects are equal to those for the model with interaction in Table 7.3. The analysis of variance table for the additive model is given in Table 7.5. The conclusions concerning the significance of the main

	Df	Sum Sq	Mean Sq	F value	Pr(>F)
city	1	522.2	522.2	3.0499	0.08623
class	2	29827.3	14913.7	87.1106	< 2e-16
residuals	56	9587.4	171.2		

Table 7.5. R-output containing the analysis of variance table for testing the main effects for the data from Table 7.2 in an additive analysis of variance model with two factors.

effects are the same as in the model with interaction. Note that the residual sum of squares and the corresponding number of degrees of freedom are greater than in the (more elaborate) model with interaction. The p-values of both tests are therefore different from those in the model with interaction.

7.4 Nonlinear and Nonparametric Regression

In the linear regression model, the conditional expectation of the dependent variable Y given the predictor variable $X = x$, that is, $x \mapsto E(Y \mid X = x)$, is a linear function of the parameter β. We can replace this regression function by a more general function,

$$E(Y \mid X = x) = f(x).$$

This equation is also called the *regression equation*.

If $f = f_\theta$ is known up to the parameter θ and f_θ is a nonlinear function of θ, then we speak of *nonlinear regression*. As with linear regression, the parameter θ can be estimated using the least-squares method. The least-squares estimator for θ minimizes the criterion

$$\theta \mapsto \sum_{i=1}^{n} \left(Y_i - f_\theta(x_i)\right)^2.$$

Often, a numeric algorithm is necessary to determine the least-squares estimate. If the measurement errors in the model

$$Y = f_\theta(x) + e$$

are normally distributed, then the least-squares estimator for θ is also the maximum likelihood estimator. An example of a parameterized nonlinear regression model is the time curve

$$E(Y \mid X = x) = g_\theta(x) = \theta_0 + \theta_1 x + \theta_2 e^{-\theta_3 x}.$$

For observations $y_1, \ldots, y_n$ at times $x_1, \ldots, x_n$, we find the least-squares estimator $\theta = (\theta_0, \theta_1, \theta_2, \theta_3)$ by minimizing

$$\sum_{i=1}^{n} \left(y_i - \theta_0 - \theta_1 x_i - \theta_2 e^{-\theta_3 x_i}\right)^2$$

283

with respect to θ.

If the form of the function f is not specified beforehand, then we speak of *nonparametric regression*. We then use the observations to determine a suitable type of function. We can use different methods for approximating functions, including Fourier series, wavelets, spline functions, and neural networks. An (almost) arbitrary function $f(x) = E(Y \mid X = x)$ on the interval $[0, 2\pi]$ can, for example, be represented as a *Fourier series*

$$f(x) = \sum_{n=0}^{\infty} \big(a_n \cos(nx) + b_n \sin(nx) \big),$$

for certain constants a_n and b_n that depend on f. By estimating these constants from the data and substituting the estimates in the formula (where we truncate the sum after a finite number of terms), we find an estimator for f. So-called wavelets are similar series expansions with appealing features.

An intermediate form consists of the *additive models*, in which the conditional expectation $E(Y \mid X = x)$ with a vector-valued predictor variable $X = (X_1, \ldots, X_k)$ is modeled as

$$E(Y \mid X = x) = f_1(x_1) + f_2(x_2) + \cdots + f_k(x_k),$$

for functions $f_1, \ldots, f_k$ that are not specified beforehand. Choosing linear functions f_i would lead back to the linear regression model.

Figures 7.3 and 7.4 illustrate the possibilities with a model that is partially additive and partially nonparametric. We have fitted the model $E(Y \mid X = x) = f(x_1) + g(x_2, x_3)$, where Y is the ozone concentration and $X = (X_1, X_2, X_3)$ contains the predictor variables radiation, temperature, and wind speed. Figure 7.3 gives the estimate of f, and Figure 7.4 give the estimate of g. This model will be discussed further in the application following Chapter 8.

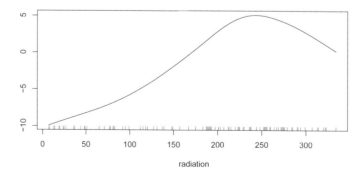

Figure 7.3. The estimate of f in the model $E(Y \mid X = x) = f(x_1) + g(x_2, x_3)$.

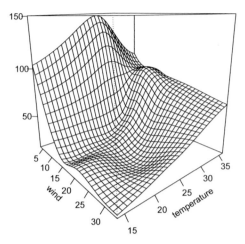

Figure 7.4. The estimate of g in the model $E(Y \mid X = x) = f(x_1) + g(x_2, x_3)$.

The dimensionality of the predictor variable plays a major role in regression problems. Without prior information or huge amounts of data, we can barely determine the correlation between a variable Y and a multidimensional predictor variable X. Even "estimating," for example, 10 unknown parameters $\beta_1, \ldots, \beta_{10}$ in a linear regression model can, in practice, cause problems. Unless the number of data points is large with respect to the number of unknown parameters, it is impossible to estimate the unknown parameters reliably. In a linear regression model, the number of unknown parameters is relatively small (the number of β_i and the unknown σ^2), but in a nonparametric model, the number of unknown parameters is theoretically infinitely large. A certain prior restriction on the model is therefore necessary.

We are dealing with a trade-off, which we have also encountered in another context. If we use a small model with few unknown parameters (for example, a linear model), then we can determine these parameters reasonably well based on the available data. We run a high risk, however, that the model is incorrect, because of which applying it (for example as a prediction) could have disastrous consequences. A large model (for example, a nonparametric model or a linear model that includes, in addition to each predictor variable X_i, the powers $X_i^2, X_i^3, \ldots$ and mixed products $X_i X_j, X_i X_j^2, \ldots$) can potentially describe the reality better, but the amount of available data may be too small to accurately estimate the parameters reliably.

7.5 Classification

To determine the "credit risk" of clients, insurance companies use background variables such as age, living conditions, income, number of claims in the past year, size of the claims, etc. Based on this data, the insurance company wants to estimate whether the client will file a substantial claim.

This is an example of a *classification* problem. Based on the data $x = (x_1, \ldots, x_m)$, we want to predict whether a certain event will take place. If we encode these two possibilities as $Y = 1$ and $Y = 0$, respectively, then the problem is to predict the Y-value of an individual based on the measured "input" x. These values x are often called *covariates*. In contrast to the assumption in Section 7.2, in this section, we take a random predictor variable X and view the observed (x, y) as a realization of a random vector (X, Y). We are now looking for the conditional distribution of Y given $X = x$. In most cases, there will not exist a perfect correlation between x on the one hand and Y given $X = x$ on the other. It seems natural to use a model expressed in probabilities, and therefore we are interested in the conditional probability

$$P(Y = 1 | X = x) = 1 - P(Y = 0 | X = x)$$

that an arbitrary event for which the value of the predictor variable X is known has a Y-value equal to 1.

In this set-up, a statistical model consists of a detailed description of the probability distribution of (X, Y). In particular, we need a detailed description of the conditional probabilities mentioned above. A classical model is the *logistic regression model*, in which

$$(7.12) \qquad P(Y = 1 | X = x) = \frac{1}{1 + e^{-\sum_{j=1}^{m} \beta_j x_j}}.$$

In this model, the value of the predictor variable $x = (x_1, \ldots, x_m)$ influences the probability distribution of the dependent variable Y through a linear combination $\sum_{j=1}^{m} \beta_j x_j$, for certain coefficients $\beta_1, \ldots, \beta_m$ that express the relative importance of the different x_j. The function $x \mapsto \Psi(x) = (1 + e^{-x})^{-1}$ is the logistic distribution function and maps the real numbers $\sum_{j=1}^{m} \beta_j x_j$ onto the interval $[0, 1]$, so that the function values may have the character of a probability. Choosing this function seems useful for computations, but other than that, it is rather arbitrary. The normal distribution function is also often used, in which case we speak of *probit regression*. In that case,

$$P(Y = 1 | X = x) = \Phi \left(\sum_{j=1}^{m} \beta_j x_j \right).$$

Figure 7.5 shows that $\Psi(x) \approx \Phi(x/1.8)$, so that the two functions lead to almost identical results, up to a scaling constant of β.

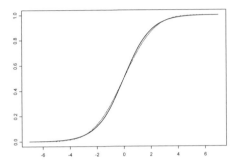

Figure 7.5. The logistic function (solid line) and the normal distribution with standard deviation 1.8 (dotted).

The variables $X_1, \ldots, X_m$ can measure more or less isolated aspects of the individuals, but can also be related. If the correlation between the independent variable Y and the predictor variable X_1 is quadratic instead of linear (for example, if both small and large values of X_1 lead to a high probability that $Y = 1$, while intermediate values usually give $Y = 0$), then it is wise to include x_1^2 in addition to x_1 in the model of the conditional probability. As in the case of linear regression, we then, for example, take $x_2 = x_1^2$. Logarithmic and exponential transformations are also often applied. Interactions between predictor variables can be modeled in the regression model by also including products like $x_1 x_2$. Categorical predictor variables that are classified in finitely many classes, like regions, can be included in the model using dummy variables, as was done for linear regression (see Section 7.2.2.1).

In addition to polynomials, there are many other possibilities to model the probabilities $P(Y = 1 | X = x)$, for example using wavelets or neural networks. The underlying idea remains the same: in a class of possibilities for the probabilities $P(Y = 1 | X = x)$ we determine the one that fits best with the observed data $(x_1, y_1), \ldots, (x_n, y_n)$ and then use this for the classification of new cases. In the machine learning literature, the full observed data is also called the *training sample* and finding the best-fitting model is called *training* or *learning*. Here, we will keep the statistical terms "observed data" and "estimating."

7.5.1 Estimation

In this section, we restrict ourselves to a one-dimensional variable X and deduce equations for the maximum likelihood estimators for the classification problem

$$P_{\alpha,\beta}(Y = 1 | X = x) = \frac{1}{1 + e^{-\alpha - \beta x}},$$

where α and β are the unknown parameters. The parameters can be estimated based on a sample of values $(x_1, y_1), \ldots, (x_n, y_n)$. Then, we can use the estimated model to predict Y based on a new value of x in the future.

Suppose that $Y_1, \ldots, Y_n$ given $X_1, \ldots, X_n$ are independent random variables with values in $\{0, 1\}$, distributed following the probability distribution above. The probability distribution of Y_i given X_i can then also be written as

$$P_{\alpha,\beta}(Y_i = y_i | X_i = x_i) = \left(\frac{1}{1 + e^{-\alpha - \beta x_i}}\right)^{y_i} \left(1 - \frac{1}{1 + e^{-\alpha - \beta x_i}}\right)^{1 - y_i},$$

for $i = 1, \ldots, n$. In other words, Y_i given $X_i = x_i$ has the Bernoulli distribution, and the probability of "success" is some function of the value x_i.

The likelihood function is given by

$$L(\alpha, \beta; Y_1, \ldots, Y_n) = \prod_{i=1}^{n} \left(\frac{1}{1 + e^{-\alpha - \beta x_i}}\right)^{Y_i} \left(1 - \frac{1}{1 + e^{-\alpha - \beta x_i}}\right)^{1 - Y_i}.$$

After some computation, setting the partial derivatives of the log-likelihood function with respect to α and β equal to 0 leads to the following equations:

$$\sum_{i=1}^{n} \frac{Y_i - \Psi(\hat{\alpha} + \hat{\beta} x_i)}{\Psi(\hat{\alpha} + \hat{\beta} x_i)\left(1 - \Psi(\hat{\alpha} + \hat{\beta} x_i)\right)} \Psi'(\hat{\alpha} + \hat{\beta} x_i) = 0,$$

$$\sum_{i=1}^{n} \frac{Y_i - \Psi(\hat{\alpha} + \hat{\beta} x_i)}{\Psi(\hat{\alpha} + \hat{\beta} x_i)\left(1 - \Psi(\hat{\alpha} + \hat{\beta} x_i)\right)} \Psi'(\hat{\alpha} + \hat{\beta} x_i) x_i = 0,$$

where $\Psi(x) = (1 + e^{-x})^{-1}$. We cannot solve $\hat{\alpha}$ and $\hat{\beta}$ explicitly from these equations. In practice this is not a problem because we can easily solve the equations numerically, using an iterative algorithm. This also holds in the case of a multidimensional predictor variable X.

The likelihood equations given above also hold when we use another "linking" function than the logistic function Ψ. In probit regression, for example, where instead of Ψ we use the normal distribution function Φ, we find the same equations with Ψ replaced by Φ. The logistic distribution function Ψ has the computational advantage that $\Psi' = \Psi(1 - \Psi)$, so that the likelihood equations can be greatly simplified. They do, however, remain nonlinear.

7.5.2 Tests

In the logistic regression model with a multidimensional predictor variable X, it is interesting to test whether a certain component of the predictor variable influences the response. This is, for example, an interesting hypothesis when the dependent variable Y stands for filing or not filing a claim with an insurance company in the past two years and the first coordinate X_1 indicates the age of the insured person. When the insurance company wants to know whether the age indeed has a predictive value for filing claims, the null hypothesis $H_0: \beta_1 = 0$ that the first coordinate of the parameter vector $\beta = (\beta_1, \ldots, \beta_m)$ is 0 can be tested against the alternative $H_1: \beta_1 \neq 0$.

To test a null hypothesis of the form $H_0: (\beta_j: j \in J) = 0$, the likelihood ratio test is often used. Although there is no analytic expression for the likelihood ratio statistic, the value of the likelihood ratio statistic can easily be determined numerically using an iterative algorithm. For observed values $(x_{1,1}, \ldots, x_{1,m}, y_1)$, $\ldots, (x_{n,1}, \ldots, x_{n,m}, y_n)$, the value of the likelihood can be computed as

$$L(\beta, y_1, \ldots, y_n) = \prod_{i=1}^{n} \left(\frac{1}{1 + e^{-\sum_{j=1}^{m} \beta_j x_{i,j}}} \right)^{y_i} \left(1 - \frac{1}{1 + e^{-\sum_{j=1}^{m} \beta_j x_{i,j}}} \right)^{1-y_i};$$

see Section 7.5.1. To compute the likelihood ratio statistic, it therefore suffices to determine the maximum likelihood estimators under the full model and under the null hypothesis. In Section 7.5.1, we discuss a particular case of this computation.

Standard computer software packages often do not report the likelihood ratio statistic directly; they rather give it through a so-called (residual) *deviance*. This is equal to twice the log-likelihood ratio statistic, $2 \log \lambda_n$, for testing the null hypothesis that the model is correct, that is, that there exists a vector $\beta \in \mathbb{R}^m$ such that (7.12) holds, inside the full model that $Y_1, \ldots, Y_n$ are independent Bernoulli variables with possibly different probabilities of success $p_i = P(Y_i = 1)$. The deviance therefore gives a measure for the fit of the logistic regression model. Moreover, the difference between the deviance for the full model (7.12) and the deviance for a submodel (such as the model under $H_0: (\beta_j: j \in J) = 0$) is equal to twice the log-likelihood ratio statistic for testing the submodel.

An alternative for the likelihood ratio test is the Wald test (see Section 4.8), whose p-values are often reported in computer outputs.

7.5.3 Confidence Regions

We can determine confidence regions for the parameter β using the Fisher information matrix. In this section, we again restrict ourselves to a one-dimensional predictor variable, as in Section 7.5.1,

$$P_{\alpha,\beta}(Y = 1 | X = x) = \frac{1}{1 + e^{-\alpha - \beta x}}.$$

Let $(X_1, Y_1), \ldots, (X_n, Y_n)$ be independent, identically distributed random vectors with $Y_i \in \{0, 1\}$, and assume that the joint probability distribution of X and Y is given by

$$P_{\alpha,\beta}(X = x, Y = y) = P_{\alpha,\beta}(Y = y | X = x) p^X(x)$$
$$= \left(\frac{1}{1 + e^{-\alpha - \beta x}} \right)^y \left(1 - \frac{1}{1 + e^{-\alpha - \beta x}} \right)^{1-y} p^X(x),$$

for unknown parameters (α, β). Here, p^X is the marginal density (or probability mass function) of X. We can estimate the parameters α and β using their maximum likelihood estimators (see Section 7.5.1). The score function of the model is given by

$$\dot{\ell}_{\alpha,\beta}(x, y) = \frac{y - \Psi(\alpha + \beta x)}{\Psi(\alpha + \beta x)(1 - \Psi(\alpha + \beta x))} \Psi'(\alpha + \beta x) \binom{1}{x}.$$

The Fisher information matrix is therefore given by (see Section 5.4.2)

$$i_{\alpha,\beta} = \int \frac{\Psi'(\alpha + \beta x)^2}{\Psi(\alpha + \beta x)\big(1 - \Psi(\alpha + \beta x)\big)} \begin{pmatrix} 1 & x \\ x & x^2 \end{pmatrix} p^X(x)\,dx,$$

where the integral is taken over all possible outcomes of X. We can estimate this matrix by replacing the integral by a sum over the observations and replacing the marginal density p^X by $1/n$ for every observation. In other words, we estimate the marginal distribution of X using the empirical marginal distribution of X. We moreover use the idea of a plug-in estimator for the Fisher information and replace α and β by their maximum likelihood estimators. This gives

$$\widehat{i_{\alpha,\beta}} = \frac{1}{n}\sum_{i=1}^{n} \frac{\Psi'(\hat\alpha + \hat\beta x_i)^2}{\Psi(\hat\alpha + \hat\beta x_i)\big(1 - \Psi(\hat\alpha + \hat\beta x_i)\big)} \begin{pmatrix} 1 & x_i \\ x_i & x_i^2 \end{pmatrix}.$$

An approximate confidence region for (α, β) of size α_0 is given by the set

$$\left\{ (\alpha, \beta) : \begin{pmatrix} \alpha - \hat\alpha & \beta - \hat\beta \end{pmatrix} n\,\widehat{i_{\alpha,\beta}} \begin{pmatrix} \alpha - \hat\alpha \\ \beta - \hat\beta \end{pmatrix} \leq \chi^2_{2,1-\alpha_0} \right\}.$$

* 7.6 Cox Regression Model

In *survival analysis*, we are interested in the probability distribution of time spans. You can think of the life span of a device, the incubation time of an illness, the time before dying after a serious operation, the time before an ex-convict commits a new crime (see the application after Chapter 1), but also the time before the next bug occurs in a computer program ("reliability analysis").

Models in survival analysis are often given in terms of the *risk function* or the *hazard function*. The hazard function associated with a probability density f is defined as

$$\lambda(t) = \frac{f(t)}{1 - F(t)},$$

where F is the corresponding distribution function. If we view $f(t)\,dt$ as the probability that a life span T lies in the interval $[t, t + dt)$, then $\lambda(t)\,dt$ has the interpretation

$$\lambda(t)\,dt \approx \frac{\mathrm{P}(t \leq T < t + dt)}{\mathrm{P}(T \geq t)} = \mathrm{P}(t \leq T < t + dt \mid T \geq t).$$

The value $\lambda(t)$ is therefore the conditional probability of "dying" right after time t given that at time t, the person or product is still "alive." It is this interpretation as an "instantaneous probability" that makes the hazard function appealing as a tool for modeling.

The hazard function $t \mapsto \lambda(t)$ is the derivative of $t \mapsto -\log(1 - F(t))$ with respect to t. Given the hazard function λ, we can recover the distribution function F using the formula $F(t) = 1 - e^{-\Lambda(t)}$, for Λ the cumulative hazard function, that is, the primitive function of λ with $\Lambda(0) = 0$ (if $F(0) = 0$). The density f is then equal to $f(t) = \lambda(t)e^{-\Lambda(t)}$.

A popular model from medical statistics is the *Cox model*, proposed by Cox in the 1970s. In this model, the life span T (the dependent variable) is correlated to a vector X of predictor variables such as age, weight, blood pressure, prognosis, etc. The Cox model postulates that the hazard function of a patient with "covariate vector" x is equal to

$$\lambda^{T|X=x}(t) = e^{\beta^T x} \lambda(t).$$

Here, λ is the hazard function of a patient with predictor variable $x = 0$, the so-called "baseline hazard." Thus, the Cox model postulates that the hazard functions of two patients with predictor variables x_1 and x_2 are proportional,

$$\frac{\lambda^{T|X=x_1}(t)}{\lambda^{T|X=x_2}(t)} = e^{\beta^T (x_1 - x_2)},$$

independently of t. This gives a simple interpretation for the parameter β: it determines the size of the relative risks attached to certain predictor variables. For example, if x is body weight, T is the age at time of death, and $\beta = 1.4$, then the risk of dying now for someone of weight 120 kg is a factor $e^{1.4*(120-90)}$ times as great as that for someone who weights 90 kg. That this relative risk is independent of time (and therefore of age in this example) simplifies the interpretation, but is not always a reasonable assumption. That is why there are many variations of the Cox model.

In the Cox model, the hazard function λ is not specified. The model therefore has as parameter the pair $\theta = (\beta, \lambda)$ consisting of a vector β and a function λ. Both parameters are estimated from the available data, for example a sample $(T_1, X_1), \ldots, (T_n, X_n)$ of life spans and predictor variables.

We can also set as model that λ has a certain form. The assumption that the function $t \mapsto \lambda(t)$ is constant, for example, corresponds to the assumption that if the variable x is equal to 0, then the survival distribution is the exponential distribution. This assumption, which has the interpretation that "new is just as good (or bad) as used," is in general unrealistic in medical statistics, but can (unfortunately) be realistic for the number of remaining bugs in reliability theory. The *Weibull* family, for which the function λ is a power function $\lambda(t) = \beta t^\alpha$, is another possible family. The advantage of the Cox model without specification of the hazard function λ over these possibilities is that it avoids the randomness of the choice of a particular type of function, so that the parameters that are estimated using the Cox model will more frequently give a good approximation for the data. On the other hand, if there are reasonable grounds to expect a certain form for the hazard function, then it is better not to use the Cox model because it contains more prior uncertainty.

A difficult aspect of survival analysis is that, often, not all life spans are observed. At the moment when we want to draw conclusions from the data, for example, not all individuals have "died," and we only know a lower bound for those life spans. In medical applications, it frequently occurs that patients cannot be followed until their death, for example because they have moved away or because they have died from another cause than the one being studied. In those cases, too, we only observe a lower bound for the life spans. We then speak of *censored data*. Long life spans are more frequently censored than short ones. The reason is that moving away, death from another cause, or the end of the study have a greater probability of occurring in a long time interval than in a short one. It would therefore be wrong to ignore censored data because we would then ignore relatively many long life spans. This could lead to an underestimate of the life span distribution. A correct approach is to use a statistical model for all observations.

7.6.1 Estimation

In an uncensored Cox model with one-dimensional predictor variable X, we assume that we have a sample $(T_1, X_1), \ldots, (T_n, X_n)$. This model is specified by the conditional hazard function

$$\lambda^{T|X=x}(t) = e^{\beta x}\lambda(t).$$

This corresponds to a conditional density of the form

$$f^{T|X=x}(t) = e^{\beta x}\lambda(t)e^{-e^{\beta x}\Lambda(t)}$$

and a conditional distribution function

$$F^{T|X=x}(t) = 1 - e^{-e^{\beta x}\Lambda(t)},$$

with Λ the cumulative hazard function. For the maximum likelihood estimator for the parameter (β, λ), we need the likelihood function. This is given by

$$(\beta, \lambda) \mapsto \prod_{i=1}^{n} f^{T|X_i}(T_i)p_X(X_i) = \prod_{i=1}^{n} e^{\beta X_i}\lambda(T_i)e^{-e^{\beta X_i}\Lambda(T_i)}p_X(X_i)$$

$$= \prod_{i=1}^{n} e^{\beta X_i}\lambda(T_i)e^{-e^{\beta X_i}\Lambda(T_i)} \prod_{j=1}^{n} p_X(X_j),$$

where p_X is the marginal density of the predictor variable. Because it is not obvious that this distribution contains information on the parameters, we can disregard the term $\prod_{j=1}^{n} p_X(X_j)$ in the likelihood when we maximize with respect to (β, λ). The maximum likelihood estimator for (β, λ) is therefore the value that maximizes the function

$$(\beta, \lambda) \mapsto \prod_{i=1}^{n} e^{\beta X_i}\lambda(T_i)e^{-e^{\beta x}\Lambda(T_i)}$$

over all possible parameter values (β, λ).

Unfortunately, this problem has no solution (a point of maximum does not exist, and the supremum over all possible parameter values is infinite) because the parameter space for λ, the set of all hazard functions, is too large. The most common modification of the problem is to define it in terms of (β, Λ) instead of (β, λ). We then replace the factor $\lambda(T_i)$ by the jump $\Delta\Lambda(T_i)$ in the cumulative hazard function in T_i. In other words, we are looking for the pair $(\hat{\beta}, \hat{\Lambda})$ that maximizes the function

$$(\beta, \Lambda) \mapsto \prod_{i=1}^{n} e^{\beta X_i} \Delta\Lambda(T_i) e^{-e^{\beta X_i}\Lambda(T_i)}$$

over all possible parameter values (β, Λ) consisting of a scalar β and a right-continuous, monotonically increasing function $\Lambda\colon [0, \infty) \mapsto [0, \infty)$ with $\Lambda(0) = 0$. This problem does have a solution, known as the *Cox estimator*. In practice, to compute this estimator we need an iterative algorithm. This is standard in computer packages for survival analysis.

An estimate for β is a number, while an estimate for Λ is a function. Often, the computer output does not list the estimate for Λ itself; rather, it lists the corresponding baseline survival function, $1 - F^{T|X=0}(t) = e^{-\Lambda(t)}$. Figure 7.6 shows an example.

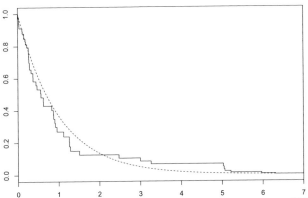

Figure 7.6. Estimate of the baseline survival function in the Cox model $\lambda^{T|X=x}(t) = e^{\beta x}\lambda(t)$ based on 50 observations, generated using the Weibull baseline hazard λ, a one-dimensional predictor variable x from the uniform distribution on $[-5, 5]$, and the parameter $\beta = -0.3$. The dashed curve is the true baseline survival function $1 - F^{T|X=0}(t) = e^{-\Lambda(t)}$.

7.6.2 Tests and Confidence Intervals

Suppose that we want to test the null hypothesis $H_0\colon \beta = \beta_0$ against the alternative $H_1\colon \beta \neq \beta_0$. We can use the likelihood ratio test. In the numerator of this test, the likelihood is maximized over the full parameter space. In the denominator, on the other hand, the maximization is restricted to the smaller parameter space with $\beta = \beta_0$. In

293

the previous subsection, we gave an (adapted) likelihood function of the general Cox model for survival analysis:

$$(\beta, \Lambda) \mapsto \prod_{i=1}^{n} e^{\beta X_i} \Delta \Lambda(T_i) e^{-e^{\beta X_i} \Lambda(T_i)} \prod_{j=1}^{n} p_X(X_j).$$

The factor $\prod_{j=1}^{n} p_X(X_j)$ occurs in both the numerator and the denominator of the likelihood ratio statistic. If the distribution p_X does not depend on the parameter (β, λ), then the products in the numerator and denominator cancel each other out. Hence, it again suffices to maximize the function

$$L(\beta, \lambda; T_1, X_1, \ldots, T_n, X_n) = \prod_{i=1}^{n} e^{\beta X_i} \Delta \Lambda(T_i) e^{-e^{\beta X_i} \Lambda(T_i)}.$$

with respect to (β, λ) under H_0 and under H_1.

The likelihood function given above is an example of a "semiparametric likelihood function" because the parameter Λ does not vary over a finite-dimensional space, but over a function space. We can, however, show that the likelihood ratio statistic for testing $H_0: \beta = \beta_0$ asymptotically has a chi-square distribution with one degree of freedom. If β is multidimensional, then the number of degrees of freedom of the chi-square distribution is equal to the dimension of β. The profile likelihood can therefore be used to construct a confidence interval for β in exactly the same way as for parametric models, that is, models in which all parameters are finite-dimensional.

Although there does not exist an analytic expression for the maximum likelihood estimator $\hat{\beta}$, we can compute the profile likelihood exactly (see Section 5.6). We first verify that for fixed β, the likelihood $\Lambda \mapsto L(\beta, \Lambda; T_1, X_1, \ldots, T_n, X_n)$ is maximized by a function Λ that has jumps at every one of the points T_i and is constant on each of the intervals $[T_{(i-1)}, T_{(i)})$, where $T_{(i)}$ is the ith order statistic. Given the jump sizes $\lambda_i = \Delta \Lambda(T_i)$, the likelihood function is given by

$$(\beta, \lambda_1, \ldots, \lambda_n) \mapsto \prod_{i=1}^{n} e^{\beta X_i} \lambda_i e^{-e^{\beta X_i} \sum_{j:T_j \leq T_i} \lambda_j}.$$

We can maximize this expression with respect to $(\lambda_1, \ldots, \lambda_n)$ in $[0, \infty)^n$ in the usual way, by first taking the logarithm and then setting the partial derivatives with respect to λ_i equal to 0. We can write the resulting likelihood equations in the form

$$\frac{1}{\lambda_i} = \sum_{k:T_k \geq T_i} e^{\beta X_k}.$$

The profile likelihood is obtained by substituting these values in the likelihood and is therefore equal to

$$L_1(\beta; T_1, X_1, \ldots, T_n, X_n) = \sup_{\Lambda} L(\beta, \Lambda; T_1, X_1, \ldots, T_n, X_n)$$

$$= \prod_{i=1}^{n} \frac{e^{\beta X_i}}{\sum_{j:T_j \geq T_i} e^{\beta X_j}} e^{-1}.$$

This expression follows from

$$\prod_{i=1}^{n} e^{-e^{\beta X_i} \sum_{j:T_j \leq T_i} \lambda_j} = e^{-\sum_{i=1}^{n} \sum_{j:T_j \leq T_i} e^{\beta X_i} \lambda_j}$$

$$= e^{-\sum_{j=1}^{n} \sum_{i:T_i \geq T_j} e^{\beta X_i} \lambda_j} = \prod_{j=1}^{n} e^{-1},$$

where the last equality follows by substituting the expression for λ_j.

We can maximize the profile likelihood with respect to β using a numeric algorithm, to determine the maximum likelihood estimator $\hat{\beta}$. The values β for which the logarithm of the profile likelihood is closer than $\frac{1}{2}\chi^2_{1,1-\alpha_0}$ to the maximum value of the logarithm of the profile likelihood form a confidence region for β of size approximately $1 - \alpha_0$.

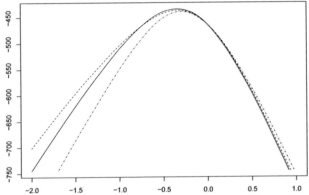

Figure 7.7. Three realizations of the log-profile likelihood for β for the Cox model $\lambda^{T|X=x}(t) = e^{\beta x} \lambda(t)$ based on 100 observations, generated using the standard Weibull baseline hazard λ, the one-dimensional predictor variable x generated from the uniform distribution on $[-5, 5]$, and the true parameter $\beta = -0.3$.

7.7 Mixed Models

A *mixed model* can be viewed as a regression model in which some of the parameters have been replaced by random variables, called random effects. Such a model is useful to model dependence between the observations or heterogeneities in the data.

In the *linear mixed model*, the observation is a vector $Y = (Y_1, \ldots, Y_n)^T$ satisfying

$$(7.13) \qquad\qquad Y = X\beta + Z\gamma + e.$$

295

Here, X and Z are known $(n \times p)$- and $(n \times q)$-matrices, $\beta \in \mathbb{R}^p$ is an unknown parameter vector, and γ and e are Gaussian random vectors in $\mathbb{R}^q$ and $\mathbb{R}^n$, respectively. The vector e is an error vector with the same interpretation as in the multivariate linear regression model (7.3) and is usually modeled as having i.i.d. coordinates $e_1, \ldots, e_n$ with mean 0. The vector $X\beta$ is also borrowed from the linear regression model. The novelty is the vector $Z\gamma$, which is random and typically creates dependencies between the $Y_1, \ldots, Y_n$. As the error vector e, the vector of *random effects* γ is not observed. We can view this vector as a device to describe the distribution of the observation Y through the structural equation (7.13) or as a way to model real world variation between observational units at a deeper level. Such an unobserved random vector is also called a *latent variable*.

Example 7.10 Variance components

In the ANOVA model (7.8), the parameters α_i, β_j and γ_{ij} model the effects of factors that influence the outcomes Y_{ijk}. When measured at factor levels i and j, the deviation of the mean value of Y_{ijk} from the overall average is given by $\alpha_i + \beta_j + \gamma_{ij}$. It may be that the levels of these factors can be considered a sample from a bigger set of possible levels, and interest is in generalizing the conclusions to this bigger population. Then, it makes sense to model one or both factors as random variables.

As an example, suppose that an experiment to measure the effect α_i of a factor is carried out in a number of different laboratories. Care has been taken to align the experimental procedures, but it is suspected that the measurements are still dependent on the laboratories. We can model this by the ANOVA model (7.8), with β_j referring to the jth laboratory, but we are not particularly interested in the laboratories and view the laboratories that participate in the experiment as exchangeable with other ones. Then, we might model the β_j, which describe the main effects of the laboratories, as a sample from a distribution. If we stick to the α_i as parameters, we then have a mixed effects model, with *fixed effects* α_i and random effects β_j. If we include interaction effects γ_{ij}, then naturally these will also be modeled as random effects.

In the general notation (7.13), the vector β now corresponds to the vector of parameters μ and α_i, and the vector γ to the vector of all β_j and all γ_{ij}. Once the variables Y_{ijk} are ordered as coordinates of a vector, the matrices X and Z can be determined so that (7.13) is valid.

Example 7.11 Repeated measures

Mixed models are frequently used when experimental units are measured more than once, such as in *longitudinal analysis*, where subjects are followed over time. If there is variation between the units, then an ordinary regression model is inappropriate because this models all observations as independent, while measurements on the same subject are clearly related. Random effects can be used to model the dependence.

Repeated measurements were also discussed in Example 3.45 on generalized estimated equations. We can view the present random effects as a way to model the error vector $(e_{i,t})$ in the GEE model structurally through the vector $Z\gamma + e$.

We consider one concrete example as illustration. Suppose that we follow S subjects, indexed $s = 1, \ldots, S$, over time, and measure each subject at times $t = 1, \ldots, T$. The observational vector Y can then best be described by two indices, for subject and time, and be partitioned in blocks as $(Y_{1,1}, \ldots, Y_{1,T}, Y_{2,1}, \ldots, Y_{2,T}, \ldots, Y_{S,1}, \ldots, Y_{S,T})^T$, where $Y_{s,1}, \ldots, Y_{s,T}$ are the consecutive measurements on individual s. A simple model would be

$$Y_{s,t} = \beta_0 + \beta_1 t + \gamma_{s,0} + \gamma_{s,1} t + e_{s,t},$$

where $e_{s,t}$ are i.i.d. univariate normal variables. The idea of this model is that every individual s follows a linear model in time t, but the intercept and slopes vary over the individuals: for individual s these are $\beta_0 + \gamma_{s,0}$ and $\beta_1 + \gamma_{s,1}$. The parameters β_0 and β_1 are the "population intercept and slope," while the parameters $\gamma_{s,0}$ and $\gamma_{s,1}$ are the deviations of individual s from the average. If the data consists of measurements on individuals sampled from some population, then it makes sense to think of the pairs $(\gamma_{s,0}, \gamma_{s,1})$ as a sample from a distribution. The vector $\gamma = (\gamma_{1,0}, \gamma_{1,1}, \ldots, \gamma_{S,0}, \gamma_{S,1})^T$ is then also random.

We define the parameter $\beta = (\beta_0, \beta_1)^T$ by collecting the two fixed effects parameters. The model can next be written in matrix form (7.13) by defining the appropriate matrices X and Z, so that the equations in the preceding display are valid. Since the pairs $(\gamma_{1,0}, \gamma_{1,1}), \ldots, (\gamma_{S,0}, \gamma_{S,1})$ correspond to different individuals, it is natural to restrict the covariance matrix Λ of γ to matrices with block structure, expressing the independence of these S pairs. For identically distributed pairs this gives 3 free parameters. Together with the parameters β_0, β_1, and σ^2, the model then has 6 free parameters. ▭

If the random effects and error vectors are distributed as $\gamma \sim N_q(0, \sigma^2 \Lambda)$ and $e \sim N_n(0, \sigma^2 I)$, then the observation Y in (7.13) possesses an $N_n(X\beta, \sigma^2 Z\Lambda Z^T + \sigma^2 I)$-distribution, where I is the $(n \times n)$-identity matrix. The parameter value is $(\beta, \sigma^2, \Lambda)$, where $\beta \in \mathbb{R}^p$, $\sigma^2 > 0$, and Λ is a positive-definite matrix, which is usually restricted to a subset of such matrices through further specification of the model. The likelihood function is given by the multivariate normal density, and its logarithm is, up to a constant, equal to

$$(\beta, \sigma^2, \Lambda) \mapsto -n \log \sigma - \frac{1}{2} \log \det(Z\Lambda Z^T + I)$$
$$- \frac{1}{2\sigma^2}(Y - X\beta)^T (Z\Lambda Z^T + I)^{-1}(Y - X\beta).$$

This may be maximized with respect to the parameter to find the maximum likelihood estimator, and likelihood ratio tests or confidence sets can be obtained as usual. A difficulty is that many situations lack an analytic expression for the maximum likelihood estimator. A numerical routine, such as Fisher scoring, is necessary to approximate the solution. The presence of the matrices Λ can make this nontrivial, but several packages are available.

It turns out that the maximum likelihood estimators are somewhat biased, similarly to how the maximum likelihood estimator of the variance σ^2 in an ordinary regression model is biased. This is often solved by *restricted maximum likelihood estimation* or *REML*, which separates the estimation of the parameter β from the estimation of the variances. The idea is to remove the parameter β from the model by premultiplying the equation (7.13) by a matrix P whose rows are orthogonal to the column space of X, so that $PX = 0$. Then, the distribution of $PY \sim N(0, \sigma^2 PZ\Lambda Z^T P^T + \sigma^2 PP^T)$ is free of β and is used to estimate (σ^2, Λ). This removes the bias resulting from the "loss of degrees of freedom" by estimating the parameter β. In a second step, the parameter β is estimated using the conditional likelihood of Y given PY, in which the unknown parameters (σ^2, Λ) are fixed at the estimates from the first step (which depend on PY only). A "maximal" matrix P with the desired properties is given by the orthogonal projection $P = I - X(X^TX)^{-1}X^T$ onto the orthocomplement of the range of X.

Because the random effects γ can be viewed as "random parameters," a Bayesian description of the model is natural. The distribution of γ is then equated to a prior distribution on these parameters. Indeed, the general Bayesian framework makes little difference between latent variables and parameters, and even data is special only in that it is observed, all quantities of interest being random. This is in contrast to the non-Bayesian view, which considers the parameter $(\beta, \sigma^2, \Lambda)$ as unknown but having a true value independent of the statistician, whereas the latent vector γ is truly random. Bayesian methods for mixed models are typically also of a numerical nature, but several standard software solutions are available (e.g. MCMC). One advantage offered by the Bayesian setup is that we can speak of the posterior distribution of the random effects γ given the data Y, which is an ordinary conditional distribution, given by Bayes's rule. Non-Bayesians often also like to reconstruct the random effects from the data, but as these are random this does not fit naturally in the framework of "parameter estimation."

Just as the linear regression model, other models can be similarly augmented with random effects. For instance, the logistic regression model (7.12) can be augmented to

$$
\mathrm{P}(Y = 1 \mid X = x, Z = z, \gamma) = \frac{1}{1 + e^{-\sum_{j=1}^{m} \beta_j x_j - \sum_{j=1}^{k} \gamma_j z_j}}.
$$

As the random effects γ are not observed, the distribution of Y that is relevant for forming a likelihood is its marginal distribution. This is obtained by multiplying the probability in the display by the marginal density of γ, which could be multivariate Gaussian, and then integrating over γ. A difference with the linear mixed model is that this cannot be achieved analytically, and hence implementation of general linear mixed models requires advanced numerical routines.

A Cox model that is augmented with a random effect is known as a *frailty model*. In a survival setting, the random effect then often models an unmeasured risk factor for death, which explains the name "frailty."

7.8 Summary

For a given *dependent variable* Y and *predictor variables* $x_1, \ldots, x_p$, the relations among the dependent and predictor variables can be described in a *regression model*. Here is an overview of common regression models:

- In a *linear regression model*, we assume that Y is a continuous random variable. The relation between Y and $(x_1, \ldots, x_p)$ is given by

$$Y = \beta_0 + \beta_1 x_1 + \ldots + \beta_p x_p + e,$$

  a linear function of the unknown regression parameters $\beta_0, \ldots, \beta_p$, with e a normally distributed error term with expectation 0 and unknown variance σ^2.

- In *analysis of variance*, the dependent variable Y is continuous, and the predictor variables ("factors") are categorical. In a model with two factors with interactions, the relation is modeled as

$$Y_{ij} = \mu + \alpha_i + \beta_j + \gamma_{ij} + e,$$

  where Y_{ij} is the dependent variable for the categories i and j, respectively, of the two factors, μ is the grand mean, α_i and β_j are the unknown main effects of the two factors, γ_{ij} is the unknown interaction between the two factors, and e is a normally distributed error term with expectation 0 and unknown variance σ^2.

- In a *nonlinear model*, we have

$$Y = f_\theta(x) + e,$$

  where f_θ is a nonlinear function in θ and e is an error term, which is often assumed to be normally distributed with expectation 0 and unknown variance σ^2.

- In a *logistic regression model*, the dependent variable Y is binary, so that we are dealing with a classification problem. The relation between Y and the vector $x = (x_1, \ldots, x_p)$ is given by

$$P(Y = 1 | X = x) = \frac{1}{1 + e^{-\sum_{j=1}^{p} \beta_j x_j}}.$$

- In a *Cox regression model*, the dependent variable is the time up to a certain event and is often denoted by T instead of Y. The relation between T and the predictor vector $x = (x_1, \ldots, x_p)$ is described using the hazard function

$$\lambda^{T|x}(t) = \frac{f^{T|x}(t)}{1 - F^{T|x}(t)} = e^{\sum_{j=1}^{p} \beta_j x_j} \lambda(t),$$

  with λ the hazard function for the individual or object with $x_1 = \ldots = x_p = 0$, which is completely unknown.

There exist different methods to estimate or test the unknown parameters in the models described above.

Exercises

1. We want to research whether there is a correlation between the average grade for the final exams of the last year of high school (on a scale from 1 to 10) and the duration of university studies (in months). We have the following data, which refer to a sample under students working toward (the equivalent of) a master's degree in mathematics who arrived at the University of Amsterdam in 1970.

grade	8	7	7	7	7.5	7	6.5	9	7	9	8	7.5
duration	82	80	66	77	79	75	58	46	58	56	70	55

(i) Determine estimates for the regression coefficients in the linear regression model using the method of least squares.
(ii) Compute the explained variance.
(iii) Make a graph showing the observations and the fitted straight line.

2. We want to include the predictor variable "education" in a regression model. There are three categories: low, middle, high. We include the variable X in the model as follows. We define $X = 1$ if the person is low skilled, $X = 2$ if the person has an average education level, and $X = 3$ if the person is highly educated. Is this a good choice, or would you choose differently? Why?

3. Consider the simple linear regression model. Based on observations $Y_1, \ldots, Y_n$ and $x_1, \ldots, x_n$, we want to predict the expected value of Y corresponding to a given x. Determine an unbiased estimator for this value.

4. Consider the linear regression model of Section 7.2.
(i) Show that the estimators for the regression coefficients are linear combinations $\hat{\alpha} = \sum \mu_i Y_i$ and $\hat{\beta} = \sum \lambda_i Y_i$ of the observations.
(ii) Show that $\hat{\alpha}$ and $\hat{\beta}$ are unbiased estimators for α and β.
(iii) Show that $\mathrm{MSE}(\alpha, \beta, \sigma^2; \hat{\alpha}) = \sigma^2/n + \bar{x}^2 \sigma^2 / \left((n-1)s_x^2\right)$.
(iv) Show that $\mathrm{MSE}(\alpha, \beta, \sigma^2; \hat{\beta}) = \sigma^2 / \left((n-1)s_x^2\right)$.
(v) Suppose that the values $x_1, \ldots, x_n$ can be fixed by the researcher within a given interval $[a, b]$. What is the optimal choice to obtain the best estimate of β?
(vi) Is there a practical reason not to make this optimal choice?

5. Consider the simple linear regression model, but suppose that we know beforehand that $\alpha = 0$.
(i) Determine the least-squares estimator for β.
(ii) Is this estimator unbiased?
(iii) Determine the variance of this estimator.
(iv) Suppose that the values $x_1, \ldots, x_n$ can be fixed by the researcher within a given interval $[0, b]$. What is the optimal choice to obtain the best estimate of β?
(v) Is there a practical reason not to make this optimal choice?

6. Consider the simple regression model $Y_i = \alpha + \beta x_i + e_i$, where α and β are unknown parameters, $x_1, \ldots, x_n$ are known constants, and $e_1, \ldots, e_n$ are independent, normally distributed with $\mathrm{E}e_i = 0$ and $\mathrm{E}e_i^2 = z_i \sigma^2$ for known positive numbers $z_1, \ldots, z_n$. Such a model with errors of different levels of accuracy is called *heteroscedastic*. Determine the maximum likelihood estimators for α and β.

7. Consider the simple linear regression model of Section 7.2.
(i) Find a sufficient vector for $(\alpha, \beta, \sigma^2)$ that is as small as possible.
(ii) Show that this model forms an exponential family.
(iii) Show that the least-squares estimators $\hat{\alpha}$ and $\hat{\beta}$ are UMVU.
(vi) Find a UMVU estimator for σ^2.

8. Suppose that we want to determine the gravitational constant using a pendulum test. We attach a massive ball to a thin cord. We then swing the ball and measure the period of oscillation. The period of oscillation is the time it takes the ball to make one complete cycle. If the amplitude of the ball is not too great, then the period of oscillation is constant and therefore independent of the amplitude. By Newton's second law of motion, we have $T = 2\pi\sqrt{l}/\sqrt{g}$, where T (in s) is the period of oscillation, g (in m/s^2) is the gravitational constant, and l (in m) is the length of the cord. We assume that in our set-up, Newton's second law holds. We carry out the pendulum test n times with cord lengths $l_1, \ldots, l_n$. This gives observations $(l_1, T_1), \ldots, (l_n, T_n)$. We make small measurement errors, so that the measurements do not follow Newton's second law exactly, but only approximately.
(i) Describe a suitable simple linear regression model.
(ii) Determine the least-squares estimator for the gravitational constant g.
(iii) Determine the variance of the least-squares estimator for the regression parameter $2\pi/\sqrt{g}$ that represents the slope.
(iv) How would you choose the cord lengths in the experiment to minimize the variance in question (iii)?
(v) Explain (or compute) how you would choose the cord lengths in the experiment to minimize the variance of the least-squares estimator for the gravitational constant.

9. In a study on the influence of caffeine on memory, 18 students are asked to study a book for 8 hours. Every hour they have a break and are given coffee with little, an average amount of, or much caffeine. The coffee is sweetened with only sugar, half sugar and half sweetener, or only sweetener. Other food or drinks are not allowed. In all, there are 3×3 possible drinks. Two students are arbitrarily chosen for each combination, which they then get during each break. After 8 hours of study, the students are given a test with 100 questions on the material they have studied. The numbers of wrong answers are registered.

	little	average	much
only sugar	12 10	7 12	19 21
half sugar, half sweetener	6 3	10 16	21 29
only sweetener	15 12	8 6	24 13

Table 7.3. Number of wrong answers made by the students. Per combination of amount of caffeine and type of sweetener, the numbers of mistakes made by two students have been registered.

(i) To begin simply, we first disregard the factor "sugar/sweetener" and the interaction term with caffeine in the model. Set up a suitable analysis of variance model.
(ii) Test whether the amount of caffeine in the coffee influences the number of wrong answers. Give the null and alternative hypotheses, the test statistic, and the p-value. What is your conclusion?
(iii) Now, consider the full model, including interaction terms, and model this. Again, test whether the amount of caffeine in the coffee influences the number of wrong answers. What is your conclusion?

    (iv) Would you still use this model if the students could themselves choose how much sugar they want in their coffee?

10. A farmer wants to study which combination of wheat species (2 species), fertilizer (3 types), and method of irrigation (2 methods) produces the highest yield. Describe an experiment to study this, and give the corresponding analysis of variance model with the yield as dependent variable.

11. Suppose that the probability of contracting disease Z is influenced by a single gene. Suppose, moreover, that the gene is biallelic with alleles A_1 and A_2. A person can then have genotype (A_1, A_1), (A_1, A_2), or (A_2, A_2). The variables blood pressure and gender also influence the probability of contracting disease Z. We have observed the occurrence of the disease, genotype, gender, and blood pressure of 100 individuals.

    (i) Suppose that A_2 is dominant over A_1, which means that if we disregard the other factors, then a person with genotype (A_1, A_2) has the same risk of contracting disease Z as a person with genotype (A_2, A_2). Describe a suitable logistic regression model for the data.

    (ii) Suppose that a person with genotype (A_1, A_2) has a smaller risk of contracting disease Z than a person with genotype (A_1, A_1), but a greater risk than someone with genotype (A_2, A_2). Describe a suitable logistic regression model.

12. Obese people who are put on a diet exhibit an exponential decrease of their adipose tissue during the period of the diet. To study how long someone should be put on a specific diet, the weight loss of n subjects is measured daily for a month while they follow the diet. Formulate a suitable statistical model for the weight loss. (Several models are possible!)

13. People with heart valve disease often get a new heart valve. There are two types of replacement heart valves: biological and mechanical. In a study on the life span of the biological valve, n patients are followed from their operation until the valve fails. We suspect that the age, weight, and gender of the patient influence the life span of the valve. Formulate a suitable statistical model to investigate whether mechanical heart valves last longer or shorter than biological heart valves.

14. Suppose that the random variable T has probability density f and hazard function λ. Show that $f(t) = \lambda(t)e^{-\Lambda(t)}$.

15. Show that the hazard function of a random variable X is constant if and only if X has an exponential distribution.

16. In a study carried out to compare two new methods for teaching arithmetic, 30 schools are divided arbitrarily into two groups. Each group is assigned one method. After two years, all students at the 30 schools are given an arithmetic test (adapted to the age of the child). Formulate a suitable model to study whether there is a correlation between the method that is used and the test scores.

17. A researcher studies the effect of exercise on the BMI of overweight children. In all, 25 children participate in the study. These children do cardio workouts every week and are weighed monthly. Formulate a suitable model to estimate the effect of working out.

Regression models quantify the correlation between an output variable Y and an input variable X and can be used to predict Y from X. Sometimes we would also like to use the regression model to "explain" Y from X. We speak of a causal explanation when X can be seen as a cause of Y. A change in the value of X is then necessarily followed by a change in Y; the size of the change is determined by the regression model.

Certainly not all regression models may be interpreted in a causal sense. The "price of a house" can, for example, be predicted in part from the income of the residents of the neighborhood; however, "income" is certainly not a cause for the price. Likewise, the "interest rates in the last few weeks" cannot be seen as cause for the "interest rates tomorrow," although the historical interest rates can certainly be used for predictions. On the other hand, we can probably view "temperature" as a cause for the "speed of a chemical reaction." Suppose that we use regression to explain "number of days unemployed" from, among other things, "highest level of education," where the regression coefficient turns out to be negative. Does this mean that a higher level of education is a cause for a shorter unemployment period?

The concept of causality belongs to philosophy rather than statistics, but it is of great significance for the interpretation of regression models. If we only want to use the regression model to predict Y based on an x observed under the same conditions as the data on which the regression model was fit, then causality is not very important. It is different if we want to use the outcomes of a regression analysis for an "intervention." If a higher level of education does lead to a shorter unemployment period, then it makes sense socio-economically to offer people more training since this must lead to lower unemployment. Causality can, in general, be studied this way by looking at the effects of an intervention. Suppose that we could change the x-value of an object while keeping the other relevant factors equal, does substituting the new x-value in the regression model give a correct prediction for the Y-value of the object? If this is the case, then it is justified to view X as a cause for Y.

In addition to the practical situation, the possibility of a causal interpretation greatly depends on the way the data are collected. Suppose, for example, that we want to know whether schools outside the Randstad (megalopolis in the Netherlands) offer better education than schools inside the Randstad. To study this, we take a sample from pupils in Group 8, the final year of elementary school (6th grade), in the Dutch population and compare the average CITO exam scores of the students at schools in the four large cities in the Randstad with the average score of pupils at schools outside these cities.[‡] We observe a small difference (see the first column in Table 7.3). A causal explanation of the observed difference from a difference in the quality of education between schools in and outside the Randstad seems logical but is certainly not justifiable. It is, for instance, quite conceivable that pupils at schools in and outside

---

[‡] The CITO exam is a national exam used to help decide which type of secondary school a child is admitted to (there are three types, preparing for a vocational, polytechnic, or university education). The score is an integer between 500 and 550.

303

the Randstad are not comparable and that other differences are responsible for the measured difference in CITO scores. Children at schools outside the Randstad may, for example, more often have highly-educated and/or Dutch-speaking parents than those in the Randstad. To rule out the influence of such confounding factors, we should set up a randomized trial: we randomly select a group of four-year-old children from the population and then determine by a randomized test to which school each child will go. (In the Netherlands, the choice of school is not determined by the location of your family residence.) In that case, the two groups of children would be comparable, and a possible difference in CITO scores could reasonably only be explained by the quality of the education. Unfortunately, we cannot carry out such an experiment. Our data will necessarily be observational data: the location of the school is determined by factors outside of our control, and we take a sample from the population as it has formed without our interference. In this type of observational studies, we are thus dealing with unintended (and statistically undesirable) selection.

Although a causal interpretation based on an observed correlation between CITO score and location is incorrect within our observational context, the estimated regression model can have a predictive value. Suppose that the correlation, quantified by the regression model, is indeed present. We must see the predictive values as follows: If we were to select a new pupil in the same way as our original sample, then the regression model gives a reasonable prediction of their CITO score. If, on the other hand, as an intervention, we were to close the schools whose pupils score low (in or outside the Randstad), then there is no guarantee at all that the school results will improve in general.

The realization that other factors, such as the level of education and ethnic background of the parents, could explain the measured difference in CITO scores motivates us to try to correct ("control") for these results in our analysis. Indeed, effects observed in an overall analysis can disappear or even reverse in an analysis on subgroups. This remarkable fact is known as Simpson's paradox. Consider, for example, Table 7.3, which gives the average scores on CITO exams for pupils in Group 8 in 2005, split into 7 strata determined by the proportion of pupils of non-Dutch background, for all of the Netherlands (first row) and only for the four largest cities (second row). According to the table, the pupils in the large cities obtain a score that is on average 2.3 points lower (first column). If we look in the different strata, however, we see that they are on average better and at most 1.4 point worse (in stratum 4). Is there an error in the table? Not at all. The somewhat surprising result is a consequence of the fact that in large cities, there are many more children from the higher strata $7, 6, \ldots$, which clearly have a lower average score. The grand mean is a weighted mean of the 7 scores, weighted according to the number of children per stratum. The grand mean in the large cities is therefore lower than the national average. A naive statistician would conclude from the lower grand mean in the large cities that the education in those cities is worse than in other schools in the country. However, the stratification rather suggests the opposite. But this stratification only partially solves the problem of observational data. There may be other factors involved that are not taken into account in the table. We cannot defend a causal explanation until we have

	mean	1	2	3	4	5	6	7
national	534.5	537.5	536.0	534.0	531.8	532.1	529.6	528.4
C4	532.2	541.0	537.3	534.0	530.4	531.4	529.6	528.5

Table 7.3. Final CITO exam for pupils in Group 8 in 2005. The column "mean" gives the grand mean. The columns 1 through 7 gives the average scores for schools with an increasing proportion of non-Dutch background. The rows "national" and "C4" gives the scores in, respectively, the whole country and the four large cities.

a better understanding of this situation, and perhaps not at all.

To correct for alternative explanations, we often set up a regression model including possible "confounding variables." In our example, for example, we carry out a regression of the CITO score Y on a vector $(X, Z) = (X, Z_1, \ldots, Z_k)$ in which X is "school location," Z_1 is "educational level parents," Z_2 is "living conditions," etc. The regression equation is then $E(Y|X = x, Z = z) = f(x, z)$. This is a step in the right direction but is certainly not yet the solution to all problems. First, we have the question of how the function f should look, that is, how the different variables Z_i should be included in the regression equations: additively? with interactions? etc. Second, it is not clear that the problem with observational data is solved using this approach.

There exist many tools for the first problem (setting up a correct regression model), including some discussed in this chapter. We assume that we have a well-fitting model. We can sharpen the second problem (observational data) by introducing so-called counterfactual variables. Every person can, in principle, go to school in ("0") or outside ("1") the Randstad. In practice, only one of the two possibilities can be realized, but let us denote by Y_0 and Y_1 the CITO scores that would be obtained by using the locations "0" and "1", respectively. The school choice itself will be encoded in the variable X, and possible confounding variables are encoded in the vector Z. The observed CITO score Y is

$$Y = \begin{cases} Y_0 & \text{if } X = 0, \\ Y_1 & \text{if } X = 1. \end{cases}$$

We want to answer the question whether the location has a causal effect on the CITO score. In the previous notation, the effect of location "1" instead of "0" can be measured by the difference $Y_1 - Y_0$. Of the pair (Y_0, Y_1), we always observe only one, namely the variable realized in Y. In this sense, the pair (Y_0, Y_1) is "counterfactual" and seems useless for solving the problem.

The counterfactual variables, however, supply a useful framework, provided that we assume more structure. We first consider the simple case where we assume that every person is randomly assigned to one of the locations, as in a randomized study. In terms of the counterfactual variables, this corresponds to the assumption that the location indicator X is independent of the pair (Y_0, Y_1). For every individual, at birth two possible outcomes are available, Y_0 and Y_1, and a randomized test ("generate X") determines which of the two is realized: Y_0 if $X = 0$ and Y_1 if $X = 1$. The expectation of Y_i is then also independent of the location: at birth, there exists an expectation of the CITO score for both locations, independently of which location will *be chosen. (This assumption is not realistic because there are confounding factors Z that influence both the choice of location and the height of the CITO score, but the computations for this simple case serve as a basis for the computations in which we correct for the information in Z.)*

For the simple case, we have

(7.14)
$$\mathrm{E}Y_i = \mathrm{E}(Y_i | X = i) = \mathrm{E}(Y | X = i), \qquad i = 0, 1,$$

where we use the independence for the first equality and the correlation between Y and Y_0, Y_1, X for the second. The expected location effect $\mathrm{E}Y_1 - \mathrm{E}Y_0$ is therefore given by

(7.15)
$$\mathrm{E}Y_1 - \mathrm{E}Y_0 = \mathrm{E}(Y | X = 1) - \mathrm{E}(Y | X = 0).$$

This can be estimated from the data by (for example) subtracting the average CITO score of pupils at location "0" from the average CITO score of pupils at location "1". This estimate is only correct if the experiment is randomized, without influence of confounding variables. It is clear that we cannot carry out such an experiment concerning school choice ourselves, and it is unrealistic to hope that the "natural life" would orchestrate such an experiment itself. To correct for the confounding variables in Z, we can again use the idea of subgroups. The subgroups are determined by equal values of the vector Z. If Z contains all relevant background information for the outcome of the CITO score and the school choice, then it is reasonable to assume that given Z, the variable X is independent of (Y_0, Y_1). Conditioning on Z corresponds to making the independence assumption for each subgroup. Suppose that the only variable in Z is the level of education of the parents. Then this assumption means that within a level of education, the school indicator X is independent of the vector (Y_0, Y_1). In other words, within the group with same level of education, at birth one has two possible outcomes Y_0 and Y_1, and the randomized test ("generate X") again determines which of the two outcomes is realized. Both Y_i and the randomization X depend on the level of education of the parents, but within the subgroups, this influence is the same for all individuals. When Z contains all relevant background information, the independence assumption should hold; the variation in CITO scores between individuals with a fixed value of Z is then based on "irrelevant" factors, and must be the product of "background noise". Let us assume that we have found a suitable Z. The equation (7.14) no longer holds because it is based on the assumption of unconditional independence (the case where Z is empty). Instead, we have

(7.16)
$$\mathrm{E}Y_i = \int \mathrm{E}(Y_i | Z = z) \, p^Z(z) \, dz = \int \mathrm{E}(Y_i | X = i, Z = z) \, p^Z(z) \, dz$$
$$= \int \mathrm{E}(Y | X = i, Z = z) \, p^Z(z) \, dz.$$

The first equality is the general expression of an expectation in conditional expec-
tations (with p^Z the marginal density of Z); for the second equality, we use the
assumption of conditional independence of (Y_0, Y_1) and X given Z, and for the third
we use the correlation between Y and Y_0, Y_1, X. The average school effect is therefore
given by

$$(7.17) \quad \mathrm{E}Y_1 - \mathrm{E}Y_0 = \int \big(\mathrm{E}(Y \,|\, X = 1, Z = z) - \mathrm{E}(Y \,|\, X = 0, Z = z)\big)p^Z(z)\,dz.$$

A first conclusion is that this effect can, in principle, be estimated from the
observed data (Y, X, Z): although we have unobservable "counterfactual" variables
Y_0, Y_1 on the left-hand side, the expressions on the right-hand side concern only
observable data. A second conclusion is that the expression in (7.15) for the causal
effect $\mathrm{E}Y_1 - \mathrm{E}Y_0$ no longer holds. We assumed that we had a randomized trial in which
the average CITO scores of pupils at school locations "1" and "0" were compared. In
the current subgroup analysis, the expected average CITO scores per school location
are of the form

$$\mathrm{E}(Y \,|\, X = i) = \int \mathrm{E}(Y \,|\, X = i, Z = z)p^{Z|X}(z|\,i)\,dz.$$

If X and Z are independent, then $p^{Z|X} = p^Z$, and this reduces to the last expression
in (7.16). In that case (for example, if Z is empty), the two expressions for the causal
effect are equal. If X and Z are not independent, then in general the two expressions
in (7.15) and (7.17) are not equal.

The difference between the first and second analyses is only the conditioning on
the vector Z. In fact, the second analysis is exactly equal to the first on every one of
the subgroups defined by a fixed value of Z, after which the estimated effects in (7.17)
are averaged over the subgroups, weighted with the marginal density p^Z of Z. The
estimated effect $\mathrm{E}(Y \,|\, X = 1, Z) - \mathrm{E}(Y \,|\, X = 0, Z)$ in the subgroups is given by a
joint regression model of Y on (X, Z). As long as we have a correct regression model
and a confounding variable Z for which the assumption of conditional independence
of X and (Y_0, Y_1) is correct, estimating a causal effect is not difficult. Unfortunately,
these assumptions are less innocent than they seem.

We have illustrated the problem of observational data in the context of CITO
scores. The problem of unintentional selection from a population in observational
studies also plays a major role in statistical research into medical treatment.
Medical treatment (for example, the administration of a drug or an operation) is
an intervention, and we would like to give a causal interpretation of the effect. A
regression model based on a sample of treated and untreated persons can, however,
give a wrong impression if the sample does not take other possible predictor variables
into account. For example, if we compare the remaining life span of treated and
untreated patients, but people with a poorer general health had a higher probability of
being treated (for example, because they went to the doctor for another condition and
the doctor performed a general health check), then the results of a regression analysis
will be misleading. Treated persons will live less long, but that is because they are less
healthy in general, not because they received treatment.

8 Model Selection

8.1 Introduction

In a regression analysis we try and explain a dependent variable using a number of independent (predictor) variables. It is often unwise to include all measured predictor variables in the regression model. Not only can a model with many variables be more difficult to interpret, but it often also has less predictive value. Model selection is aimed at finding the best-fitting model from a given collection of models.

In this chapter, we discuss the most important methods for choosing a suitable model in a general framework: methods based on tests, penalty methods (including AIC), the Bayesian approach, and cross-validation. Choosing suitable variables in a regression model is an important example. Within (nonparametric) estimation theory, model selection is also applied to find a model of a complexity that fits the data.

8.2 Goal of Model Selection

Suppose that for given data X, several statistical models $\mathcal{P}_d$ seem reasonable. We index the models with a parameter d and write the models as $\mathcal{P}_d = \{P_{d,\theta} : \theta \in \Theta_d\}$. The dth model $\mathcal{P}_d$ thus has parameter space Θ_d. The index d can, for example, indicate the number of "free" parameters in the model or the number of indices in a group of predictor variables; we can then choose between a "large model" (large d) or a "small model" (small d).

Every model $\mathcal{P}_d$ is a collection of possible probability distributions for X, and, of course, $\cup_d \mathcal{P}_d$ is also such a collection and therefore also a statistical model. At first glance, the statistical problem is choosing the "best-fitting" probability distribution from $\cup_d \mathcal{P}_d$, and we have only complicated the notation by introducing the extra symbol d. However, the models $\mathcal{P}_d$ will generally differ greatly from one another, and the extra notation allows us to further define the notion of "best-fit model." Moreover, the methods to estimate d are of a different nature than the previously described methods.

Example 8.1 Regression

In the multiple regression problem of Chapter 7, a "dependent variable" $Y = (Y_1, \ldots, Y_n)$ is modeled as a linear regression on a number of "predictor variables" $x_{i,j}$ using the equation

$$Y_i = \sum_{j=1}^{p} \beta_j x_{i,j} + e_i, \qquad i = 1, \ldots, n.$$

Here, $e_1, \ldots, e_n$ are random variables with expectation 0 and finite variance σ^2. The independent variables $x_{i,1}, \ldots, x_{i,p}$ are usually the values of background variables measured on the ith object, and Y_i is a "response" measured on the same object.

Within the context of model selection, it is useful to consider the model as a specification of an outcome Y as a function of input vectors $X_1, \ldots, X_p$. To emphasize this, we can write the displayed equation in vector form,

$$Y = \sum_{j=1}^{p} \beta_j X_j + e = X\beta + e.$$

Here, the symbols Y and $X_1, \ldots, X_p$ are vectors of length n whose ith coordinate refers to the ith object, and e is an error vector. In matrix notation, X is an $(n \times p)$-matrix with columns $X_1, \ldots, X_p$, and $\beta = (\beta_1, \ldots, \beta_p)^T$ is the vector of parameters.

In Chapter 7, we discussed how the unknown parameters $\beta_1, \ldots, \beta_p$ ("regression coefficients") can be estimated from the data. In practice, however, there are often many possible candidates X_j. Does it make sense to involve the postal code in the equation, in addition to gender, age, and level of education? Do neighborhood or volume give extra information on the price of a property when included in addition to surface area, year of building, and subdistrict? Do two specific genes give a good description of the genetic component of a disease, or is it wise to involve all 40 000 genes in the regression equation?

These questions are more complex than they seem. Not only do the questions relate to very diverse situations, they also have different purposes. The question concerning the price of a property is likely to come from the IRS or a broker, who is looking for a simple formula to determine a price objectively. The "best fit" is the model that supplies the best predictions. The question from genetics is probably meant causally: certain genes will influence the occurrence of a disease through biochemical processes, while others will not. If we are only interested in predicting a disease, it does not matter if we also involve a few genes from the second group in the regression equation (they may be linked to genes that do have a direct influence and therefore have predictive value), but to understand the development of the disease (causal explanation), it is essential to make a sharp distinction between the two groups of genes.

As can be seen in the previous example, model selection can focus both on predicting and on explaining. For predicting, the structure of the model does not matter. The model may be arbitrarily complicated as long as the predictions are good. For explaining, we want to include in the model only variables (parameters) that have a causal relation with the studied outcome. This distinction between predicting and explaining can be recognized in part in the different methods of model selection.

Neither of these objectives focuses on a model that approximates the *observed* data as closely as possible. Such a model is usually the most complex type of model, with the most parameters, which tries to follow every aspect of the data. A good model abstracts only those elements that are systematic. We say that a model that is too large *overfits* the given data, with as a result that it does not *generalize* to new data. We illustrate this in the following example.

Example 8.2 Overfitting

Suppose that we want to describe the data $(x_1, y_1), \ldots, (x_n, y_n)$ in the plane using a $(d+1)$-dimensional regression model $y = \beta_0 + \beta_1 x + \beta_2 x^2 + \cdots + \beta_d x^d$. If we assume that the observations (x_i, y_i) follow this curve up to additive errors with expectation 0, then the least squares method is a reasonable method to estimate the coefficients $\beta_0, \ldots, \beta_d$.

The dimension $d + 1$ is the essential parameter of the model. A large value of d gives more possible regression curves $y = \beta_0 + \beta_1 x + \cdots + \beta_d x^d$, but also requires that we estimate more unknown parameters from the available data. Figure 8.1 gives the least squares curves for $d = 1$ (the straight line) and $d = 10$ (the dashed curve, partially outside the frame of the figure) based on $n = 11$ observations. The linear and 11-dimensional fit are extremely different. The true regression curve in this simulated example is the parabola given in the figure (the solid curve). The straight line is a better approximation of it than the dashed curve and is therefore preferable.

The true curve can be given explicitly by a curve of type $y = \beta_0 + \beta_1 x + \cdots + \beta_d x^d$ with $d = 10$, by taking $\beta_3 = \cdots = \beta_{10}$ equal to 0. This model was therefore not wrong. However, the least squares method did not "know" that the true curve only had degree 2, and fit a degree 10 curve with nonzero coefficients $\beta_3, \ldots, \beta_{10}$. This degree 10 curve has an excellent fit with the actual observations $(x_1, y_1), \ldots, (x_{11}, y_{11})$, but gives a bad impression of the mechanism according to which these data were simulated. It "overfits" the observed data.

Conversely, the linear model $y = \beta_0 + \beta_1 x$ cannot describe the true curve perfectly but does give a usable approximation.

The difference in the success of the two models can be understood as a consequence of a different trade-off between bias and variance, the two ingredients of the mean square error. The least-squares estimators for the model with $d = 10$ are unbiased, and therefore the prediction $\hat{\beta}_0^{10} + \hat{\beta}_1^{10} x + \cdots + \hat{\beta}_{10}^{10} x^{10}$ is also unbiased. However, this equation contains 11 estimators, each with a variance (or estimation inaccuracy). These combine to an unbiased estimator for the regression function with a large variance. On the other hand, the estimator $\hat{\beta}_0^1 + \hat{\beta}_1^1 x$ is biased, because the linear model cannot explain the quadratic function, but has a small variance. Of the two methods, the second works best, though a quadratic model would probably have been better.

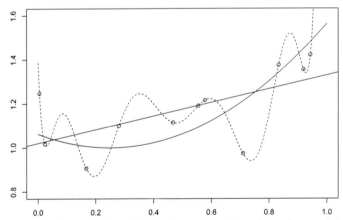

Figure 8.1. Regression analysis applied to observations $(x_1, y_1), \ldots, (x_{11}, y_{11})$ from the model $Y_i = (x_i - 0.25)^2 + 1 + e_i$ for a sample $e_1, \ldots, e_{11}$ from the $N(0, 0.1^2)$-distribution. Plotted are the parabola $y = (x - 0.25)^2 + 1$ and the least squares curve for the linear model $y = \beta_0 + \beta_1 x$ and the polynomial model $y = \beta_0 + \beta_1 x + \cdots + \beta_{10} x^{10}$. The points $x_1, \ldots, x_{11}$ were simulated from the uniform distribution on $[0, 1]$.

8.3 Test Methods

Hypothesis tests, as described in Chapter 4, can also be used for model selection. In a linear regression model, for example, we test whether the coefficient of a certain variable differs significantly from 0. If not, then this variable is not included in the model. More generally, we take a model with many parameters in mind and use tests to determine whether a certain parameter value differs significantly from 0. Next, insignificant parameters are set equal to 0 and only significant parameters are left free in the model and estimated from the data.

The relevant tests can be carried out in several ways. To obtain initial insight, one often tests every parameter separately in the model with only that parameter (and all other parameters equal to 0). Popular overall methods are the step-down and step-up methods, which can be applied to every (finite) parameter set $(\theta_i : i \in I)$.

The *step-down method* begins with testing all null hypotheses $H_0 : \theta_i = 0$, for every $i \in I$, separately, each time within the model with all parameters θ_j for $j \in I$. If all null hypotheses are rejected (at a given threshold, for example 5%), then we choose the model with all parameters. In the other case, the parameter θ_i with the greatest p-value is removed, after which we repeat the procedure with the model without this parameter. (By removing the parameter, the p-values of the null hypotheses $H_0 : \theta_j = 0$ for the remaining parameters change; they need to be computed again!) Continuing in this manner, we remove one parameter at a time, until the p-values for all remaining parameters are less than the chosen threshold.

The *step-up method*, on the other hand, begins with the model without parameters (or the model with only an intercept) and first tests the null hypothesis $H_0 : \theta_i = 0$ for all parameters separately, within the model with only the parameter θ_i. If none of the null hypotheses are rejected, then we choose the model without parameters. In the other case, we add the parameter with the smallest p-value and then test the null hypotheses $H_0 : \theta_i = 0$ for each of the remaining parameters within the model consisting of this parameter and the previously chosen ones. If none of the p-values is less than the threshold, then we choose the model consisting of the previously added parameters; otherwise, we repeat the procedure.

These two methods can be varied in numerous ways. We can, for example, test several parameters at a time, alternate step-up and step-down, or vary the thresholds we use.

One disadvantage of all test methods is that after repeatedly carrying out a test (at a given level), the overall size is unclear. Another disadvantage is that the final model will depend on the order in which the parameters are tested. In general, the step-up and step-down methods will not lead to the same model, and certain models will never be taken into consideration.

The simplicity of the test method is its main merit.

8.4 Penalty Methods

The maximum likelihood and least squares methods, and more general estimation methods that minimize or maximize a criterion function, generally choose the largest model when several are available. After all, a larger model gives a larger space in which to optimize the criterion function, and therefore always has a better value for the criterion. We only obtain a smaller model if the optimum over the larger model belongs to the smaller model. This appears to be a rather fortuitous situation, so the naive application of these methods generally leads to overfitting.

Adding a *penalty* to the criterion can correct this. Suppose that we obtain a suitable parameter for the model $\mathcal{P}_d = (P_{d,\theta} : \theta \in \Theta_d)$ by maximizing a criterion $\theta \mapsto L(\theta; X)$ over Θ_d. We then determine a suitable parameter and suitable model by maximizing the function

$$D \times \Theta_d \ni (d, \theta) \mapsto L(\theta; X) - \mathrm{pen}(d)$$

with respect to $\theta \in \Theta_d$ and $d \in D$. The term $\mathrm{pen}(d)$ in this expression is the penalty; its purpose is to decrease the criterion as the model increases in size. This discourages large models.

The *Akaike information criterion (AIC)* is one of the best-known model selection criteria. In the case of the log-likelihood for observing a sample $X_1, \ldots, X_n$ from a density $p_{d,\theta}$ indexed by a $|d|$-dimensional parameter, the AIC penalty is equal to the dimension of the parameter. The best model is thus determined by maximizing

$$d \mapsto \log \prod_{i=1}^{n} p_{d,\hat{\theta}_d}(X_i) - |d|$$

with respect to d, where $|d|$ is the number of parameters of the model $\mathcal{P}_d$ and $\hat{\theta}_d$ is the maximum likelihood estimator for the parameter using this model. It is common practice to multiply this expression by -2 and then define the AIC penalty as $2|d|$, but this leads to the same optimum.

Example 8.3 Regression

The log-likelihood function for the parameters $\beta_1, \ldots, \beta_d$ in the regression model $Y = \beta_1 X_1 + \cdots + \beta_d X_d + e$ with normally distributed errors e is equal to $-n \log \sigma - \|Y - \beta_1 X_1 - \cdots - \beta_d X_d\|^2/(2\sigma^2)$, up to a constant. For given σ^2, the AIC criterion is therefore equal to minimizing

$$(\beta_1, \beta_2, \ldots, \beta_d, d) \mapsto \|Y - \beta_1 X_1 - \cdots - \beta_d X_d\|^2 + 2\sigma^2 d.$$

The penalty thus increases the sum of squares by $2\sigma^2$ times the number of parameters of the model. The minimum over $(\beta_1, \ldots, \beta_d)$ is achieved by the least-squares estimator, for which the expression is equal to the residual sum of squares plus $2\sigma^2 d$. The best model according to AIC is the model for which this sum is minimal.

Generally, the variance σ^2 will be unknown and also estimated from the data. In that case, the corresponding AIC criterion is equivalent to minimizing

$$(\beta_1, \beta_2, \ldots, \beta_d, \sigma^2, d) \mapsto n \log \sigma + \frac{1}{2\sigma^2} \| Y - \beta_1 X_1 - \cdots - \beta_d X_d \|^2 + d + 1.$$

Minimizing for σ^2 gives $\hat{\sigma}^2(\beta_1, \ldots, \beta_d) = n^{-1} \| Y - \beta_1 X_1 - \cdots - \beta_d X_d \|^2$. Substituting this gives an AIC criterion for only $(\beta_1, \ldots, \beta_d)$ equal to the expression

$$\frac{n}{2} \log \left(\frac{1}{n} \| Y - \beta_1 X_1 - \cdots - \beta_d X_d \|^2 \right) + \frac{n}{2} + d + 1.$$

The factor $1/n$ inside the logarithm, the term $+n/2$, and the $+1$ in the last term can be left out. The minimum over $(\beta_1, \ldots, \beta_d)$ is achieved by the least-squares estimator. We substitute these in the equation. AIC chooses the model for which the resulting value is minimal. ▭

AIC aims at choosing the model that gives the best estimate (or "prediction") of the density of the observation. The quality is measured by the *Kullback–Leibler divergence* between the chosen density $p_{d,\hat{\theta}_d}$ (where $\hat{\theta}_d$ is the maximum likelihood estimator when the model $\mathcal{P}_d$ is assumed to be correct) and the true density p. The Kullback–Leibler divergence is a measure for the difference between two probability densities, closely related to the maximum likelihood method, defined as $K(p; q) = E_p(\log(p/q)(X))$, where E_p means that the quantity X has density p. The relevant divergence in this case is therefore

$$(8.1) \qquad K(p; p_{d,\hat{\theta}_d}) = E_p \log \frac{p(X)}{p_{d,\theta}(X)} \bigg|_{\theta = \hat{\theta}_d}.$$

This divergence depends on the unknown density p and therefore cannot simply be minimized with respect to the model index d. The probability density in the numerator of the expression can be avoided, because $d \mapsto -E_p \log p_{d,\theta}(X)|_{\theta = \hat{\theta}_d}$ has the same point of minimum. We could replace the expectation E_p by an average over a sample of n observations $X_1, \ldots, X_n$ from the density p, which leads to the criterion

$$-\frac{1}{n} \sum_{i=1}^{n} \log p_{d,\hat{\theta}_d}(X_i).$$

This estimator for $-E_p \log p_{d,\theta}(X)|_{\theta = \hat{\theta}_d}$ turns out to be biased and systematically underestimates its estimand. An intuitive explanation is that the same data $X_1, \ldots, X_n$ is used twice: first, to determine the maximum likelihood estimator $\hat{\theta}_d$ and then, to replace the expectation E_p in (8.1) by an empirical mean (a form of overfitting). Akaike made it seem plausible that the bias is approximately equal to n^{-1} times the number parameters $|d|$, which leads to the AIC penalty.

Lemma 8.4

Suppose $\Theta_d \subset \mathbb{R}^{|d|}$ and that the function $\theta \mapsto \log p_{d,\theta}(x)$ is twice continuously differentiable with derivatives $\dot\ell_{d,\theta}(x)$ and $\ddot\ell_{d,\theta}(x)$, and let θ_d be the point of minimum of $\theta \mapsto K(p; p_{d,\theta})$ over $\theta \in \Theta_d$. Then, under certain regularity conditions, we have

$$\mathrm{E}_p\left(-\frac{1}{n}\sum_{i=1}^{n}\log p_{d,\hat\theta_d}(X_i) + \frac{\tilde d}{n}\right) - \left(-\mathrm{E}_p\log p_{d,\hat\theta_d}(X_1)\right) = o(n^{-1}),$$

for $\tilde d$ the trace of the matrix $\left(\mathrm{E}_p\dot\ell_{d,\theta_d}\dot\ell_{d,\theta_d}^T(X_1)\right)\left(-\mathrm{E}_p\ddot\ell_{d,\theta_d}(X_1)\right)^{-1}$. If $p \in \mathcal{P}_d$ (and $p = p_{d,\theta_d}$), then $\tilde d = |d|$.

Proof. (Sketch). Write $\mathbb{P}_n f$ and Pf for, respectively, $n^{-1}\sum_{i=1}^{n}f(X_i)$ and $\mathrm{E}_p f(X_1)$. By the definitions of $\hat\theta_d$ and θ_d, we have $\mathbb{P}_n\dot\ell_{d,\hat\theta_d} = 0$ and $P\dot\ell_{d,\theta_d} = 0$. Second-order Taylor expansions about $\hat\theta_d$ and θ_d, respectively, therefore give

$$\mathbb{P}_n\log\frac{p_{d,\theta_d}}{p_{d,\hat\theta_d}} = \tfrac{1}{2}(\hat\theta_d - \theta_d)^T\mathbb{P}_n\ddot\ell_{d,\hat\theta_d}(\hat\theta_d - \theta_d) + \cdots,$$

$$-P\log\frac{p_{d,\theta_d}}{p_{d,\hat\theta_d}} = \tfrac{1}{2}(\hat\theta_d - \theta_d)^T P\ddot\ell_{d,\theta_d}(\hat\theta_d - \theta_d) + \cdots.$$

On the right-hand side of the first equation, we replace $\mathbb{P}_n$ by P, as a further approximation. We then use that, for large n, $\sqrt{n}(\hat\theta_d - \theta_d)$ is approximately distributed as a multivariate normal vector V with expectation 0 and covariance matrix $\Sigma = (P\ddot\ell_{d,\theta_d})^{-1}(P\dot\ell_{d,\theta_d}\dot\ell_{d,\theta_d}^T)(P\ddot\ell_{d,\theta_d})^{-1}$. (This follows from an extension of Theorem 5.9 to the case where the density p of the data does not belong to the model.) From $\mathrm{E}V^T AV = \mathrm{tr}(A\Sigma)$, we then deduce that the expectations of the two right-hand sides are asymptotically equal to $-\tilde d/(2n)$. Finally, we compute the expected values of the two equations, use $\mathrm{E}_p\mathbb{P}_n\log p_{d,\theta_d} = \mathrm{E}_p\log p_{d,\theta_d}$, and take the sum of the two equations.

If $p = p_{d,\theta_d}$, then both matrices $P\ddot\ell_{d,\theta_d}$ and $-P\dot\ell_{d,\theta_d}\dot\ell_{d,\theta_d}^T$ reduce to the Fisher information matrix. The relevant matrix is then the identity, which has trace $|d|$. ∎

Lemma 8.4 shows that Akaike's correction for the bias is correct only if the density p of the data belongs to the model $\mathcal{P}_d$ (in which case $\tilde d = |d|$). Unfortunately, the value $\tilde d$ depends on the unknown parameter θ_d and therefore cannot be used as penalty. (We could replace $\tilde d$ by an estimate; this gives the "Takeuchi method.")

A correct expectation of a criterion is no guarantee for good behavior, but the AIC criterion proves to have good properties. This follows from so-called *oracle inequalities*, of the form

$$\mathrm{E}_p K(p; p_{\hat d,\hat\theta_{\hat d}}) \le C \inf_{d \in D}\left(K(p; \mathcal{P}_d) + \frac{|d|}{n}\right).$$

Here, $\hat{d}$ is the index of the model chosen by AIC, and $\hat{\theta}_{\hat{d}}$ is the maximum likelihood estimator for the parameter in this model. Furthermore, $K(p; \mathcal{P}_d) = K(p; p_{d,\theta_d})$ is the Kullback–Leibler divergence between the model $\mathcal{P}_d$ and the true density p, and $|d|$ is the number of parameters in the model $\mathcal{P}_d$. The left-hand side of the inequality gives the expected Kullback–Leibler distance between the density $p_{\hat{d},\hat{\theta}_{\hat{d}}}$ chosen by AIC and the true density p. This is bounded above (up to a multiplicative constant) by the minimum over all models $\mathcal{P}_d$ of the sum $K(p; \mathcal{P}_d) + |d|/n$ of the distance to the model and the AIC penalty divided by n. The term $K(p; \mathcal{P}_d)$ can be viewed as a necessary, minimal (square) bias when using the model $\mathcal{P}_d$. Every estimator in this model will have at least this "distance" to the true density p. The term $|d|/n$ can be viewed as a variance term, or estimation inaccuracy, as a result of estimating $|d|$ unknown parameters when using the model $\mathcal{P}_d$. Such a variance term is also necessary.

When using the model $\mathcal{P}_d$, we could hope to make an error of order $K(p; \mathcal{P}_d) + |d|/n$. The infimum over d in the inequality shows that the model chosen by AIC makes a smaller error than every other model, at least up to a multiplicative constant C (which must usually be taken greater than 1). AIC therefore works as an "oracle" that knows the best model.

Without further specification of the constant C, the oracle inequality is not very interesting for a small number of low-dimensional models. After all, the largest model then has the smallest bias and also gives order $1/n$. For models with very different dimensions $|d|$, however, the inequality suggests a high potential gain for model selection methods.

For a precise mathematical formulation, we consider the linear regression model $Y = \beta_1 X_1 + \cdots + \beta_p X_p + e = X\beta + e$, with models defined by subcollections of the predictor variables. For $d \subset \{1, 2, \ldots, p\}$, define the model $\mathcal{P}_d$ as the linear regression model $Y = \sum_{j \in d} \beta_j X_j + e$ with coefficients $\beta_d = (\beta_j : j \in d)$. The least-squares estimator in this model is $\hat{\beta}_d = (X_d^T X_d)^{-1} X_d Y$, for X_d the matrix with columns X_j for $j \in d$, and the least squares prediction for EY is $X_d \hat{\beta}_d = P_d Y = P_d X\beta + P_d e$, for P_d the orthogonal projection on the column space of X_d. The mean square error of this prediction is

$$\mathrm{E}_\beta \|X_d \hat{\beta}_d - X\beta\|^2 = \|P_d X\beta - X\beta\|^2 + \mathrm{E}\|P_d e\|^2 = \|P_d X\beta - X\beta\|^2 + |d|\sigma^2.$$

The last equality follows from the fact that $\|P_d e\|^2/\sigma^2$ has a chi-square distribution with d degrees of freedom. The first term on the right-hand side is the square of the bias of the estimator. Since $\|P_d X\beta - X\beta\| \leq \|X_d \beta_d - X\beta\|$, this term is certainly equal to 0 if the true coefficients β_j for $j \notin d$ are equal to 0. The fact that this is not known beforehand advocates the use of larger models. However, the second term $|d|\sigma^2$ is then large, and it is better to balance the two terms. This is a "bias-variance trade-off" in the form of model selection.

AIC is capable of making this "trade-off."[b]

---

[b] Proofs of the following two results can be found in, e.g., L. Birg and P. Massart, Gaussian model selection. *J. Eur. Math. Soc.* (2001), no. 3, 203-268; and R. Nishii, Asymptotic properties of criteria for selection of variables in multiple regression, *Ann. Statist.* 12 (1984), no. 2, 758-765.

Theorem 8.5 AIC

Let $Y = \beta_1 X_1 + \cdots + \beta_p X_p + e = X\beta + e$, with e a vector of independent, normally distributed errors with expectation 0 and variance σ^2. Let $\hat{d}$ be the model chosen by AIC from an arbitrary set of models D given by subsets $(X_j: j \in d)$. Then the least-squares estimator $\hat{\beta}_d$ satisfies

$$\mathrm{E}_\beta \|X_{\hat{d}}\hat{\beta}_{\hat{d}} - X\beta\|^2 \leq C \min_d \left(\|X_d\beta_d - X\beta\|^2 + |d|\sigma^2 \right).$$

The theorem is restricted to finitely many models, but AIC can also be applied to infinitely many models. The regression model $y = \beta_0 + \beta_1 x + \beta_2 x^2 + \cdots + \beta_d x^d + e$ from Example 8.2 can, for example, be applied with arbitrary, unknown $d \in \mathbb{N}$, to approximate a true model $y = f(x) + e$ as well as possible, where the function f is not necessarily a polynomial.

The quality of AIC is in the prediction. Somewhat unexpectedly, AIC is not capable of finding the right model. In the regression example, this is the model d_0 consisting of all regression coefficients β_j that are nonzero. For large n, AIC never chooses a small model; rather, it adds every zero coefficient to the model with positive probability.

Theorem 8.6 AIC

In the situation of Theorem 8.5, as $n \to \infty$, the model $\hat{d}$ converges in distribution to a random variable that allocates a positive probability to every element d of the set $\{d: d \supset d_0\}$, where $d_0 = \{j: \beta_j \neq 0\}$.

8.5 Bayesian Model Selection

In Bayesian statistics, model selection does not give any conceptual problems. The additional parameter d is simply given a prior distribution, just like the parameters in the models. Bayes's rule then gives a posterior distribution for all parameters, in particular for the model index d. The latter is a vector of "probabilities" of the different models, given the data. Although we could choose the most probable model, the Bayesian approach does not lead to model *selection*. Instead, we speak of *model averaging*.

For a more precise description, we denote the prior probabilities of the models P_d, for $d \in D$, by a probability vector $(p_d: d \in D)$ (so $0 \leq p_d \leq 1$ for every d and $\sum_d p_d = 1$). We, moreover, denote by $\theta \mapsto \pi_d(\theta)$ a prior density on the parameter set Θ_d, for every d. If the probability density of the data X given (d, θ) is written as $x \mapsto p_{d,\theta}(x)$, then Bayes's rule tells us that the conditional density of (d, θ) satisfies

$$(8.2) \qquad \pi(d, \theta \mid X) \propto p_d \, \pi_d(\theta) \, p_{d,\theta}(X).$$

This depends on both the prior distribution $(p_d: d \in D)$ on the models and the relative probabilities $\pi_d(\theta)$ of the parameters in the dth model. For a small number of models of not too different dimensions, the first is often chosen uniformly: $p_d = 1/\#D$ for every d. In nonstandard situations, however, the posterior distribution can depend strongly on the choice. The necessity to choose prior distributions is seen both as a strength and as a weakness of the Bayesian method, depending on the statistician.

The joint posterior distribution for (d, θ) can be marginalized over θ to give the posterior distribution of the model index as

$$(8.3) \qquad \pi(d \mid X) \propto \int p_d \, \pi_d(\theta) \, p_{d,\theta}(X) \, d\theta.$$

Instead of this full posterior distribution, often only the Bayes factors are reported. The *Bayes factor* for the pair of models with indices d_1 and d_2 is given by

$$\mathrm{BF}(d_1, d_2) = \frac{\int \pi_{d_1}(\theta) \, p_{d_1,\theta}(x) \, d\theta}{\int \pi_{d_2}(\theta) \, p_{d_2,\theta}(x) \, d\theta}.$$

When the prior weights of the models are equal ($p_{d_1} = p_{d_2}$), this is exactly the ratio of the posterior probabilities, also called "posterior odds ratio." More generally, we have the rule that " the posterior odds are the prior odds times the Bayes factor":

$$\frac{\pi(d_1 \mid X)}{\pi(d_2 \mid X)} = \frac{p_{d_1}}{p_{d_2}} \, \mathrm{BF}(d_1, d_2).$$

A value greater than 1 means that model P_{d_1} has a greater posterior probability than model P_{d_2}.

Prior probabilities of models can, in particular, be sensitive to the variance of the prior density. In many cases, we want to choose this large, as an expression of uncertainty over the parameter (so-called *vague prior distribution*). To avoid this sensitivity, the *method of the intrinsic Bayes factor* proposes to split the data into two parts. The first is used to compute a posterior distribution for every parameter and the second is used to compute (8.2), where the prior densities π_d are replaced by the posterior distributions from the first step.

The following simple example shows that, in practice, Bayesian model selection is not that simple.

318

Example 8.7 Bayes factors for a vague prior distribution

Suppose that the data $X_1, \ldots, X_n$ form a sample from either a $N(0, 1)$-distribution or a $N(\theta, 1)$-distribution for an unknown $\theta \in \mathbb{R}$. For an $N(0, \tau^2)$ prior density on the parameter θ, the Bayes factor between the two models is equal to

$$\frac{\int \phi(\theta/\tau)/\tau \prod_{i=1}^{n} \phi(X_i - \theta) \, d\theta}{\prod_{i=1}^{n} \phi(X_i)} = \sqrt{\frac{1}{1 + n\tau^2}} e^{\frac{1}{2} n \overline{X}^2 / (1 + (n\tau^2)^{-1})}.$$

For large n and fixed τ this expression has the same order of magnitude as $(\tau^2 n)^{-1/2} \exp(n\frac{1}{2}\overline{X}^2)$. The prior standard deviation τ therefore plays a crucial role. This is usually not seen as an advantage because $\tau \to \infty$ expresses the full uncertainty of θ.

An alternative for a normal distribution with large variance is the "uniform distribution" on $\mathbb{R}$, the distribution with density 1. This is not a probability distribution (and is therefore called an "improper" prior distribution), but the relevant Bayes factor is well defined and is equal to

$$\frac{\int \prod_{i=1}^{n} \phi(X_i - \theta) \, d\theta}{\prod_{i=1}^{n} \phi(X_i)} = \sqrt{\frac{2\pi}{n}} e^{\frac{1}{2} n \overline{X}^2}.$$

As for the normal prior distribution, this is an increasing function of $|\overline{X}|$. This is reasonable, because the choice between the two models in fact comes down to testing the hypothesis that the expectation θ is 0. The Bayes factor is greater than a certain value c when $\sqrt{n}|\overline{X}_n| > \sqrt{\log(c^2/2\pi) + \log n}$. For large n, the presence of the term $\log n$ makes the Bayes factor more conservative than the Gauss test, which rejects the null hypothesis for values greater than quantiles from the standard normal distribution. Within the context of model selection, this is probably wise. The theory of tests deals asymmetrically with the two hypotheses, and always works with a fixed type I error, regardless of the number of observations. On the other hand, the choice of the prior density equal to 1 is somewhat arbitrary, which also makes the choice of the constant c arbitrary. Since the "uniform distribution on $\mathbb{R}$" has infinite mass, there is no canonical normalization, and every other constant is just as logical.

To compute an intrinsic Bayes factor, we could split the data in $X_1, \ldots, X_m$ and $X_{m+1}, \ldots, X_n$, for a given m. The posterior distribution given the data $X_1, \ldots, X_m$ and an $N(0, \tau^2)$ prior distribution is the normal distribution with expectation $\nu_m := m/(m + \tau^{-2})\overline{X}_m$ and variance $\tau_m^2 := 1/(m + \tau^{-2})$. We take this as prior distribution for computing the Bayes factor given the data $X_{m+1}, \ldots, X_n$ and find

$$\frac{\int \phi((\theta - \nu_m)/\tau_m)/\tau_m \prod_{i=m+1}^{n} \phi(X_i - \theta) \, d\theta}{\prod_{i=m+1}^{n} \phi(X_i)}.$$

The dependence on m and the asymmetry in the observations are unpleasant. We could choose $m = 1$ and take the average over the n expressions obtained by replacing X_1 by X_i.

In addition to the posterior probabilities of the different models, the full posterior distribution (8.2) also gives the posterior distribution of the parameter. For a given function $g: \cup_d \Theta_d \to \mathbb{R}^d$, we have

$$P\big(g(\theta) \in B \mid X\big) = \sum_d P\big(g(\theta) \in B \mid d, X\big)\pi(d \mid X)$$

$$= \frac{\sum_d p_d \int_{\theta:g(\theta)\in B} \pi_d(\theta)\, p_{d,\theta}(X)\, d\theta}{\sum_d p_d \int \pi_d(\theta)\, p_{d,\theta}(X)\, d\theta}.$$

This shows that we obtain the best prediction of $g(\theta)$ by combining the different models. The Bayesian approach does not select one model, but averages the predictions of all models, weighed according to their posterior probability.

Nevertheless, in many cases, we want to indicate one model as best fitting. Then the value d that maximizes (8.3) is frequently natural, that is, the mode of the posterior distribution of the model index. In standard situations, with a large number of observations, this model index turns out to correspond to a penalized maximum likelihood estimator for d, with penalty equal to half the logarithm of the number of observations times the AIC penalty, the so-called BIC penalty. We consider this for the situation that the observation forms a sample $X_1, \ldots, X_n$ with (marginal) probability density p, and take $n \to \infty$.[♯]

Theorem 8.8 BIC

Consider a model $\mathcal{P}_d$ with parameter $\theta \in \Theta_d \subset \mathbb{R}^{|d|}$ given by probability densities $p_{d,\theta}$ that satisfy the conditions of Theorems 4.50 and 5.26 and such that the maximum likelihood estimators $\hat{\theta}_d$ are consistent for the point of minimum of $\theta \mapsto K(p; p_{\theta,d})$. Then

$$\log \int \pi_d(\theta) \prod_{i=1}^n p_{d,\theta}(X_i)\, d\theta = \log \prod_{i=1}^n p_{d,\hat{\theta}_d}(X_i) - \tfrac{1}{2}|d| \log n + O_P(1).$$

Furthermore, $\mathrm{BF}(d, d_0) \to 0$ in probability as $n \to \infty$ whenever p belongs to $\mathcal{P}_d \cap \mathcal{P}_{d_0}$ and $|d| > |d_0|$, and, under certain regularity conditions, the same is true for every model $\mathcal{P}_d$ with $|d| < |d_0|$ such that $K(p; \mathcal{P}_d) > 0$.

In addition to establishing a link between the Bayesian method for model selection and the penalized maximum likelihood, the theorem also shows that, ultimately, the Bayesian methods choose (with high probability) the correct model, namely the smallest model that contains the true distribution of the data. We say that these methods are *consistent* for model selection.

[♯] A proof of the following theorem for exponential families can be found in the original paper: Gideon Schwarz, Estimating the dimension of a model. *Ann. Statist.* 6 (1978), no. 2, 461464.

The penalty $|d| \log n$ is known as the *BIC penalty*. This is a factor $\frac{1}{2} \log n$ times the AIC penalty $2|d|$. Since $\log n$ increases slowly with n (for example, $\log(1000) = 7$), this does not make a great difference for small values of n. For larger n, however, the BIC penalty is significantly greater, which leads to the choice for smaller models than those given by AIC.

The higher penalty for larger models is necessary for the consistency of the method. The AIC method overestimates the dimension (see Theorem 8.6) and is not consistent. On the other hand, we have seen that AIC gives good estimates for the distribution of the observations, in the sense of the Kullback–Leibler divergence between estimate and true density. It is somewhat paradoxical that for BIC, and consistent model selection methods in general, oracle inequalities such as in Theorem 8.5 do not hold. The problem is that consistent model selection methods necessarily reject larger models with one or more parameters that are almost 0 in favor of the smaller model in which these parameters are 0. Thus, consistent model selection methods "shrink" the parameters too much for good predictions. A larger model may well be closer to the truth.

Theorem 8.8 is restricted to finitely many models of finite dimensions. However, this is not a requirement for the Bayesian approach, which conceptually allows infinitely many models, possibly of infinite dimensions. This leads to a very different situation than BIC, which unfortunately is not easy to summarize. The choice of the priors $(p_d : d \in D)$ plays a much more important role when there are more models, in particular when the dimensions of these models differ greatly.

8.6 Cross-Validation

The idea behind cross-validation is to base the choice of a model and the estimation of its parameters on independent observations. The problem of "overfitting," as identified in Example 8.2, is thereby avoided.

The method can be applied to every estimation method that is defined to optimize a criterion. Consider, as an example, the least squares method for linear regression. Given a response vector Y and a corresponding collection $X_1, \ldots, X_p$ of predictor variables, we determine a suitable parameter $\hat{\beta}_d$ for a regression model $Y = X_d \beta_d + e$ by minimizing the sum of squares

$$\|Y - X_d \beta_d\|^2 = \sum_{i=1}^{n} (Y_i - \sum_{j \in d} \beta_{d,j} x_{i,j})^2.$$

We know from Chapter 7 that this least squares method provides unbiased estimators as long as the errors e_i have expectation 0. If, however, we choose from a given number of models the model with predictor variables $(X_j : j \in d)$ with minimal residual sum of squares $\|Y - X_d \hat{\beta}_d\|^2$, then we are overfitting the model. After all, the largest model gives the smallest residues.

Suppose that a second data set $(\tilde{X}, \tilde{Y})$ is available, with the same statistical properties as the first data set (X, Y), and statistically independent of it. We base the estimate $\hat{\beta}_d = \hat{\beta}_d(X, Y)$ on the first data set, but evaluate the quality of the different models using the second data set: for every model index d, we compute the residual sum of squares $\|\tilde{Y} - \tilde{X}_d \hat{\beta}_d\|^2$ on the new data and then choose the model with index d for which this is minimal. The data in the second set is called the *validation data*, and the corresponding sum of squares is called the validation sum; the data in the first set (X, Y) is called the *training data*.

Because we use new data, the residual sum of squares $\|\tilde{Y} - \tilde{X}_d \hat{\beta}_d\|^2$ gives an honest measure for the quality of the regression model d. Indeed, this is exactly the error we would make by using this model for predicting a future response $\tilde{Y}$ from the corresponding independent variables $\tilde{X}$. There is no question of overfitting (provided that there is sufficient validation data to give a good estimate of the future error).

In practice, we do not have a second data set, but we can split a given data set (arbitrarily). In the procedure described above, we would then use the two half data sets asymmetrically. It would make more sense to use both data sets in both roles: both in the role of training set and in the role of validation set. We then measure the quality of a model by taking the sum of the two residual sums of squares computed on one half of the data with parameter estimates based on the other half. This is the principle of *cross-validation*.

Instead of two equal parts, we can also use other partitions. We speak of *k-fold cross-validation* if the data set is split into k parts of equal size, after which each part is used as validation set and the parameter estimates are computed on the union of the remaining $k-1$ parts of the data. The best model minimizes the sum of the k validation sums. When $k = n$, where n is the number of data points, this process is called the *leave-one-out cross-validation*. Estimates $\hat{\beta}_d^{-i}$ are then computed on all data points except (X_i, Y_i) and validated with score $(Y_i - x^i \hat{\beta}_d^{-i})^2$, where $x^i = (x_{i,1}, \ldots, x_{i,p})$ is the row vector of predictor variables for the ith observation. The best model minimizes the sum over all these scores.

Cross-validation is simple to implement and reasonably accurate if we have sufficient data. However, partitioning the data leads to a loss in efficiency, because estimates and model choice are, in fact, based on only part of the data. This method is less suitable for small sets of data because in that case, the validation sum is not an accurate estimate of the actual error.

Cross-validation has a goal similar to that of AIC, namely finding the model that provides the best prediction. AIC uses the entire data set but adds a penalty to the criterion to correct for "using the data twice" (for estimating the parameters and selecting the model). This penalty is based on theoretical analysis and depends on the structure of the data and the models. The justification of cross-validation is simpler, and this method can therefore be used in more situations.

8.7 Post-Model Selection Analysis

Suppose that we have used one of the methods described above to select a suitable model $\mathcal{P}_d$ from a given collection. The second step is to determine a suitable parameter θ_d within this model, possibly with a corresponding confidence region. This is called *post-model selection estimation*.

In practice, in this second step, one often uses the standard methods for the model $\mathcal{P}_d$ without taking into account that this model was itself estimated. This can give misleading results, in particular for confidence regions. After all, the uncertainty holds for both the model and the parameter value, and a standard method for the model $\mathcal{P}_d$ will ignore the first. Confidence regions will be too small, in particular if the model choice and parameter estimate are based on the same data.

Unfortunately, there is no simple or generally accepted solution for this dilemma: confidence statements are difficult to reconcile with model selection. For fair confidence regions, as defined in Chapter 5, we will generally need to use the *largest* model.

8.8 Summary

Let $X_1, \ldots, X_n$ be a sample from an unknown distribution P with density p. Model selection methods select the best model from a set $\cup_d \mathcal{P}_d$ with $\mathcal{P}_d = \{P_{d,\theta} : \theta \in \Theta_d\}$. This "best" model is often not the model that fits best with the observed data; the best-fitting model for the data suffers from overfitting.

Four methods for selecting a model:

- *Test methods*. We first test which parameter values in the model differ significantly from 0. The selected model only involves the significant parameters. This model selection method is often used with regression models to select predictor variables for the model.
- *Penalty methods*. Many estimation methods optimize a criterion function. Penalty methods add a term to the criterion function that "penalizes" more as the model grows. The *Akaike information criterion (AIC)* chooses the best model by maximizing

$$d \mapsto \log \prod_{i=1}^{n} p_{d,\hat{\theta}_d}(X_i) - |d|$$

(the log-likelihood minus the number of estimated model parameters in $\mathcal{P}_d$), where $\hat{\theta}_d$ is the maximum likelihood estimator for $\theta \in \Theta_d$ under the assumption that $\mathcal{P}_d$ is the right model. This choice of penalty is justified by using the *Kullback–Leibler divergence* as distance measure between the chosen density and the true density p.
- *Bayesian model selection*. In the Bayesian approach, the different models $\mathcal{P}_d$ and the parameter $\theta_d \in \Theta_d$ are assigned prior probabilities. Bayesian estimation takes the average of the predictions concerning θ over the different models, weighted by their posterior distribution. When we nevertheless want to select a single model, we choose the model d with the maximal mode of the posterior density $d \mapsto \pi(d|X_1, \ldots, X_n)$. Under certain conditions, this corresponds asymptotically to maximizing the function

$$d \mapsto \log \prod_{i=1}^{n} p_{d,\hat{\theta}_d}(X_i) - \frac{1}{2}|d| \log n.$$

The term $\frac{1}{2}|d| \log n$ is called the *BIC penalty* (where BIC stands for Bayesian information criterion).
- *Cross-validation*. In cross-validation, the choice of model and the parameter estimate are based on independent parts of the data. In the simplest case, the data is split into two equal parts. The first part (the *training data*) is used to estimate the model parameters, and the second part (the *validation data*) is used to determine the fit of the models quantitatively using a validation sum. The roles are then reversed. This method can be generalized to k-fold cross-validation, where the data is split into k equal parts. The best model minimizes the sum of the validation sums.

In New York, in the summer of 1973, daily measurements were carried out to study the relation between the ozone concentration and a number of other meteorological variables. In addition to the ozone concentration, the solar radiation, the wind speed, and the temperature were measured. Ozone is an important component of smog. It forms from nitrogen dioxides and volatile hydrocarbons under the influence of sunlight. In warm weather with little wind, the ozone concentration can rise sharply and cause respiratory problems.

The ozone concentration is measured in parts per billion (ppb); 1 ppb corresponds to 2μ g/m^3. The unit for solar radiation is langley (Ly). It expresses the amount of solar energy per surface unit, $1\ Ly = 41840\ J\ /\ m^2$. Wind speed is measured in miles per hour (mph) and the temperature in degrees Fahrenheit. In this application, we have converted the latter two to km/h and degrees Celsius, respectively.[†]

We consider four regression models to explain the ozone concentration based on the three background variables wind speed, temperature, and solar radiation. We will then use leave-one-out cross-validation to select the best model. The four regression models we consider are

> *model 1:* $Y = \beta_0 + \beta_1 X_1 + \beta_2 X_2 + \beta_3 X_3 + e$
> *model 2:* $Y = \beta_0 + \beta_1 X_1 + \beta_2 X_2 + \beta_3 X_3 + \beta_4 X_1 X_2 + e$
> *model 3:* $Y = f_1(X_1) + f_2(X_2) + f_3(X_3) + e$
> *model 4:* $Y = g(X_1, X_2) + h(X_3) + e.$

Here, Y is the ozone concentration, X_1 is the wind speed, X_2 is the temperature, and X_3 is the solar radiation. Models 1 and 2 are linear regression models. In addition to a term for each background variable, model 2 has a term $X_1 X_2$ for the interaction between wind and temperature. Models 3 and 4 are nonparametric additive models. Model 3 is a generalization of model 1: the linear functions $\beta_i X_i$ are replaced by nonspecified functions f_i. In model 4, we have again included the interaction between X_1 (wind speed) and X_2 (temperature), in the function g, in addition to a nonparametric function h for the influence of the solar radiation. In all models, we assume that the measurement errors e are normally distributed.

The estimated parameters of model 1 are in Table 8.1. In this model, the estimated parameter value $\hat{\beta}_1$ (wind speed) is negative, while $\hat{\beta}_2$ (temperature) and $\hat{\beta}_3$ (solar radiation) are both positive. This corresponds to the increased smog in warm weather with little wind. The last column in the table gives the p-value of the t-tests for H_0: $\beta_i = 0$ based on the T-statistic from (7.4). The three parameters β_1, β_2, and β_3 all differ significantly from 0.

Table 8.2 gives the results of model 2. In this model, the estimated parameters $\hat{\beta}_1$, $\hat{\beta}_2$, and $\hat{\beta}_3$ are all positive. This is, however, compensated by a negative estimate for the interaction parameter β_4.

[†] The data can be found on the book's webpage at `http://www.aup.nl` under ozone.

```
Coefficients:
              Estimate Std. Error t value  Pr(>|t|)
(Intercept) -11.47511   15.64553  -0.733    0.4649
wind         -2.07184    0.40672  -5.094  1.52e-06
temperature   2.97377    0.45635   6.516  2.42e-09
radiation     0.05982    0.02319   2.580    0.0112
```

Table 8.1. R-output containing the parameter estimates and p-values for model 1. The numbers under Std. Error are the standard errors corresponding to the estimates. The numbers under t value are the values of the T-statistic from (7.4), and those under Pr(>|t|) are the corresponding p-values.

```
Coefficients:
                   Estimate Std. Error t value  Pr(>|t|)
(Intercept)      -119.84436   28.89624  -4.147  6.80e-05
wind                4.40662    1.54137   2.859   0.00512
temperature         7.04471    1.02990   6.840  5.26e-10
radiation           0.06599    0.02152   3.067   0.00274
wind:temperature   -0.25501    0.05883  -4.334  3.34e-05
```

Table 8.2. R-output containing the parameter estimates and p-values for model 2. The numbers under Std. Error are the standard errors corresponding to the estimates. The numbers under t value are the values of the T-statistic from (7.4), and those under Pr(>|t|) are the corresponding p-values.

The fitted functions for models 3 and 4 were computed using the function mgcv::gam from package mgcv version 1.8-6 in the R-language for statistical computing, with default choices of the parameters. This function iteratively determines functions for the additive terms by fitting penalized regression splines, with the smoothing parameter determined by generalized cross validation (GCV). See the R-documentation for details. Figure 8.2 shows the estimates of the three functions f_1, f_2, and f_3 in model 3. The estimated function $\hat{f}_1$ for the influence of the wind speed on the formation of ozone is chiefly decreasing, whereas $\hat{f}_2$ (temperature) and $\hat{f}_3$ (solar radiation) are increasing. This agrees with model 1, where $\hat{\beta}_1$ is negative and $\hat{\beta}_2$ and $\hat{\beta}_3$ are positive. The nonlinearity of the estimated functions in clearly visible in Figure 8.2, which shows that model 3 is more general than model 1.

Figure 8.3 shows the estimate of the function g in model 4. The figure on the left gives a perspective view of the estimate $\hat{g}$, and the figure on the right shows the estimate using contour lines. The estimated form of g is more complicated than the function $\beta_1 X_1 + \beta_2 X_2 + \beta_4 X_1 X_2$ used in model 2. Figure 8.4 shows the estimate of the function h in model 4.

To determine the best model, we use cross-validation. In this method, we split the data in a part used to estimate the model and a part used to evaluate the estimated model (the validation data). In leave-one-out cross-validation, each individual observation is used once as validation data. The validation sum of leave-

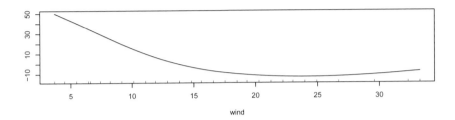

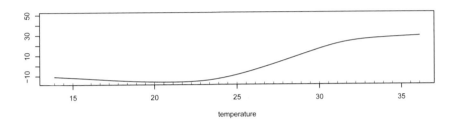

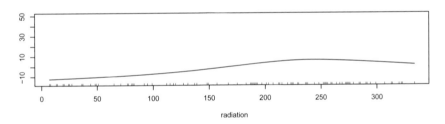

Figure 8.2. Estimates of the functions f_1, f_2, and f_3 in model 3. The short vertical line segments on the horizontal axis indicate the measured values of the three background variables.

one-out cross-validation is

$$\sum_{i=1}^{n}(Y_i - \hat{l}^{-i}(x^i))^2,$$

where $\hat{l}^{-i}$ is the estimated regression function, estimated on all data points except the ith observation, and x^i contains the values of the predictor variables of the ith observation. Table 8.3 gives the values of these validation sums for the four models. It shows that model 4 is the best. Apparently, for this situation, it is important that the interaction between the variables wind speed and temperature is included in the model, and in a more general way than taking their product as in model 2.

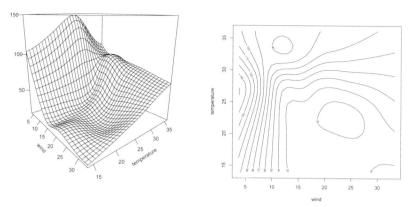

Figure 8.3. Estimate of the function g in model 4: a perspective view on the left and contour lines on the right.

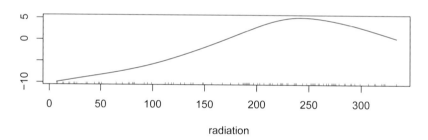

radiation

Figure 8.4. Estimate of the function h in model 4. The short vertical line segments on the horizontal axis indicate the measured value of the solar radiation.

model	1	2	3	4
validation sum	52038.87	44454.35	40635.28	32770.13

Table 8.3. R-output containing the value of the validation sum for the four models.

A Probability Theory

A.1 Introduction

This appendix contains some definitions and results from probability theory that are relevant when reading this book. The aim is to present this matter briefly. For further information, including proofs of theorems, examples, and applications, we refer to textbooks on probability theory, such as S. Ross, *A First Course in Probability*, Prentice-Hall and R. Meester, *A Natural Introduction to Probability Theory*, Birkhäuser.

A.2 Distributions

The basis of all statistical procedures is an observation X influenced by uncertainty, chance, or another form of randomness. As in probability theory, in statistics the uncertainty is translated mathematically by assigning a *probability distribution* to X.

Definition A.1 Random variable

A *random variable is an observation subject to uncertainty, described by a probability distribution.*

The set of all possible outcomes of X is called the *sample space* and denoted by Ω. A probability distribution is a function that gives the probability of finding the observation X in A for (almost) all subsets $A \subseteq \Omega$,

$$P(X \in A).$$

Probability distributions have three properties:
  (i) $P(X \in \Omega) = 1$.
  (ii) $0 \le P(X \in A) \le 1$ for all $A \subseteq \Omega$.
  (iii) (*σ-additivity*) For *disjoint* sets $A_1, A_2, \ldots \subseteq \Omega$ ($A_i \cap A_j = \emptyset$ if $i \ne j$), we have

$$P\left(X \in \bigcup_{i=1}^{\infty} A_i\right) = \sum_{i=1}^{\infty} P(X \in A_i).$$

All other general properties of a probability distribution can be deduced directly from these three; for example, $P(X \in A) \le P(X \in B)$ for $A, B \subseteq \Omega$ with $A \subseteq B$. For other properties of probability distributions, we refer to books on probability theory.

In some cases, we are not interested in the random variable X itself, but in a function g of this variable, for example $g(X) = X^2$. In many cases, the function g is defined on the entire real line, and $g(X)$ is therefore defined for all real-valued random quantities X.[‡] Suppose that the probability distribution of X is known; the distribution of $Y = g(X)$ is then determined by

$$P(Y \in A) = P(g(X) \in A) = P(X \in g^{-1}(A)),$$

where $g^{-1}(A)$ is called the *inverse image* of A under g:

$$g^{-1}(A) = \{x \colon g(x) \in A\}.$$

(The notation g^{-1} might suggest that we need the inverse of g for the definition of the inverse image. This is not the case; even if g is not invertible, the right-hand side is well defined.)

A.2.1 Discrete and Continuous Distributions

There are two basic types of probability distributions: discrete distributions and continuous distributions. A *discrete distribution* is characterized by a finite or countable set of possible outcomes of the random variable, while a random variable with a *continuous distribution* can have outcomes in an interval of the real line. To every discrete and every continuous distribution correspond a probability density (or density for short) and a distribution function.

[‡] The condition that X lie in the domain of g with probability 1 ensures that, in general, $g(X)$ is well defined. Strictly speaking, there is a condition on g (measurability), but in this book we do not discuss nonmeasurable functions.

For a discrete distribution, the probability *density* assigns a weight (probability mass) to every possible outcome, equal to the probability of that outcome. These weights are nonnegative and add up to 1. The probability of an outcome in a subset A of the sample space Ω is equal to

$$P(X \in A) = \sum_{\omega \in A} p(\omega),$$

where $p(\omega) = P(X = \omega)$. Some examples of common discrete distributions are the Bernoulli, binomial, Poisson, geometric, hypergeometric, and negative binomial distributions.

If X has a continuous distribution over (part of) the real line, we use a *probability density function* $f \colon \mathbb{R} \to \mathbb{R}$, which we also call probability density for short. The summation in discrete distributions is replaced by integration in continuous distributions. The probability of an outcome in $A \subseteq \mathbb{R}$ for the continuous random variable X with probability density f is given by

$$P(X \in A) = \int_A f(x)\, dx.$$

Some examples of common continuous distributions are the uniform, normal, exponential, Cauchy, chi-square, t, gamma, and beta distributions.

A.2.2 Distribution Functions

Probability densities are a way to specify distributions. Another, equivalent, way is by using so-called distribution functions.

Definition A.2 Distribution function

Let X be a random variable following some probability distribution. The cumulative distribution function or distribution function F corresponding to the probability distribution is defined by
$$F(x) = P(X \le x).$$
The distribution function is a monotonically increasing function, that is, if $x \le y$, then $F(x) \le F(y)$. The definition of the distribution function in this form holds for both discrete and continuous random variables that are real valued.

For a real-valued, discrete random variable X, the relation between the probability density p and the distribution function F can be expressed as follows:

$$F(x) = P(X \le x) = \sum_{\omega \le x} p(\omega).$$

The distribution function has jumps at all points that are possible outcomes of X. Between these jumps, the distribution function is constant. The size of the jump at the point w is equal to $p(w)$. Hence, discrete distributions can be specified in two ways: either using the probability density p (the distribution function F can be found using summation) or using the distribution function F (the probability density p follows from the size of the jumps).

For a continuous random variable X, the relation between the probability density f and distribution function F can be expressed as follows:

$$F(x) = P(X \leq x) = \int_{-\infty}^{x} f(u)\, du.$$

The distribution function F can therefore be seen as the primitive function of the probability density f. Conversely, f is the derivative of F,

$$f(x) = F'(x).$$

It follows that continuous distributions can also be specified in two ways: either using the probability density f (the distribution function F follows by integration) or using the distribution function F (the probability density follows by differentiation).

Using the distribution function, it is easy to compute the probability for intervals of the form $(c, d]$:

$$P(c < X \leq d) = P(X \leq d) - P(X \leq c) = F(d) - F(c).$$

For discrete distributions, it matters whether the interval is open, closed, or half closed. For example, the probability $P(c \leq X \leq d)$ is greater than $P(c < X \leq d)$ when $P(X = c) > 0$, because $P(c \leq X \leq d) = P(c < X \leq d) + P(X = c)$. For continuous random variables we have $P(X = c) = 0$ for all c; hence the choice of an open or closed interval is of no consequence.

A.3 Expectation and Variance

The expectation and variance of a distribution are properties that reflect the location and dispersion of the distribution, respectively. The *location* is a point around which the distribution is centered, while the *dispersion* is a measure for the width of the distribution around its location. Several properties can act as location or dispersion. Expectation and variance are examples that are often used.

The *expected value* or *expectation* $E(X)$ of a random variable X corresponds to the notion of weighted average. The weights are based on the probability density. When there is no risk of confusion, we write simply EX for the expectation. For a discrete random variable X with probability density p, the expectation EX is given by

$$EX = \sum_{w \in \Omega} w\, p(w),$$

For a continuous random variable X with probability density f, the expectation is defined by

$$EX = \int_{-\infty}^{\infty} x\, f(x)\, dx.$$

For a function g, the expectation of $g(X)$ is

$$E(g(X)) = \int_{-\infty}^{\infty} g(x)\, f(x)\, dx.$$

The expected value of X or $g(X)$ is not always well defined. The integral or sum may not converge. For example, the expectation of a random variable with Cauchy distribution does not exist. In this book, we assume that all integrals we use exist.

The *variance* is the expectation of the square of the distance from X to its expected value,

$$\operatorname{var} X = E(X - EX)^2.$$

We can easily check that the variance can be written as

$$\operatorname{var} X = E(X^2) - (EX)^2.$$

This way of writing it is often useful when computing the variance of a random variable. The expectation $E(X^2)$ is computed as $E(g(X))$ with $g(X) = X^2$. The variance is large if the probability that X is found at a considerable distance from EX is relatively large. This property characterizes the notion of dispersion.

The *covariance* of two random variables X and Y is equal to

$$\operatorname{cov}(X, Y) = E\big((X - EX)(Y - EY)\big) = E(XY) - EX\,EY.$$

The following identities can be deduced from the definitions of the expected value and variance:

$$E(a + bX) = a + b\,EX$$
$$\operatorname{var}(a + bX) = b^2\,\operatorname{var}(X)$$
$$E(X + Y) = EX + EY$$
$$\operatorname{var}(X + Y) = \operatorname{var}(X) + \operatorname{var} Y + 2\operatorname{cov}(X, Y).$$

A.4 Standard Distributions

In this section, we give examples of common discrete and continuous distributions.

A.4.1 Discrete Distributions

Example A.3 Bernoulli distribution

The random variable X has a *Bernoulli distribution* with parameter $p \in [0, 1]$ if

$$P(X = 0) = 1 - p \quad \text{and} \quad P(X = 1) = p.$$

This probability mass function can also be written as

$$P(X = x) = p^x (1 - p)^{1-x}, \quad x \in \{0, 1\}.$$

The expected value and variance are then equal to $EX = p$ and $\text{var } X = p(1 - p)$. If $X_1, \ldots, X_n$ are independent Bernoulli random variables with parameter p, then $X_1 + \ldots + X_n$ is binomially distributed with parameters n and p. ⬜

Example A.4 Binomial distribution

The random variable X has a *binomial distribution* with parameters $n \in \mathbb{N}$ and $p \in [0, 1]$ if

$$P(X = k) = \binom{n}{k} p^k (1 - p)^{n-k}$$

for $k \in \{0, 1, \ldots, n\}$. The expected value and variance are then equal to $EX = np$ and $\text{var } X = np(1 - p)$. The binomial distribution with parameters $n = 1$ and $p \in [0, 1]$ is equal to the Bernoulli distribution with parameter p. If X_1 and X_2 are two independent binomial random variables with parameters (n, p) and (m, p), respectively, then $X_1 + X_2$ is again binomially distributed, with parameters $(n + m, p)$. ⬜

Example A.5 Multinomial distribution

The random variable $X = (X_1, \ldots, X_r)$ has a *multinomial distribution* with parameters $(n, p_1, \ldots, p_r)$, where $n \in \mathbb{N}$ and $p_i \in [0, 1]$ for $i = 1, \ldots, r$ with $\sum_{i=1}^{r} p_i = 1$, if

$$P(X_1 = k_1, \ldots, X_r = k_r) = \binom{n}{k_1 \cdots k_r} p_1^{k_1} \cdots p_r^{k_r}$$

for $k_i \in \{0, 1, \ldots, n\}$ for $i = 1, \ldots, r$ with $\sum_{i=1}^{r} k_i = n$, where

$$\binom{n}{k_1 \cdots k_r} = \frac{n!}{k_1! \cdots k_r!}.$$

In the case $r = 2$, the multinomial distribution reduces to the binomial distribution with parameters n and p_1. ⬜

Example A.6 Poisson distribution

The random variable X has a *Poisson distribution* with parameter $\lambda > 0$, denoted by Poisson(λ), if

$$P(X = k) = \frac{\lambda^k e^{-\lambda}}{k!}$$

for $k \in \{0, 1, \ldots\}$. The expected value and variance are then equal to $EX = \lambda$ and var $X = \lambda$. If X_1 and X_2 are two independent Poisson random variables with means λ and μ, respectively, then $X_1 + X_2$ again has a Poisson distribution, with parameter $\lambda + \mu$.

Example A.7 Geometric distribution

The random variable X has a *geometric distribution* with parameter $p \in (0, 1]$ if

$$P(X = k) = p(1 - p)^{k-1}$$

for $k \in \{1, 2, \ldots\}$. The expected value and variance are then equal to $EX = 1/p$ and var $X = (1 - p)/p^2$. If $X_1, \ldots, X_r$ are independent, geometrically distributed random variables with parameter[nope: means?] p, then $X_1 + \ldots + X_r$ has a negative binomial distribution with parameters r and p.

Example A.8 Negative binomial distribution

The random variable X has a *negative binomial distribution* with parameters $r \in \mathbb{N}$ and $p \in (0, 1]$ if

$$P(X = k) = \binom{k-1}{r-1} p^r (1 - p)^{k-r}$$

for $k \in \{r, r + 1, \ldots\}$. The expected value and variance are then equal to $EX = r/p$ and var $X = r(1 - p)/p^2$. The negative binomial distribution with parameters $r = 1$ and $p \in (0, 1]$ is equal to the geometric distribution with parameter p.

Example A.9 Hypergeometric distribution

The random variable X has a *hypergeometric distribution* with parameters $N, m, n \in \mathbb{N}$, where $n, m < N$, if

$$P(X = k) = \frac{\binom{m}{k}\binom{N-m}{n-k}}{\binom{N}{n}}$$

for $k \in \{0, 1, \ldots, n\}$. The expected value and variance are then equal to $EX = nm/N$ and var $X = n(m/N)(1 - m/N)(N - n)/(N - 1)$.

A.4.2 Continuous Distributions

Example A.10 Uniform distribution

The random variable X has a *uniform distribution* $U[a, b]$ on the interval $[a, b]$ if the density of X is equal to

$$f(x) = \frac{1}{b-a} 1_{[a,b]}(x).$$

The indicator function $1_{[a,b]}(x)$ takes on the value 1 for $x \in [a, b]$ and 0 elsewhere. The expected value and variance are then equal to $EX = (a + b)/2$ and $\operatorname{var} X = (b - a)^2/12$. When $a = 0$ and $b = 1$, the density is equal to $f(x) = 1_{[0,1]}(x)$, the expected value is $1/2$, and the variance is $1/12$.

Example A.11 Normal distribution

The random variable X has a *normal distribution* with parameters $\mu \in \mathbb{R}$ and $\sigma^2 > 0$ if the density of X is equal to

$$f(x) = \frac{1}{\sqrt{2\pi\sigma^2}} e^{-\frac{1}{2}\frac{(x-\mu)^2}{\sigma^2}}.$$

The expected value and variance are then equal to $EX = \mu$ and $\operatorname{var}(X) = \sigma^2$. The *standard normal distribution* is the normal distribution with parameters $\mu = 0$ and $\sigma^2 = 1$. The density and distribution function of the standard normal distribution are denoted by ϕ and Φ, respectively. If X_1 and X_2 are two independent normal random variables, then $X_1 + X_2$ is again normally distributed, with parameters $(\mu + \nu, \sigma^2 + \tau^2)$.

Example A.12 Exponential distribution

The random variable X has an *exponential distribution* with parameter $\lambda > 0$ if the density of X is equal to

$$f(x) = \lambda e^{-\lambda x}, \qquad x \geq 0.$$

The expected value and variance are then equal to $EX = 1/\lambda$ and $\operatorname{var} X = 1/\lambda^2$. If $X_1, \ldots, X_n$ are independent, exponentially distributed random variables with parameter λ, then the sum $X_1 + \ldots + X_n$ has a gamma distribution with shape parameter n and inverse scale parameter or rate parameter λ.

Example A.13 Gamma distribution

The random variable X has a *gamma distribution* with shape parameter n and inverse scale parameter $\lambda > 0$ (or scale parameter $1/\lambda$) if the density of X is equal to

$$f(x) = \frac{x^{\alpha-1}\lambda^{\alpha}e^{-\lambda x}}{\Gamma(\alpha)}, \qquad x \geq 0,$$

where Γ is the so-called *gamma function*

$$\Gamma(\alpha) = \int_0^{\infty} x^{\alpha-1}e^{-x}dx.$$

When $\alpha \in \mathbb{N}$, we have $\Gamma(\alpha) = (\alpha - 1)!$. The expected value and variance of X are given by $EX = \alpha/\lambda$ and $\operatorname{var} X = \alpha/\lambda^2$. The gamma distribution with parameters $\alpha = 1$ and $\lambda > 0$ is equal to the exponential distribution with parameter λ. If X_1 and X_2 are two independent random variables with gamma distributions with parameters (α, λ) and (β, λ), respectively, then $X_1 + X_2$ again has a gamma distribution, with parameters $\alpha + \beta$ and λ.

Example A.14 Beta distribution

The random variable X has a *beta distribution* met parameters $\alpha > 0$ and $\beta > 0$ if the density of X is equal to

$$f(x) = \frac{x^{\alpha-1}(1-x)^{\beta-1}}{B(\alpha, \beta)}, \qquad x \in [0, 1],$$

where B is the so-called *beta function*

$$B(\alpha, \beta) = \int_0^1 x^{\alpha-1}(1-x)^{\beta-1}dx.$$

We have $B(\alpha, \beta) = \Gamma(\alpha)\Gamma(\beta)/\Gamma(\alpha + \beta)$ and $B(\alpha + 1, \beta)/B(\alpha, \beta) = \alpha/(\alpha + \beta)$. The expectation of X can be calculated as follows:

$$\int_0^1 x\frac{x^{\alpha-1}(1-x)^{\beta-1}}{B(\alpha, \beta)}dx = \frac{B(\alpha + 1, \beta)}{B(\alpha, \beta)}\int_0^1 \frac{x^{(\alpha+1)-1}(1-x)^{\beta-1}}{B(\alpha + 1, \beta)}dx$$

$$= \frac{\alpha}{\alpha + \beta},$$

where the last equality follows from the fact that the integral in the second-to-last expression is equal to 1. The variance of X is given by $\operatorname{var} X = \alpha\beta/((\alpha + \beta)^2(\alpha + \beta + 1))$.

Example A.15 Cauchy distribution

The random variable X has a *Cauchy distribution* with location parameter θ if the density of X is equal to

$$f(x) = \frac{1}{\pi(1 + (x - \theta)^2)}.$$

When $\theta = 0$, this is called the standard Cauchy distribution. The expected value and variance of the Cauchy distribution do not exist.

Example A.16 Chi-square distribution

The random variable X has a *chi-square distribution* with n degrees of freedom if X has the same distribution as $\sum_{i=1}^{n} Z_i^2$ for $Z_1, \ldots, Z_n$ independent standard normal random variables. The expected value and variance of X are given by $\mathrm{E}X = n$ and $\mathrm{var}\, X = 2n$. The chi-square distribution with n degrees of freedom is denoted by χ_n^2. This distribution is identical to the gamma distribution with parameters $(n/2, 1/2)$.

Example A.17 t-distribution

The random variable X has a *t-distribution* (or *Student's t-distribution*) with n degrees of freedom if X has the same distribution as $Z/\sqrt{Y/n}$, where Y and Z are independent random variables, Z has a standard normal distribution, and Y has a χ_n^2-distribution. The t-distribution with n degrees of freedom is denoted by t_n. An expression for the density is given in Exercise 4.44.

Example A.18 F-distribution

A random variable X has an *F-distribution* with m and n degrees of freedom if X has the same distribution as $(U/m)/(V/n)$, where U and V are independent, chi-square distributed random variables with m and n degrees of freedom, respectively. The F-distribution with m and n degrees of freedom is denoted by $F_{m,n}$.

A.5 Multivariate and Marginal Distributions

In many cases, we are not interested in a single observation but want to consider several measured quantities at the same time. In probability theory, random vectors are used in such cases. The simplest case is that of two random variables X and Y that are combined into a vector (X, Y). Let Ω_X and Ω_Y be the sample spaces of X and Y, respectively. The possible outcomes of (X, Y) are the points $(x, y) \in \Omega = \Omega_X \times \Omega_Y$. When X and Y are real valued, the combined sample space Ω is equal to (part of) the real plane; that is, $\Omega \subseteq \mathbb{R}^2$. The *joint distribution* of X and Y describes the probabilities of the form

$$\mathrm{P}\big((X, Y) \in A\big),$$

where A is a subset of Ω.

For random vectors, too, there is a distinction between discrete and continuous distributions. If the vector (X,Y) has a discrete distribution, then the distribution is determined by the *multivariate probability density* $p(\omega_1, \omega_2)$, where $p(\omega_1, \omega_2) = P((X,Y) = (\omega_1, \omega_2))$ for all possible outcomes $(\omega_1, \omega_2) \in \Omega$. In this case, the probability of an outcome in a subset A of Ω is equal to the sum

$$P((X,Y) \in A) = \sum_{(\omega_1, \omega_2) \in A} p(\omega_1, \omega_2).$$

When the vector (X,Y) has a continuous distribution, we use a *multivariate probability density function* $f : \mathbb{R}^2 \to \mathbb{R}$, also called a multivariate density for short. The probability of an outcome in a set A is then given by the integral

$$P((X,Y) \in A) = \int_A f(x,y)\, dx\, dy.$$

Examples of joint probability densities and computations involving them can be found in textbooks on probability theory.

From the multivariate distribution of the random vector (X,Y), we can deduce the marginal distributions of X and Y. A *marginal distribution* is determined by the corresponding *marginal probability density*. When the random variables are discrete, the marginal probability density p_X of X is deduced as follows from the multivariate probability density p:

$$p_X(\omega) = \sum_{\omega_2 \in \Omega_Y} p(\omega, \omega_2).$$

For continuous random variables, the summation is replaced by integration, and the *marginal probability density function* f_X of X is deduced from the multivariate probability density f,

$$f_X(x) = \int_{-\infty}^{\infty} f(x,y)\, dy.$$

Analogous formulas hold for the marginal distribution of Y. For random vectors $(X_1, \ldots, X_n)$ with $n > 2$, we can easily generalize the above. In that case, the marginal density of X_1, for example, can be deduced by integrating the multivariate density over $x_2, \ldots, x_n$. In Appendix B, we discuss the multivariate normal distribution.

A.6 Independence and Conditioning

Independence and the conditioning of random variables play an important role in statistics.

Definition A.19 Independent random variables

Two random variables X and Y are called independent if for all events $A \subseteq \Omega_X$ and $B \subseteq \Omega_Y$, we have

$$P(X \in A, Y \in B) = P(X \in A) P(Y \in B).$$

The following theorem shows how the independence of two random variables is reflected in the joint distribution. Proofs of this theorem and Theorem A.22 can be found in textbooks on probability theory.

Theorem A.20

If the random variables X and Y have a discrete joint distribution with probability density p, then X and Y are independent if and only if

$$p(\omega_1, \omega_2) = p_X(\omega_1) p_Y(\omega_2) \quad \text{for all } \omega_1, \omega_2.$$

If (X, Y) has a continuous joint distribution with probability density p, then X and Y are independent if and only if

$$f(x, y) = f_X(x) f_Y(y) \quad \text{for all } x, y.$$

The independence of X and Y means that information on the realization of Y does not influence the distribution of X and vice versa. This heuristic interpretation can be substantiated by considering conditional probabilities.

Definition A.21 Conditional probability

For random variables X and Y, the conditional probability of $X \in A$ given $Y \in B$ is defined by

$$P(X \in A | Y \in B) = \frac{P(X \in A, Y \in B)}{P(Y \in B)}$$

for $A \subseteq \Omega_X$, $B \subseteq \Omega_Y$, and $P(Y \in B) > 0$.

For independent random variables X and Y, the probability of $X \in A$ given $Y \in B$ reduces to

$$P(X \in A | Y \in B) = P(X \in A)$$

for all $A \subseteq \Omega_X$ and $B \subseteq \Omega_Y$ because the product form from Definition A.19 then holds. This calculation shows that if using additional information on Y by conditioning does not influence the distribution of X, then X and Y are independent.

Theorem A.22 Bayes's rule

Let $A_1, \ldots, A_n$ be a partition of Ω, that is, $A_i \cap A_j = \emptyset$ for $i \neq j$ and $\bigcup_{i=1}^{n} A_i = \Omega$, and assume $\mathrm{P}(X \in A_i) > 0$ for all i. Then for an arbitrary B with $\mathrm{P}(Y \in B) > 0$, we have

$$\mathrm{P}(X \in A_i \mid Y \in B) = \frac{\mathrm{P}(Y \in B \mid X \in A_i)\mathrm{P}(X \in A_i)}{\sum_{j=1}^{n} \mathrm{P}(Y \in B \mid X \in A_j)\mathrm{P}(X \in A_j)}.$$

Definition A.23 Conditional density

For continuous random variables X and Y, the conditional density of X given Y is equal to

$$f_{X|Y}(x \mid y) = \frac{f(x,y)}{f_Y(y)}.$$

In fact, the conditional density is an application of Bayes's rule. Here, too, if X and Y are independent, the conditional density of X given Y has a simpler form: $f_{X|Y}(x \mid y) = f_X(x)$ for arbitrary y.

The expected value and variance of the sum of two independent random variables are equal to

$$\mathrm{E}(X + Y) = \mathrm{E}X + \mathrm{E}Y$$
$$\mathrm{var}(X + Y) = \mathrm{var}(X) + \mathrm{var}(Y)$$

because in that case $\mathrm{Cov}(X, Y) = \mathrm{E}(XY) - \mathrm{E}X\mathrm{E}Y = 0$.

The expressions above can easily be extended to sums of n random variables. Let $X_1, \ldots, X_n$ be random variables with finite expectation μ, then

$$\mathrm{E}\left(\sum_{i=1}^{n} X_i\right) = \sum_{i=1}^{n} \mathrm{E}X_i = n\mu.$$

If $X_1, \ldots, X_n$ have finite variance σ^2 and are independent, then we also have

$$\mathrm{var}\left(\sum_{i=1}^{n} X_i\right) = \sum_{i=1}^{n} \mathrm{var}\, X_i = n\sigma^2.$$

The expectation and variance of the sample mean

$$\overline{X} = \frac{1}{n}\sum_{i=1}^{n} X_i$$

are then equal to

$$\mathrm{E}\overline{X} = \frac{1}{n}\mathrm{E}\left(\sum_{i=1}^{n} X_i\right) = \mu,$$

$$\mathrm{var}\,\overline{X} = \frac{1}{n^2}\mathrm{var}\left(\sum_{i=1}^{n} X_i\right) = \frac{\sigma^2}{n}.$$

341

A.7 Limit Theorems and the Normal Approximation

For independent, identically distributed continuous random variables $X_1, X_2, \ldots, X_n$ with marginal probability density f_X, the joint density f is equal to the product of the marginal densities:

$$f(x_1, \ldots, x_n) = \prod_{i=1}^{n} f_X(x_i).$$

The joint density comes up frequently in statistical problems. The following important theorems from probability theory apply to sequences of independent and identically distributed random variables. Since the limit theorems concern the limit for $n \to \infty$, in this section the sample mean $\overline{X}$ will be denoted by $\overline{X}_n$ to highlight the dependence on n.

Definition A.24 Convergence in probability

A sequence of random vectors T_n converges in probability to the random vector T, denoted by $T_n \overset{P}{\to} T$, if $P(\|T_n - T\| > \varepsilon) \to 0$ as $n \to \infty$ for all $\varepsilon > 0$.

Definition A.25 Convergence in distribution

A sequence of random vectors T_n converges in distribution to the random vector T, denoted by $T_n \rightsquigarrow T$, if $P(T_n \leq x) \to P(T \leq x)$ as $n \to \infty$ for all x for which the map $x \mapsto P(T \leq x)$ is continuous.

Theorem A.26 Weak law of large numbers

Let $X_1, X_2, \ldots$ be independent, identically distributed random variables with a marginal distribution with finite expected value μ. Then for every $\varepsilon > 0$,

$$\lim_{n \to \infty} P\left(|\overline{X}_n - \mu| > \varepsilon\right) = 0.$$

In other words, $\overline{X}_n$ converges in probability to μ.

Theorem A.27 Strong law of large numbers

Let $X_1, X_2, \ldots$ be independent, identically distributed random variables with a marginal distribution with finite expected value μ. Then

$$P\left(\lim_{n \to \infty} \overline{X}_n = \mu\right) = 1.$$

Theorem A.28 Central limit theorem

Let $X_1, X_2, \ldots$ be independent, identically distributed random variables with a marginal distribution with finite expected value μ and finite variance σ^2. Then

$$\lim_{n \to \infty} P\left(\frac{\sqrt{n}(\overline{X}_n - \mu)}{\sqrt{\sigma^2}} \leq z \right) = \Phi(z),$$

where Φ is the distribution function of the standard normal distribution. In other words, the standardized sample mean converges in distribution to a standard normal random variable.

The central limit theorem as such cannot be applied in practice because we never have an infinite amount of data. However, for large values of n, the probability on the left-hand side is approximately equal to the probability on the right-hand side. How large n must be exactly for a reasonable approximation depends, among other things, on the skewness of the marginal distribution.

The random variable on the left-hand side can also be written as

$$\frac{\sqrt{n}(\overline{X}_n - \mu)}{\sqrt{\sigma^2}} = \frac{\overline{X}_n - \mu}{\sqrt{\sigma^2/n}} = \frac{\overline{X}_n - E\overline{X}_n}{\sqrt{\operatorname{var} \overline{X}_n}}.$$

Thus, the central limit theorem implies that the standardized sample mean approximately follows the standard normal distribution when the number of observations is large.

Example A.29 Normal approximation of the binomial distribution

Let $X_1, \ldots, X_n$ be a sample from the Bernoulli distribution with parameter p. The corresponding expected value and variance are both finite, respectively p and $p(1-p)$. The central limit theorem implies that for large values of n,

$$\frac{\overline{X}_n - E\overline{X}_n}{\sqrt{\operatorname{var} \overline{X}_n}} = \frac{\overline{X}_n - p}{\sqrt{p(1-p)/n}}$$

approximately has the standard normal distribution. The sample mean therefore approximately has distribution $N(p, p(1-p)/n)$. From this, we can also derive an approximation for the distribution of $Y = \sum_{i=1}^{n} X_i$, namely the binomial distribution with parameters n and p. If $\overline{X}_n$ approximately has distribution $N(p, p(1-p)/n)$, then $Y = n\overline{X}_n$ approximately has distribution $N(np, np(1-p))$. This approximation is reasonable when n is not too small and p is not too close to 0 or 1. As a rule of thumb, we use the condition $np(1-p) \geq 5$. Since the binomial distribution is discrete and the normal distribution is continuous, we generally apply a *continuity correction* to improve the approximation. The probability mass at $Y = i$ in the discrete distribution is, as it were, spread out over the interval $(i - 1/2, i + 1/2]$ in the continuous distribution, $P(Y = i) = P(i - 1/2 < Y \leq i + 1/2)$ for all $i \in \mathbb{N}$. This gives

$$P(Y \leq i) = P(Y \leq i + \tfrac{1}{2}) \quad \text{and} \quad P(Y \geq i) = P(Y > i - \tfrac{1}{2}).$$

For example, for $n = 25$ and $p = 0.4$, the combination of the normal approximation and the continuity correction gives

$$P(Y \leq 11) = P(Y \leq 11.5) = P\left(\frac{Y - 10}{\sqrt{6}} \leq \frac{11.5 - 10}{\sqrt{6}}\right) \approx \Phi\left(\frac{1.5}{\sqrt{6}}\right) = 0.730.$$

The exact probability in this case is equal to 0.732. For comparison, we also give the approximated probability without continuity correction: $\Phi(1/\sqrt{6}) = 0.658$. The correction clearly improves the approximation.

Example A.30 Normal approximation of the Poisson distribution

Let $X_1, \ldots, X_n$ be a sample from the Poisson distribution with parameter λ. The expected value and the variance of this distribution are both equal to λ and therefore finite. By the central limit theorem,

$$\frac{\overline{X}_n - E\overline{X}_n}{\sqrt{\operatorname{var} \overline{X}_n}} = \frac{\overline{X}_n - \lambda}{\sqrt{\lambda/n}}$$

approximately follows the standard normal distribution. This quantity can also be written as

$$\frac{\sum_{i=1}^n X_i - n\lambda}{\sqrt{n\lambda}}.$$

Because the sum of independent Poisson random variables again has a Poisson distribution, see Example A.6, the random variable $Y = \sum_{i=1}^n X_i$ has a Poisson distribution with parameter $\mu = n\lambda$. Consequently,

$$\frac{Y - \mu}{\sqrt{\mu}}$$

approximately follows the standard normal distribution. This is equivalent to saying that for large values of μ, the distribution Poisson(μ) can be approximated using the distribution $N(\mu, \mu)$.

Exercises

1. Compute EX^2 if X has a Poisson distribution with parameter θ.

2. Compute $EX(X - 1)(X - 2)$ if X has a Poisson distribution with parameter 1.

3. Compute Ee^X if X has a standard normal distribution.

4. Let X and Y be independent exponential random variables with expectation 1. Determine the probability density and the expectation of $\max(X, Y)$.

5. Let $X = (X_1, X_2, X_3)$ follow the multinomial distribution with parameters n and $p = (p_1, p_2, p_3)$. Show that
   (i) $EX_i = p_i$,
   (ii) $\text{var } X_i = np_i(1 - p_i)$,
   (iii) $\text{cov}(X_i, X_j) = -np_ip_j$ for $i \neq j$. *Hint: Write X_i and X_j as sums of independent random variables.*

6. Let X and Y be independent normal random variables with distributions $N(0, 1)$ and $N(1, 2)$, respectively.
   (i) Determine $P(X + Y \leq 2)$.
   (ii) Determine a number ξ such that $P(X + Y > \xi) = 0.95$.

7. Determine the distribution function and the probability density of $X^2 + Y^2$ if X and Y are independent standard normal random variables.

8. The random vector (X, Y) has an absolutely continuous distribution with density

$$f(x, y) = \begin{cases} \frac{e^{-y}}{\sqrt{\pi}} & \text{if } y > x^2, \\ 0 & \text{otherwise.} \end{cases}$$

   (i) Show that X has a normal distribution with expectation 0 and variance $1/2$.
   (ii) Show that the marginal density of Y is given by $2\sqrt{y}e^{-y}/\sqrt{\pi}$ for $y > 0$.
   (iii) Given x, determine the conditional density of Y given $X = x$.
   (iv) Given x, determine $E(Y|X = x)$.
   (v) Determine EY.

9. The random vector (X, Y) has an absolutely continuous distribution with density

$$f(x, y) = \begin{cases} e^{-x^2 y} & \text{if } x > 1, y > 0, \\ 0 & \text{if } x \leq 1 \text{ or } y \leq 0. \end{cases}$$

   (i) Show that $Z = X^2 Y$ has an exponential distribution with parameter 1.
   (ii) Determine $E(X^2 Y)^2$.
   (iii) Determine Ee^{Z-X-Y}.

10. Let $X_1, \ldots, X_n$ be independent random variables with the uniform distribution on the interval $[0, 1]$. Determine the expectation and variance of $Y = \max(X_1, \ldots, X_n)$. *Hint: Deduce the density of Y from the distribution function $P(Y \leq y)$ of Y, which can be determined using the distribution functions of $X_1, \ldots, X_n$.*

11. Let $X_1, \ldots, X_n$ be independent, identically distributed random variables with expectation μ and variance σ^2. Compute $E\overline{X}_n$, $\text{var } \overline{X}_n$, $E(\overline{X}_n)^2$, and $\text{cov}(X_i - \overline{X}_n, \overline{X}_n)$.

12. Let $X_1, \ldots, X_n$ be independent, identically distributed continuous random variables with probability density f. Let $\mathbb{F}$ be the function given by $\mathbb{F}_n(A) = (1/n)\#(X_i \in A)$ for a fixed set A (for example an interval). Show that $\mathbb{F}_n(A)$ converges in probability to a limit as $n \to \infty$. What is the limit?

13. Let X be binomially distributed with parameters 100 and $1/4$. Determine an approximation for $P(X \leq 30)$ using the central limit theorem.

14. Let $X_1, \ldots, X_{25}$ be independent random variables with distribution Poisson(5). Determine an approximation for $P(\overline{X}_n \geq 4.5)$ using the central limit theorem.

B Multivariate Normal Distribution

B.1 Introduction

The multivariate normal distribution is the generalization of the usual normal distribution to higher dimensions. The distribution is used as the basis for the definition of certain statistical models, such as the general linear regression model but also occurs in limit results related to estimating or testing vector-valued parameters. In this appendix, we discuss, as background material, the main properties of multidimensional distributions. These are useful for understanding certain parts of this book.

B.2 Covariance Matrices

The covariance of two random variables X and Y is defined as $\mathrm{cov}(X,Y) = \mathrm{E}(X - \mathrm{E}X)(Y - \mathrm{E}Y)$ (provided that these expectations exist). The variance of X is equal to $\mathrm{var}\, X = \mathrm{cov}(X, X)$. The expectation is linear: $\mathrm{E}(\alpha X + \beta Y) = \alpha \mathrm{E}X + \beta \mathrm{E}Y$. The covariance is symmetric and bilinear: $\mathrm{cov}(\alpha X + \beta Y, Z) = \alpha\, \mathrm{cov}(X, Z) + \beta\, \mathrm{cov}(Y, Z)$.

The *expectation vector* and *covariance matrix* of a random vector $(X_1, \ldots, X_k)$ are the vector and matrix

$$
EX = \begin{pmatrix} EX_1 \\ EX_2 \\ \vdots \\ EX_k \end{pmatrix}, \qquad \operatorname{Cov} X = \begin{pmatrix} \operatorname{cov}(X_1, X_1) & \cdots & \operatorname{cov}(X_1, X_k) \\ \operatorname{cov}(X_2, X_1) & \cdots & \operatorname{cov}(X_2, X_k) \\ \vdots & & \vdots \\ \operatorname{cov}(X_k, X_1) & \cdots & \operatorname{cov}(X_k, X_k) \end{pmatrix}.
$$

For $k = 1$, these reduce to the expectation and variance of the variable X_1. We conclude this section with the following lemma, which states a number of properties of random vectors.

Lemma B.1

For every matrix A, vector b, and random vector X, we have
(i) $E(AX + b) = AEX + b$,
(ii) $\operatorname{Cov}(AX) = A(\operatorname{Cov} X)A^T$,
(iii) $\operatorname{Cov} X$ is symmetric and positive definite,
(iv) $P\big(X \in EX + \operatorname{range}(\operatorname{Cov} X)\big) = 1$.

B.3 Definition and Basic Properties

For given numbers $\mu \in \mathbb{R}$ and $\sigma > 0$, we say that a random variable X has the normal distribution $N(\mu, \sigma^2)$ if X has a probability density of the form

$$
x \mapsto \frac{1}{\sqrt{2\pi\sigma^2}} e^{-\frac{1}{2}(x-\mu)^2/\sigma^2}.
$$

Furthermore, we say that X has distribution $N(\mu, 0)$ if $P(X = \mu) = 1$. This is the natural extension to the case $\sigma = 0$, because we then always have $EX = \mu$ and $\operatorname{var} X = \sigma^2$.

We now want to generalize the definition of the normal distribution to higher dimensions. Let μ and Σ be an arbitrary vector and a positive definite, symmetric $k \times k$ matrix. Every positive definite, symmetric matrix Σ can be written as

$$
\Sigma = LL^T
$$

for a $k \times k$ matrix L. The matrix L is not unique, but it does not matter which we choose. One possibility with a geometric interpretation comes from the transformation to an orthonormal basis of eigenvectors of Σ. With respect to this basis, the linear transformation Σ is represented by the diagonal matrix D whose entries are the eigenvalues of Σ (in a given order), and Σ is equal to $\Sigma = ODO^T$ for the orthogonal matrix O that represents the change of basis (orthogonal means $O^T = O^{-1}$). In the decomposition, we now choose L equal to $L = OD^{1/2}O^T$ with $D^{1/2}$ the diagonal matrix whose entries are the square roots of the eigenvalues of Σ. We then have

$$LL^T = OD^{1/2}O^T OD^{1/2}O^T = OD^{1/2}D^{1/2}O^T = \Sigma.$$

This matrix L therefore has the desired decomposition property. This L is a positive definite, symmetric matrix, just as Σ is, and is therefore also known as a "positive square root" of Σ.

Definition B.2 Multivariate normal distribution

A k-dimensional random vector X has a *multivariate normal distribution* with parameters μ and Σ, denoted by $N_k(\mu, \Sigma)$, if X has the same probability distribution as the vector $\mu + LZ$ for a $k \times k$ matrix L with $\Sigma = LL^T$ and a vector $Z = (Z_1, \ldots, Z_k)^T$ whose coordinates are independent random variables with distribution $N(0, 1)$.

The notation $N_k(\mu, \Sigma)$ suggests that the distribution of X depends only on μ and Σ. This is indeed the case, although this is not immediately clear from the definition because, at first glance, the distribution of $\mu + LZ$ appears to depend on μ and L. We will see further on, in Lemmas B.3 and B.4, that the distribution of the vector $\mu + LZ$ depends only on μ and $LL^T = \Sigma$.

The parameters μ and Σ are exactly the expectation and covariance matrix of the vector X because, by Lemma B.1,

$$\mathrm{E}X = \mu + L\mathrm{E}Z = \mu, \qquad \mathrm{Cov}\, X = L\,\mathrm{Cov}\, ZL^T = \Sigma.$$

The multivariate normal distribution with $\mu = 0$ and $\Sigma = I$, the identity matrix, is called the *standard normal distribution*. By Definition B.2, the coordinates of a standard normal vector X are independent variables with distribution $N(0, 1)$.

If the matrix Σ is singular, then the multivariate normal distribution $N_k(\mu, \Sigma)$ does not have a density. (This corresponds to the case $\sigma^2 = 0$ in dimension 1.) This follows from Lemma B.1, which implies that the vector $X - \mathrm{E}X$ takes on its values in the range of the matrix Σ, which is a lower-dimensional subspace of $\mathbb{R}^k$ when Σ is singular. It also immediately follows from Definition B.2 because if Σ is singular, then L is also singular, and the range of $X - \mu = LZ$ is contained in the range of L. Conversely, if Σ is nonsingular, then the multivariate normal distribution $N_k(\mu, \Sigma)$ is continuous. The following lemma gives the probability density explicitly.

Lemma B.3

A random vector X has a multivariate normal distribution with parameters μ and a nonsingular matrix Σ if and only if X has a density of the form

$$x \mapsto \frac{1}{(2\pi)^{k/2}\sqrt{\det \Sigma}} e^{-\frac{1}{2}(x-\mu)^T \Sigma^{-1}(x-\mu)}.$$

Proof. The density of $Z = (Z_1, \ldots, Z_k)$ is the product of standard normal densities. Hence, for every vector b, we have

$$P(\mu + LZ \le b) = \int_{z:\mu+Lz \le b} \prod_{i=1}^{k} \frac{1}{\sqrt{2\pi}} e^{-\frac{1}{2}z_i^2}\, dz.$$

We apply the substitution $\mu + Lz = x$ to this integral. The Jacobian $\partial z/\partial x$ of this linear transformation is L^{-1}; it has determinant $\det L^{-1} = (\det \Sigma)^{-1/2}$. Moreover, $\sum z_i^2 = z^T z = (x-\mu)^T \Sigma^{-1}(x-\mu)$. Hence we can rewrite the integral as

$$\int_{x:x \le b} \frac{1}{(2\pi)^{k/2}} e^{-\frac{1}{2}(x-\mu)^T \Sigma^{-1}(x-\mu)} (\det \Sigma)^{-1/2}\, dx.$$

Since this is true for every b, the result follows from the definition of a probability density. ∎

It is often useful to "reduce" vectors to one-dimensional variables using linear combinations. We can prove that the distribution of a vector X is completely determined by the distributions of all linear combinations $a^T X$, in the sense that two k-dimensional random vectors X and Y are identically distributed if and only if the random variables $a^T X$ and $a^T Y$ are identically distributed for all $a \in \mathbb{R}^k$. Using this property, we can prove the following lemma concerning the normal distribution.

Lemma B.4

The random vector X has distribution $N_k(\mu, \Sigma)$ if and only if for every $a \in \mathbb{R}^k$, the one-dimensional variable $a^T X$ has distribution $N_1(a^T \mu, a^T \Sigma a)$.

Proof. When X has the normal distribution $N_k(\mu, \Sigma)$, the parameters $a^T \mu$ and $a^T \Sigma a$ are correct, because they are exactly the expectation and covariance of the variable $a^T X$. It therefore suffices to prove that $a^T X$ is normally distributed. Because X has the same distribution as $\mu + LZ$, the variable $a^T X$ has the same distribution as $a^T \mu + (L^T a)^T Z$. The latter is a constant plus a linear combination $b^T Z$ of independent variables with distribution $N(0, 1)$ (for $b = L^T a$). We know from probability theory that such a linear combination is normally distributed.

350

Conversely, when $a^T X$ has the normal distribution $N_1(a^T \mu, a^T \Sigma a)$, by the argument we just gave, $a^T X$ has the same distribution as $a^T Y$ for an $N_k(\mu, \Sigma)$-distributed vector Y. If this holds for every a, then X and Y have the same distribution because of the property mentioned before this lemma; hence X has distribution $N_k(\mu, \Sigma)$. ■

Corollary B.5

If the vector $X = (X_1, \ldots, X_k)$ has distribution $N_k(\mu, \Sigma)$ and $A : \mathbb{R}^k \to \mathbb{R}^m$ is an arbitrary matrix, then AX has distribution $N_m(A\mu, A\Sigma A^T)$.

Proof. The parameters $A\mu$ and $A\Sigma A^T$ are correct because they are the expectation and covariance matrix of AX. It suffices to prove that AX is normally distributed. For every vector a, we have $a^T(AX) = (A^T a)^T X$. This variable has a one-dimensional normal distribution by Lemma B.4. Hence AX has a multivariate normal distribution by Lemma B.4, now applied in the other direction. ■

Lemma B.4 and Corollary B.5 imply that the marginal distributions of a multivariate normal distribution are again normal. Indeed, they are the distributions of the linear combinations $e_i^T X$ for the unit vectors $e_1, \ldots, e_k$. The converse is false: it is possible that each of the variables $X_1, \ell s, X_k$ has the normal distribution, while the vector $(X_1, \ldots, X_k)$ does not have the multivariate normal distribution.

A commonly used application of Corollary B.5 is that an orthogonal transformation of a standard normal vector again has a standard normal distribution: if O is a $k \times k$ matrix with $O^T O = OO^T = I$ and Z has distribution $N_k(0, I)$, then OZ also has distribution $N_k(0, I)$ because $O0 = 0$ and $O^T IO = I$. Geometrically, this property means that the standard normal distribution is invariant under rotations.

We conclude with a surprising property of multivariate normal vectors. The random variables $X_1, \ldots, X_k$ are called uncorrelated if the covariance matrix of $(X_1, \ldots, X_k)$ is a diagonal matrix. Independent random variables are always uncorrelated, but the converse does not hold in general. If the vector $X = (X_1, \ldots, X_k)$ has a multivariate normal distribution, then the converse does hold.

Lemma B.6

The vector $X = (X_1, \ldots, X_k)$ has a multivariate normal distribution with Σ a diagonal matrix if and only if $X_1, \ldots, X_k$ are independent and have normal marginal distributions.

Proof. A symmetric, positive definite diagonal matrix Σ can be written as $\Sigma = LL^T$ for L the diagonal matrix with entries the square roots of the diagonal entries of Σ. Then, by definition, X has distribution $N_k(\mu, \Sigma)$ if it has the same distribution as $\mu + LZ = (\mu_1 + L_{11}Z_1, \ldots, \mu_k + L_{kk}Z_k)$ for independent standard normal variables $Z_1, \ldots, Z_k$. Hence the coordinates of X are independent and normally distributed.

351

Conversely, if $X_1, \ldots, X_k$ are independent and have distribution $N(\mu_i, \sigma_i^2)$, then X has the same distribution as $(\mu_1 + \sigma_1 Z_1, \ldots, \mu_k + \sigma_k Z_k) = \mu + LZ$, for L the diagonal matrix with diagonal $(\sigma_1, \ldots, \sigma_k)$. Hence X has distribution $N(\mu, LL^T)$, where LL^T is a diagonal matrix. ∎

B.4 Conditional Distributions

If (X, Y) is a random vector with density $(x, y) \mapsto f(x, y)$, then the conditional distribution of X given $Y = y$ is defined as

$$x \mapsto f_{X|Y=y}(x) = \frac{f(x, y)}{\int f(x, y)\,dx}.$$

For a multivariate normally distributed random vector, these conditional distributions are again normal.

For the sake of simplicity, we consider only two-dimensional normal distributions. The proof of the following theorem can, however, easily be extended to conditional distributions of higher-dimensional vectors. For a two-dimensional normal vector (X, Y), we write the expectation and covariance matrix as

$$(B.1) \qquad \begin{pmatrix} \mu \\ \nu \end{pmatrix}, \qquad \begin{pmatrix} \sigma^2 & \rho\sigma\tau \\ \rho\sigma\tau & \tau^2 \end{pmatrix}.$$

Then, σ^2 and τ^2 are the variances van X and Y, respectively, and ρ is the correlation coefficient of X and Y.

Theorem B.7

If (X, Y) has a two-dimensional normal distribution with expectation and covariance matrix as in (B.1), then the conditional distribution of X given $Y = y$ is equal to the normal distribution with expectation $\mu - \rho\sigma\nu/\tau + \rho\sigma y/\tau$ and variance $(1 - \rho^2)\sigma^2$.

Proof. For a given $\lambda \in \mathbb{R}$, we can write $X = X - \lambda Y + \lambda Y = Z + \lambda Y$ for $Z = X - \lambda Y$. Then (Z, Y) is a linear transformation of (X, Y) and therefore has a two-dimensional normal distribution. For $\lambda = \rho\sigma/\tau$, we have

$$\operatorname{cov}(Z, Y) = \operatorname{cov}(X - \lambda Y, Y) = \rho\sigma\tau - \lambda\tau^2 = 0.$$

By Lemma B.6 , we conclude that for the given value $\lambda = \rho\sigma/\tau$, the variables Z and Y are independent; in other words, the conditional distribution of Z given $Y = y$ is the unconditional distribution of Z. The latter is the one-dimensional normal distribution with expectation $EZ = \mu - \lambda\nu = \mu - \rho\sigma\nu/\tau$ and variance $\operatorname{var} Z = \sigma^2 + \lambda^2\tau^2 - 2\lambda\rho\sigma\tau = (1 - \rho^2)\sigma^2$. The conditional distribution of $X = Z + \lambda Y$ given $Y = y$ is then the unconditional distribution of $Z + \lambda y$, which is the normal distribution with expectation $\mu - \rho\sigma\nu/\tau + \rho\sigma y/\tau$ and variance $(1 - \rho^2)\sigma^2$. ∎

B.5 Multivariate Central Limit Theorem

The "usual" central limit theorem says that the average of a sequence of independent random variables with finite variance is approximately normally distributed. More precisely, if $Y_1, Y_2, \ldots$ is a sequence of independent, identically distributed random variables with expectation μ and finite variance σ^2, then for every $x \in \mathbb{R}$,

$$\lim_{n \to \infty} \mathrm{P}\big(\sqrt{n}(\overline{Y}_n - \mu) \le x\big) = \Phi(x/\sigma).$$

We say that the sequence $\sqrt{n}(\overline{Y}_n - \mu)$ converges in distribution to $N_1(0, \sigma^2)$.

This theorem is also true when the sequence $Y_1, Y_2, \ldots$ consists of random vectors. We then define the average $\overline{Y}_n$ as the vector whose coordinates are the averages of the coordinates of the Y_i. As parameters, we have an expectation vector μ and a covariance matrix Σ. The central limit theorem for vectors states that the sequence of vectors $\sqrt{n}(\overline{Y}_n - \mu)$ converges in distribution to $N_k(0, \Sigma)$.

B.6 Derived Distributions

The chi-square distribution with k degrees of freedom is by definition the distribution of $\sum_{i=1}^{k} Z_i^2$ for independent standard normal random variables $Z_1, \ldots, Z_k$. The sum of squares $\sum_{i=1}^{k} Z_i^2$ is exactly the Euclidean norm $\|Z\|^2$ of the vector $Z = (Z_1, \ldots, Z_k)$, which has a standard normal distribution. We conclude that the squared norm of a k-dimensional standard normal distribution has a chi-square distribution with k degrees of freedom.

If X has distribution $N_k(\mu, \Sigma)$ for a nonsingular matrix Σ and L is the symmetric, positive definite square root of Σ, so that $\Sigma = L^2$, then $L^{-1}(X - \mu)$ has the standard normal distribution. It follows that the quadratic form

$$(X - \mu)^T \Sigma^{-1} (X - \mu) = \big\| L^{-1}(X - \mu) \big\|^2$$

has a chi-square distribution with k degrees of freedom.

The chi-square distribution also occurs as the distribution of the squared norm of a projection of a multivariate normal distribution. A *projection* is a linear map $P \colon \mathbb{R}^k \to \mathbb{R}^k$ of the following form. For a given orthonormal basis $\{f_1, \ldots, f_k\}$ of $\mathbb{R}^k$ (not necessarily the standard basis!), we set $Px = \sum_{i=1}^{l} \xi_i f_i$ when $x = \sum_{i=1}^{k} \xi_i f_i$. In other words, we "forget" the component of x in the space spanned by $f_{l+1}, \ldots, f_k$. We call P the projection onto the linear space spanned by $f_1, \ldots, f_l$. The squared norm of Px is $\|Px\|^2 = \sum_{i=1}^{l} \xi_i^2$. The matrix $I - P$ gives the projection onto the space spanned by $f_{l+1}, \ldots, f_k$: $(I - P)x = \sum_{i=l+1}^{k} \xi_i f_i$.

353

If Z has distribution $N_k(0, I)$, then its coordinates $Z_1, \dots, Z_k$ with respect to the standard basis are independent and have distribution $N(0, 1)$. Because of the rotation invariance of the standard normal distribution, the coordinates $\zeta_1, \dots, \zeta_k$ with respect to an arbitrary basis are also independent and also have distribution $N(0, 1)$. For a projection P as in the previous paragraph, it follows that $\|PZ\|^2 = \sum_{i=1}^{l} \zeta_i^2$ has a chi-square distribution with l degrees of freedom. This statement is part of Cochran's theorem.

Consider a partition

$$\{f_1^1, \dots, f_{i_1}^1\}, \{f_1^2, \dots, f_{i_2}^2\}, \dots, \{f_1^r, \dots, f_{i_r}^r\}$$

of a given orthogonal basis $\{f_1, \dots, f_k\}$, with corresponding projections $P_1, \dots, P_r$. The linear subspaces $H_1, \dots, H_r$ spanned by the elements of this partition are orthogonal, and $P_1, \dots, P_r$ map $\mathbb{R}^k$ exactly onto $H_1, \dots, H_r$.

Theorem B.8 Cochran's theorem

Let $P_1, P_2, \dots, P_r$ be orthogonal projections onto orthogonal subspaces $H_1, H_2, \dots, H_r$ as defined above. If Z has distribution $N_k(0, I)$, then $P_1 Z, P_2 Z, \dots, P_r Z$ are independent random variables, and the random quantities $\|P_1 Z\|^2, \dots, \|P_r Z\|^2$ have a chi-square distribution with, respectively, $\dim(H_1), \dots, \dim(H_r)$ degrees of freedom.

It follows from this theorem that the quotients

$$\frac{\|P_j Z\|^2 / i_j}{\|P_l Z\|^2 / i_l}$$

have F-distributions with i_j and i_l degrees of freedom.

C Tables

This appendix contains some tables for the normal distribution, the t-distribution, the chi-square distribution, and the binomial distribution with $n = 10$. They are meant to be used when there is no computer available. These tables, and many more, can be computed, for example in R, in a fraction of a second and to a higher degree of precision than presented here.

C.1 Normal Distribution

	0	1	2	3	4	5	6	7	8	9
0.0	0.5000	0.5040	0.5080	0.5120	0.5160	0.5199	0.5239	0.5279	0.5319	0.5359
0.1	0.5398	0.5438	0.5478	0.5517	0.5557	0.5596	0.5636	0.5675	0.5714	0.5753
0.2	0.5793	0.5832	0.5871	0.5910	0.5948	0.5987	0.6026	0.6064	0.6103	0.6141
0.3	0.6179	0.6217	0.6255	0.6293	0.6331	0.6368	0.6406	0.6443	0.6480	0.6517
0.4	0.6554	0.6591	0.6628	0.6664	0.6700	0.6736	0.6772	0.6808	0.6844	0.6879
0.5	0.6915	0.6950	0.6985	0.7019	0.7054	0.7088	0.7123	0.7157	0.7190	0.7224
0.6	0.7257	0.7291	0.7324	0.7357	0.7389	0.7422	0.7454	0.7486	0.7517	0.7549
0.7	0.7580	0.7611	0.7642	0.7673	0.7704	0.7734	0.7764	0.7794	0.7823	0.7852
0.8	0.7881	0.7910	0.7939	0.7967	0.7995	0.8023	0.8051	0.8078	0.8106	0.8133
0.9	0.8159	0.8186	0.8212	0.8238	0.8264	0.8289	0.8315	0.8340	0.8365	0.8389
1.0	0.8413	0.8438	0.8461	0.8485	0.8508	0.8531	0.8554	0.8577	0.8599	0.8621
1.1	0.8643	0.8665	0.8686	0.8708	0.8729	0.8749	0.8770	0.8790	0.8810	0.8830
1.2	0.8849	0.8869	0.8888	0.8907	0.8925	0.8944	0.8962	0.8980	0.8997	0.9015
1.3	0.9032	0.9049	0.9066	0.9082	0.9099	0.9115	0.9131	0.9147	0.9162	0.9177
1.4	0.9192	0.9207	0.9222	0.9236	0.9251	0.9265	0.9279	0.9292	0.9306	0.9319
1.5	0.9332	0.9345	0.9357	0.9370	0.9382	0.9394	0.9406	0.9418	0.9429	0.9441
1.6	0.9452	0.9463	0.9474	0.9484	0.9495	0.9505	0.9515	0.9525	0.9535	0.9545
1.7	0.9554	0.9564	0.9573	0.9582	0.9591	0.9599	0.9608	0.9616	0.9625	0.9633
1.8	0.9641	0.9649	0.9656	0.9664	0.9671	0.9678	0.9686	0.9693	0.9699	0.9706
1.9	0.9713	0.9719	0.9726	0.9732	0.9738	0.9744	0.9750	0.9756	0.9761	0.9767
2.0	0.9772	0.9778	0.9783	0.9788	0.9793	0.9798	0.9803	0.9808	0.9812	0.9817
2.1	0.9821	0.9826	0.9830	0.9834	0.9838	0.9842	0.9846	0.9850	0.9854	0.9857
2.2	0.9861	0.9864	0.9868	0.9871	0.9875	0.9878	0.9881	0.9884	0.9887	0.9890
2.3	0.9893	0.9896	0.9898	0.9901	0.9904	0.9906	0.9909	0.9911	0.9913	0.9916
2.4	0.9918	0.9920	0.9922	0.9925	0.9927	0.9929	0.9931	0.9932	0.9934	0.9936
2.5	0.9938	0.9940	0.9941	0.9943	0.9945	0.9946	0.9948	0.9949	0.9951	0.9952
2.6	0.9953	0.9955	0.9956	0.9957	0.9959	0.9960	0.9961	0.9962	0.9963	0.9964
2.7	0.9965	0.9966	0.9967	0.9968	0.9969	0.9970	0.9971	0.9972	0.9973	0.9974
2.8	0.9974	0.9975	0.9976	0.9977	0.9977	0.9978	0.9979	0.9979	0.9980	0.9981
2.9	0.9981	0.9982	0.9982	0.9983	0.9984	0.9984	0.9985	0.9985	0.9986	0.9986
3.0	0.9987	0.9987	0.9987	0.9988	0.9988	0.9989	0.9989	0.9989	0.9990	0.9990
3.1	0.9990	0.9991	0.9991	0.9991	0.9992	0.9992	0.9992	0.9992	0.9993	0.9993
3.2	0.9993	0.9993	0.9994	0.9994	0.9994	0.9994	0.9994	0.9995	0.9995	0.9995
3.3	0.9995	0.9995	0.9995	0.9996	0.9996	0.9996	0.9996	0.9996	0.9996	0.9997
3.4	0.9997	0.9997	0.9997	0.9997	0.9997	0.9997	0.9997	0.9997	0.9997	0.9998
3.5	0.9998	0.9998	0.9998	0.9998	0.9998	0.9998	0.9998	0.9998	0.9998	0.9998
3.6	0.9998	0.9998	0.9999	0.9999	0.9999	0.9999	0.9999	0.9999	0.9999	0.9999
3.7	0.9999	0.9999	0.9999	0.9999	0.9999	0.9999	0.9999	0.9999	0.9999	0.9999
3.8	0.9999	0.9999	0.9999	0.9999	0.9999	0.9999	0.9999	0.9999	0.9999	0.9999
3.9	1.0000	1.0000	1.0000	1.0000	1.0000	1.0000	1.0000	1.0000	1.0000	1.0000

Table C.1. Distribution function of the normal distribution on the interval $[0, 4]$. The value in the table is $\Phi(x)$ for $x = a + b/100$, where a is the value in the first column and b is the value in the first row.

C.2 *t*-Distribution

df	0.6	0.7	0.75	0.8	0.85	0.9	0.925	0.95	0.975	0.98	0.99	0.995	0.999
1	0.32	0.73	1.00	1.38	1.96	3.08	4.17	6.31	12.71	15.89	31.82	63.66	318.31
2	0.29	0.62	0.82	1.06	1.39	1.89	2.28	2.92	4.30	4.85	6.96	9.92	22.33
3	0.28	0.58	0.76	0.98	1.25	1.64	1.92	2.35	3.18	3.48	4.54	5.84	10.21
4	0.27	0.57	0.74	0.94	1.19	1.53	1.78	2.13	2.78	3.00	3.75	4.60	7.17
5	0.27	0.56	0.73	0.92	1.16	1.48	1.70	2.02	2.57	2.76	3.36	4.03	5.89
6	0.26	0.55	0.72	0.91	1.13	1.44	1.65	1.94	2.45	2.61	3.14	3.71	5.21
7	0.26	0.55	0.71	0.90	1.12	1.41	1.62	1.89	2.36	2.52	3.00	3.50	4.79
8	0.26	0.55	0.71	0.89	1.11	1.40	1.59	1.86	2.31	2.45	2.90	3.36	4.50
9	0.26	0.54	0.70	0.88	1.10	1.38	1.57	1.83	2.26	2.40	2.82	3.25	4.30
10	0.26	0.54	0.70	0.88	1.09	1.37	1.56	1.81	2.23	2.36	2.76	3.17	4.14
11	0.26	0.54	0.70	0.88	1.09	1.36	1.55	1.80	2.20	2.33	2.72	3.11	4.02
12	0.26	0.54	0.70	0.87	1.08	1.36	1.54	1.78	2.18	2.30	2.68	3.05	3.93
13	0.26	0.54	0.69	0.87	1.08	1.35	1.53	1.77	2.16	2.28	2.65	3.01	3.85
14	0.26	0.54	0.69	0.87	1.08	1.35	1.52	1.76	2.14	2.26	2.62	2.98	3.79
15	0.26	0.54	0.69	0.87	1.07	1.34	1.52	1.75	2.13	2.25	2.60	2.95	3.73
16	0.26	0.54	0.69	0.86	1.07	1.34	1.51	1.75	2.12	2.24	2.58	2.92	3.69
17	0.26	0.53	0.69	0.86	1.07	1.33	1.51	1.74	2.11	2.22	2.57	2.90	3.65
18	0.26	0.53	0.69	0.86	1.07	1.33	1.50	1.73	2.10	2.21	2.55	2.88	3.61
19	0.26	0.53	0.69	0.86	1.07	1.33	1.50	1.73	2.09	2.20	2.54	2.86	3.58
20	0.26	0.53	0.69	0.86	1.06	1.33	1.50	1.72	2.09	2.20	2.53	2.85	3.55
21	0.26	0.53	0.69	0.86	1.06	1.32	1.49	1.72	2.08	2.19	2.52	2.83	3.53
22	0.26	0.53	0.69	0.86	1.06	1.32	1.49	1.72	2.07	2.18	2.51	2.82	3.50
23	0.26	0.53	0.69	0.86	1.06	1.32	1.49	1.71	2.07	2.18	2.50	2.81	3.48
24	0.26	0.53	0.68	0.86	1.06	1.32	1.49	1.71	2.06	2.17	2.49	2.80	3.47
25	0.26	0.53	0.68	0.86	1.06	1.32	1.49	1.71	2.06	2.17	2.49	2.79	3.45
26	0.26	0.53	0.68	0.86	1.06	1.31	1.48	1.71	2.06	2.16	2.48	2.78	3.43
27	0.26	0.53	0.68	0.86	1.06	1.31	1.48	1.70	2.05	2.16	2.47	2.77	3.42
28	0.26	0.53	0.68	0.85	1.06	1.31	1.48	1.70	2.05	2.15	2.47	2.76	3.41
29	0.26	0.53	0.68	0.85	1.06	1.31	1.48	1.70	2.05	2.15	2.46	2.76	3.40
30	0.26	0.53	0.68	0.85	1.05	1.31	1.48	1.70	2.04	2.15	2.46	2.75	3.39
31	0.26	0.53	0.68	0.85	1.05	1.31	1.48	1.70	2.04	2.14	2.45	2.74	3.37
32	0.26	0.53	0.68	0.85	1.05	1.31	1.47	1.69	2.04	2.14	2.45	2.74	3.37
33	0.26	0.53	0.68	0.85	1.05	1.31	1.47	1.69	2.03	2.14	2.44	2.73	3.36
34	0.26	0.53	0.68	0.85	1.05	1.31	1.47	1.69	2.03	2.14	2.44	2.73	3.35
35	0.26	0.53	0.68	0.85	1.05	1.31	1.47	1.69	2.03	2.13	2.44	2.72	3.34
36	0.26	0.53	0.68	0.85	1.05	1.31	1.47	1.69	2.03	2.13	2.43	2.72	3.33
37	0.26	0.53	0.68	0.85	1.05	1.30	1.47	1.69	2.03	2.13	2.43	2.72	3.33
38	0.26	0.53	0.68	0.85	1.05	1.30	1.47	1.69	2.02	2.13	2.43	2.71	3.32
39	0.26	0.53	0.68	0.85	1.05	1.30	1.47	1.68	2.02	2.12	2.43	2.71	3.31
40	0.26	0.53	0.68	0.85	1.05	1.30	1.47	1.68	2.02	2.12	2.42	2.70	3.31
41	0.25	0.53	0.68	0.85	1.05	1.30	1.47	1.68	2.02	2.12	2.42	2.70	3.30
42	0.25	0.53	0.68	0.85	1.05	1.30	1.47	1.68	2.02	2.12	2.42	2.70	3.30
43	0.25	0.53	0.68	0.85	1.05	1.30	1.47	1.68	2.02	2.12	2.42	2.70	3.29
44	0.25	0.53	0.68	0.85	1.05	1.30	1.47	1.68	2.02	2.12	2.41	2.69	3.29
45	0.25	0.53	0.68	0.85	1.05	1.30	1.46	1.68	2.01	2.12	2.41	2.69	3.28
46	0.25	0.53	0.68	0.85	1.05	1.30	1.46	1.68	2.01	2.11	2.41	2.69	3.28
47	0.25	0.53	0.68	0.85	1.05	1.30	1.46	1.68	2.01	2.11	2.41	2.68	3.27
48	0.25	0.53	0.68	0.85	1.05	1.30	1.46	1.68	2.01	2.11	2.41	2.68	3.27
49	0.25	0.53	0.68	0.85	1.05	1.30	1.46	1.68	2.01	2.11	2.40	2.68	3.27
50	0.25	0.53	0.68	0.85	1.05	1.30	1.46	1.68	2.01	2.11	2.40	2.68	3.26

Table C.2. Quantiles of the *t*-distribution with 1 to 50 degrees of freedom.

C.3 Chi-Square Distribution

df	0.001	0.01	0.02	0.025	0.05	0.075	0.1	0.15	0.2	0.25	0.3	0.4
1	0.00	0.00	0.00	0.00	0.00	0.01	0.02	0.04	0.06	0.10	0.15	0.27
2	0.00	0.02	0.04	0.05	0.10	0.16	0.21	0.33	0.45	0.58	0.71	1.02
3	0.02	0.11	0.18	0.22	0.35	0.47	0.58	0.80	1.01	1.21	1.42	1.87
4	0.09	0.30	0.43	0.48	0.71	0.90	1.06	1.37	1.65	1.92	2.19	2.75
5	0.21	0.55	0.75	0.83	1.15	1.39	1.61	1.99	2.34	2.67	3.00	3.66
6	0.38	0.87	1.13	1.24	1.64	1.94	2.20	2.66	3.07	3.45	3.83	4.57
7	0.60	1.24	1.56	1.69	2.17	2.53	2.83	3.36	3.82	4.25	4.67	5.49
8	0.86	1.65	2.03	2.18	2.73	3.14	3.49	4.08	4.59	5.07	5.53	6.42
9	1.15	2.09	2.53	2.70	3.33	3.78	4.17	4.82	5.38	5.90	6.39	7.36
10	1.48	2.56	3.06	3.25	3.94	4.45	4.87	5.57	6.18	6.74	7.27	8.30
11	1.83	3.05	3.61	3.82	4.57	5.12	5.58	6.34	6.99	7.58	8.15	9.24
12	2.21	3.57	4.18	4.40	5.23	5.82	6.30	7.11	7.81	8.44	9.03	10.18
13	2.62	4.11	4.77	5.01	5.89	6.52	7.04	7.90	8.63	9.30	9.93	11.13
14	3.04	4.66	5.37	5.63	6.57	7.24	7.79	8.70	9.47	10.17	10.82	12.08
15	3.48	5.23	5.98	6.26	7.26	7.97	8.55	9.50	10.31	11.04	11.72	13.03
16	3.94	5.81	6.61	6.91	7.96	8.71	9.31	10.31	11.15	11.91	12.62	13.98
17	4.42	6.41	7.26	7.56	8.67	9.45	10.09	11.12	12.00	12.79	13.53	14.94
18	4.90	7.01	7.91	8.23	9.39	10.21	10.86	11.95	12.86	13.68	14.44	15.89
19	5.41	7.63	8.57	8.91	10.12	10.97	11.65	12.77	13.72	14.56	15.35	16.85
20	5.92	8.26	9.24	9.59	10.85	11.73	12.44	13.60	14.58	15.45	16.27	17.81
21	6.45	8.90	9.91	10.28	11.59	12.5	13.24	14.44	15.44	16.34	17.18	18.77
22	6.98	9.54	10.60	10.98	12.34	13.28	14.04	15.28	16.31	17.24	18.10	19.73
23	7.53	10.20	11.29	11.69	13.09	14.06	14.85	16.12	17.19	18.14	19.02	20.69
24	8.08	10.86	11.99	12.40	13.85	14.85	15.66	16.97	18.06	19.04	19.94	21.65
25	8.65	11.52	12.70	13.12	14.61	15.64	16.47	17.82	18.94	19.94	20.87	22.62
26	9.22	12.20	13.41	13.84	15.38	16.44	17.29	18.67	19.82	20.84	21.79	23.58
27	9.80	12.88	14.13	14.57	16.15	17.24	18.11	19.53	20.70	21.75	22.72	24.54
28	10.39	13.56	14.85	15.31	16.93	18.05	18.94	20.39	21.59	22.66	23.65	25.51
29	10.99	14.26	15.57	16.05	17.71	18.85	19.77	21.25	22.48	23.57	24.58	26.48
30	11.59	14.95	16.31	16.79	18.49	19.66	20.60	22.11	23.36	24.48	25.51	27.44
31	12.20	15.66	17.04	17.54	19.28	20.48	21.43	22.98	24.26	25.39	26.44	28.41
32	12.81	16.36	17.78	18.29	20.07	21.30	22.27	23.84	25.15	26.30	27.37	29.38
33	13.43	17.07	18.53	19.05	20.87	22.12	23.11	24.71	26.04	27.22	28.31	30.34
34	14.06	17.79	19.28	19.81	21.66	22.94	23.95	25.59	26.94	28.14	29.24	31.31
35	14.69	18.51	20.03	20.57	22.47	23.76	24.80	26.46	27.84	29.05	30.18	32.28
36	15.32	19.23	20.78	21.34	23.27	24.59	25.64	27.34	28.73	29.97	31.12	33.25
37	15.97	19.96	21.54	22.11	24.07	25.42	26.49	28.21	29.64	30.89	32.05	34.22
38	16.61	20.69	22.30	22.88	24.88	26.25	27.34	29.09	30.54	31.81	32.99	35.19
39	17.26	21.43	23.07	23.65	25.70	27.09	28.20	29.97	31.44	32.74	33.93	36.16
40	17.92	22.16	23.84	24.43	26.51	27.93	29.05	30.86	32.34	33.66	34.87	37.13
41	18.58	22.91	24.61	25.21	27.33	28.76	29.91	31.74	33.25	34.58	35.81	38.11
42	19.24	23.65	25.38	26.00	28.14	29.61	30.77	32.63	34.16	35.51	36.75	39.08
43	19.91	24.40	26.16	26.79	28.96	30.45	31.63	33.51	35.07	36.44	37.70	40.05
44	20.58	25.15	26.94	27.57	29.79	31.29	32.49	34.40	35.97	37.36	38.64	41.02
45	21.25	25.90	27.72	28.37	30.61	32.14	33.35	35.29	36.88	38.29	39.58	42.00
46	21.93	26.66	28.50	29.16	31.44	32.99	34.22	36.18	37.80	39.22	40.53	42.97
47	22.61	27.42	29.29	29.96	32.27	33.84	35.08	37.07	38.71	40.15	41.47	43.94
48	23.29	28.18	30.08	30.75	33.10	34.69	35.95	37.96	39.62	41.08	42.42	44.92
49	23.98	28.94	30.87	31.55	33.93	35.54	36.82	38.86	40.53	42.01	43.37	45.89
50	24.67	29.71	31.66	32.36	34.76	36.40	37.69	39.75	41.45	42.94	44.31	46.86

Table C.3. Quantiles of the chi-square distribution with 1 to 50 degrees of freedom.

df	0.6	0.7	0.75	0.8	0.85	0.9	0.925	0.95	0.975	0.98	0.99	0.999
1	0.71	1.07	1.32	1.64	2.07	2.71	3.17	3.84	5.02	5.41	6.63	10.83
2	1.83	2.41	2.77	3.22	3.79	4.61	5.18	5.99	7.38	7.82	9.21	13.82
3	2.95	3.66	4.11	4.64	5.32	6.25	6.90	7.81	9.35	9.84	11.34	16.27
4	4.04	4.88	5.39	5.99	6.74	7.78	8.50	9.49	11.14	11.67	13.28	18.47
5	5.13	6.06	6.63	7.29	8.12	9.24	10.01	11.07	12.83	13.39	15.09	20.52
6	6.21	7.23	7.84	8.56	9.45	10.64	11.47	12.59	14.45	15.03	16.81	22.46
7	7.28	8.38	9.04	9.80	10.75	12.02	12.88	14.07	16.01	16.62	18.48	24.32
8	8.35	9.52	10.22	11.03	12.03	13.36	14.27	15.51	17.53	18.17	20.09	26.12
9	9.41	10.66	11.39	12.24	13.29	14.68	15.63	16.92	19.02	19.68	21.67	27.88
10	10.47	11.78	12.55	13.44	14.53	15.99	16.97	18.31	20.48	21.16	23.21	29.59
11	11.53	12.90	13.70	14.63	15.77	17.28	18.29	19.68	21.92	22.62	24.72	31.26
12	12.58	14.01	14.85	15.81	16.99	18.55	19.60	21.03	23.34	24.05	26.22	32.91
13	13.64	15.12	15.98	16.98	18.20	19.81	20.90	22.36	24.74	25.47	27.69	34.53
14	14.69	16.22	17.12	18.15	19.41	21.06	22.18	23.68	26.12	26.87	29.14	36.12
15	15.73	17.32	18.25	19.31	20.60	22.31	23.45	25.00	27.49	28.26	30.58	37.70
16	16.78	18.42	19.37	20.47	21.79	23.54	24.72	26.30	28.85	29.63	32.00	39.25
17	17.82	19.51	20.49	21.61	22.98	24.77	25.97	27.59	30.19	31.00	33.41	40.79
18	18.87	20.60	21.60	22.76	24.16	25.99	27.22	28.87	31.53	32.35	34.81	42.31
19	19.91	21.69	22.72	23.90	25.33	27.20	28.46	30.14	32.85	33.69	36.19	43.82
20	20.95	22.77	23.83	25.04	26.50	28.41	29.69	31.41	34.17	35.02	37.57	45.31
21	21.99	23.86	24.93	26.17	27.66	29.62	30.92	32.67	35.48	36.34	38.93	46.80
22	23.03	24.94	26.04	27.30	28.82	30.81	32.14	33.92	36.78	37.66	40.29	48.27
23	24.07	26.02	27.14	28.43	29.98	32.01	33.36	35.17	38.08	38.97	41.64	49.73
24	25.11	27.10	28.24	29.55	31.13	33.20	34.57	36.42	39.36	40.27	42.98	51.18
25	26.14	28.17	29.34	30.68	32.28	34.38	35.78	37.65	40.65	41.57	44.31	52.62
26	27.18	29.25	30.43	31.79	33.43	35.56	36.98	38.89	41.92	42.86	45.64	54.05
27	28.21	30.32	31.53	32.91	34.57	36.74	38.18	40.11	43.19	44.14	46.96	55.48
28	29.25	31.39	32.62	34.03	35.71	37.92	39.38	41.34	44.46	45.42	48.28	56.89
29	30.28	32.46	33.71	35.14	36.85	39.09	40.57	42.56	45.72	46.69	49.59	58.30
30	31.32	33.53	34.80	36.25	37.99	40.26	41.76	43.77	46.98	47.96	50.89	59.70
31	32.35	34.60	35.89	37.36	39.12	41.42	42.95	44.99	48.23	49.23	52.19	61.10
32	33.38	35.66	36.97	38.47	40.26	42.58	44.13	46.19	49.48	50.49	53.49	62.49
33	34.41	36.73	38.06	39.57	41.39	43.75	45.31	47.40	50.73	51.74	54.78	63.87
34	35.44	37.80	39.14	40.68	42.51	44.90	46.49	48.60	51.97	53.00	56.06	65.25
35	36.47	38.86	40.22	41.78	43.64	46.06	47.66	49.80	53.20	54.24	57.34	66.62
36	37.50	39.92	41.30	42.88	44.76	47.21	48.84	51.00	54.44	55.49	58.62	67.99
37	38.53	40.98	42.38	43.98	45.89	48.36	50.01	52.19	55.67	56.73	59.89	69.35
38	39.56	42.05	43.46	45.08	47.01	49.51	51.17	53.38	56.90	57.97	61.16	70.70
39	40.59	43.11	44.54	46.17	48.13	50.66	52.34	54.57	58.12	59.20	62.43	72.05
40	41.62	44.16	45.62	47.27	49.24	51.81	53.50	55.76	59.34	60.44	63.69	73.40
41	42.65	45.22	46.69	48.36	50.36	52.95	54.66	56.94	60.56	61.67	64.95	74.74
42	43.68	46.28	47.77	49.46	51.47	54.09	55.82	58.12	61.78	62.89	66.21	76.08
43	44.71	47.34	48.84	50.55	52.59	55.23	56.98	59.30	62.99	64.12	67.46	77.42
44	45.73	48.40	49.91	51.64	53.70	56.37	58.13	60.48	64.20	65.34	68.71	78.75
45	46.76	49.45	50.98	52.73	54.81	57.51	59.29	61.66	65.41	66.56	69.96	80.08
46	47.79	50.51	52.06	53.82	55.92	58.64	60.44	62.83	66.62	67.77	71.20	81.40
47	48.81	51.56	53.13	54.91	57.03	59.77	61.59	64.00	67.82	68.99	72.44	82.72
48	49.84	52.62	54.20	55.99	58.14	60.91	62.74	65.17	69.02	70.20	73.68	84.04
49	50.87	53.67	55.27	57.08	59.24	62.04	63.88	66.34	70.22	71.41	74.92	85.35
50	51.89	54.72	56.33	58.16	60.35	63.17	65.03	67.50	71.42	72.61	76.15	86.66

Table C.3. (Continued) Quantiles of the chi-square distribution with 1 to 50 degrees of freedom.

C.4 Binomial Distribution ($n = 10$)

p	0	1	2	3	4	5	6	7	8	9	10
0.01	904	996	1000	1000	1000	1000	1000	1000	1000	1000	1000
0.02	817	984	999	1000	1000	1000	1000	1000	1000	1000	1000
0.03	737	965	997	1000	1000	1000	1000	1000	1000	1000	1000
0.04	665	942	994	1000	1000	1000	1000	1000	1000	1000	1000
0.05	599	914	988	999	1000	1000	1000	1000	1000	1000	1000
0.06	539	882	981	998	1000	1000	1000	1000	1000	1000	1000
0.07	484	848	972	996	1000	1000	1000	1000	1000	1000	1000
0.08	434	812	960	994	999	1000	1000	1000	1000	1000	1000
0.09	389	775	946	991	999	1000	1000	1000	1000	1000	1000
0.1	349	736	930	987	998	1000	1000	1000	1000	1000	1000
0.11	312	697	912	982	997	1000	1000	1000	1000	1000	1000
0.12	279	658	891	976	996	1000	1000	1000	1000	1000	1000
0.13	248	620	869	969	995	999	1000	1000	1000	1000	1000
0.14	221	582	845	960	993	999	1000	1000	1000	1000	1000
0.15	197	544	820	950	990	999	1000	1000	1000	1000	1000
0.16	175	508	794	939	987	998	1000	1000	1000	1000	1000
0.17	155	473	766	926	983	997	1000	1000	1000	1000	1000
0.18	137	439	737	912	979	996	1000	1000	1000	1000	1000
0.19	122	407	708	896	973	995	999	1000	1000	1000	1000
0.2	107	376	678	879	967	994	999	1000	1000	1000	1000
0.21	95	346	647	861	960	992	999	1000	1000	1000	1000
0.22	83	318	617	841	952	990	998	1000	1000	1000	1000
0.23	73	292	586	821	943	987	998	1000	1000	1000	1000
0.24	64	267	556	799	933	984	997	1000	1000	1000	1000
0.25	56	244	526	776	922	980	996	1000	1000	1000	1000
0.26	49	222	496	752	910	976	996	999	1000	1000	1000
0.27	43	202	466	727	896	971	994	999	1000	1000	1000
0.28	37	183	438	702	882	966	993	999	1000	1000	1000
0.29	33	166	410	676	866	960	991	999	1000	1000	1000
0.3	28	149	383	650	850	953	989	998	1000	1000	1000
0.31	24	134	357	623	832	945	987	998	1000	1000	1000
0.32	21	121	331	596	813	936	984	997	1000	1000	1000
0.33	18	108	307	568	794	927	981	997	1000	1000	1000
0.34	16	96	284	541	773	916	978	996	1000	1000	1000
0.35	13	86	262	514	751	905	974	995	999	1000	1000
0.36	12	76	241	487	729	893	969	994	999	1000	1000
0.37	10	68	221	460	706	879	964	993	999	1000	1000
0.38	8	60	202	434	682	865	959	991	999	1000	1000
0.39	7	53	184	408	658	850	952	990	999	1000	1000
0.4	6	46	167	382	633	834	945	988	998	1000	1000
0.41	5	41	152	358	608	817	937	985	998	1000	1000
0.42	4	36	137	333	582	798	929	983	997	1000	1000
0.43	4	31	124	310	556	779	919	980	997	1000	1000
0.44	3	27	111	288	530	759	909	976	996	1000	1000
0.45	3	23	100	266	504	738	898	973	995	1000	1000
0.46	2	20	89	245	478	717	886	968	995	1000	1000
0.47	2	17	79	226	453	694	873	963	994	999	1000
0.48	1	15	70	207	427	671	859	958	992	999	1000
0.49	1	13	62	189	402	647	844	952	991	999	1000
0.5	1	11	55	172	377	623	828	945	989	999	1000

Table C.4. Cumulative probabilities ($\times 1000$) for the binomial distribution with parameters 10 and p going from 0.01 to 0.5.

p	0	1	2	3	4	5	6	7	8	9	10
0.5	1	11	55	172	377	623	828	945	989	999	1000
0.51	1	9	48	156	353	598	811	938	987	999	1000
0.52	1	8	42	141	329	573	793	930	985	999	1000
0.53	1	6	37	127	306	547	774	921	983	998	1000
0.54	0	5	32	114	283	522	755	911	980	998	1000
0.55	0	5	27	102	262	496	734	900	977	997	1000
0.56	0	4	24	91	241	470	712	889	973	997	1000
0.57	0	3	20	81	221	444	690	876	969	996	1000
0.58	0	3	17	71	202	418	667	863	964	996	1000
0.59	0	2	15	63	183	392	642	848	959	995	1000
0.6	0	2	12	55	166	367	618	833	954	994	1000
0.61	0	1	10	48	150	342	592	816	947	993	1000
0.62	0	1	9	41	135	318	566	798	940	992	1000
0.63	0	1	7	36	121	294	540	779	932	990	1000
0.64	0	1	6	31	107	271	513	759	924	988	1000
0.65	0	1	5	26	95	249	486	738	914	987	1000
0.66	0	0	4	22	84	227	459	716	904	984	1000
0.67	0	0	3	19	73	206	432	693	892	982	1000
0.68	0	0	3	16	64	187	404	669	879	979	1000
0.69	0	0	2	13	55	168	377	643	866	976	1000
0.7	0	0	2	11	47	150	350	617	851	972	1000
0.71	0	0	1	9	40	134	324	590	834	967	1000
0.72	0	0	1	7	34	118	298	562	817	963	1000
0.73	0	0	1	6	29	104	273	534	798	957	1000
0.74	0	0	1	4	24	90	248	504	778	951	1000
0.75	0	0	0	4	20	78	224	474	756	944	1000
0.76	0	0	0	3	16	67	201	444	733	936	1000
0.77	0	0	0	2	13	57	179	414	708	927	1000
0.78	0	0	0	2	10	48	159	383	682	917	1000
0.79	0	0	0	1	8	40	139	353	654	905	1000
0.8	0	0	0	1	6	33	121	322	624	893	1000
0.81	0	0	0	1	5	27	104	292	593	878	1000
0.82	0	0	0	0	4	21	88	263	561	863	1000
0.83	0	0	0	0	3	17	74	234	527	845	1000
0.84	0	0	0	0	2	13	61	206	492	825	1000
0.85	0	0	0	0	1	10	50	180	456	803	1000
0.86	0	0	0	0	1	7	40	155	418	779	1000
0.87	0	0	0	0	1	5	31	131	380	752	1000
0.88	0	0	0	0	0	4	24	109	342	721	1000
0.89	0	0	0	0	0	3	18	88	303	688	1000
0.9	0	0	0	0	0	2	13	70	264	651	1000
0.91	0	0	0	0	0	1	9	54	225	611	1000
0.92	0	0	0	0	0	1	6	40	188	566	1000
0.93	0	0	0	0	0	0	4	28	152	516	1000
0.94	0	0	0	0	0	0	2	19	118	461	1000
0.95	0	0	0	0	0	0	1	12	86	401	1000
0.96	0	0	0	0	0	0	0	6	58	335	1000
0.97	0	0	0	0	0	0	0	3	35	263	1000
0.98	0	0	0	0	0	0	0	1	16	183	1000
0.99	0	0	0	0	0	0	0	0	4	96	1000

Table C.4. (Continued) Cumulative probabilities ($\times 1000$) for the binomial distribution with parameters 10 and p going from 0.5 to 0.99.

D Answers to Exercises

1.1. model: $X \sim \text{bin}(n, p), p \in [0, 1]$, estimate: X/n

1.2. (i). X_i blood pressure change ith person in treatment group, Y_j blood pressure change jth person in control group, model: $X_1, \ldots, X_n, Y_1, \ldots, Y_m$ independent, $X_i \sim N(\mu_1, \sigma_1^2)$, $Y_i \sim N(\mu_2, \sigma_2^2)$

(ii). estimate: $\bar{x}_n - \bar{y}_m$

1.3. (i). model: $X \sim \text{hyp}(N, r, n)$ with $N \geq \max(r, n)$

(ii). estimate: rn/X

(iii). model: $X \sim \text{bin}(n, r/N)$ with $N \geq \max(r, n)$, estimate: rn/X

1.4. (i). model: $X \sim \text{neg} - \text{bin}(3, p)$ with $p \in (0, 1]$

(ii). estimate: $3/50$

1.5. (i). X_{ij} is the number of customers in week i on half-day j, where j has 11 possible values; model: $X_j \sim \text{Poisson}(\theta_{ij})$ independent, with $\theta_{ij} > 0$

(ii). estimate: $(1/10) \sum_{i=1}^{10} x_{ij}$, where j corresponds to Monday afternoon

1.8. too long

1.9. not good, the estimate is too low

2.3. (i). yes

(ii). if $Y \sim F_{a,b}$, then $b = \sqrt{\text{var } Y}$ and $a = EY - \sqrt{\text{var } Y}$

2.4. (i). $F(x) = ((x + 3)/5)1_{-3<x<2} + 1_{x \geq 2}$

(ii). $F^{-1}(\alpha) = -3 + 5\alpha$

2.5. (i). $F(x) = x^2/\theta^2 1_{0<x<\theta} + 1_{x \geq \theta}$

(ii). $F^{-1}(\alpha) = \theta\sqrt{\alpha}$

2.6. $y = 2 + 4x$

2.7. $y = 2 + 2x/3$

2.10. 0

2.12. $1/\sqrt{2}$

3.2. $c = (n + 2)/(n + 1)$

3.3. (i). $c = 1$

(ii). $c^2 p(1 - p)/n + p^2(c - 1)^2$

(iii). $c = pn/(1 - p + pn)$, not usable

(iv). $c \to 1$

3.4. (i). no

(ii). many possibilities, including $\overline{X}^2 - \overline{X}/n$

3.5.

(ii). $(\sum_{i=1}^{m} X_i + \sum_{j=1}^{n} Y_j)/(m+n)$

3.6. (i). $p_M = (2/3)p$ and $p_V = (4/3)p$, $\mathrm{MSE}(p,(X+Y)/100) = p/100 - p^2/90$

(ii). $\mathrm{MSE}(p, Z/100) = p/100 - p^2/100$

(iii). $(X+Y)/100$ is better

3.8. $\overline{X}$

3.9. (i). $n/\sum_{i=1}^{n} x_i^a$

(ii). $n^{-1}\sum_{i=1}^{n} x_i^a$

3.10. (i). $\theta/(\theta+1)$

(ii). $\hat{\theta} = -n/\sum_{i=1}^{n} \log(X_i)$ and $\widehat{\theta/(\theta+1)} = \hat{\theta}/(\hat{\theta}+1)$

3.11. $1/\overline{Y}$

3.12. (i). $X_{(1)}$

(ii). no

(iii). $\mathrm{MSE}(\theta; X_{(1)}) = \theta^2/((n-1)(n-2))$

3.14. $\left(\overline{X}, \overline{Y}, (m+n-2)^{-1}\left(\sum_{i=1}^{m}(X_i - \overline{X})^2 + \sum_{j=1}^{n}(Y_j - \overline{Y})^2\right)\right)$

3.16. $X_{(1)}$

3.17. (i). $\mathrm{hyp}(N, r, n)$

(ii). $\lfloor rn/x \rfloor$ (round down to the nearest integer)

3.18. (i). $\mathrm{bin}(n, F(x))$

(ii). yes

(iii). $F(x)(1 - F(x))/n$

3.20. (i). $\overline{X}^2$

(ii). $\overline{X}^2$

(iii). $(n/(n+1))\overline{X}^2$

3.21. $\hat{p} = \overline{X}$ in both cases

3.22. (i). $\hat{p} = \overline{X}$

(ii). $(n/(n-1)\overline{X}^2 - (1/(n-1))\overline{X}$, for example, is unbiased

3.23. $\hat{p} = 1/\overline{X}$

3.24. $(\overline{X} + 1)/\overline{X}$

3.25. (i). $3\overline{X}/2$

(iii). $(3\overline{X}/2)^2$

(iv). $(18n/(8n+1))\overline{X}^2$, for example, is unbiased

3.26. (i). $2\overline{X} - 1$

(ii). $X_{(n)}$

3.27. (i). $(\hat{\sigma}, \hat{\tau}) = (X_{(1)}, X_{(n)})$

(ii). $(\hat{\sigma}, \hat{\tau}) = (\overline{X} - (3\overline{X}^2/2 - 3\overline{X^2})^{1/2}, \overline{X} + (3\overline{X}^2/2 - 3\overline{X^2})^{1/2})$

3.28. (i). $\max(|X_{(1)}|, X_{(n)})$

(ii). $(3\overline{X^2})^{1/2}$

3.31. (i). θ

(ii). the sample median: $\mathrm{med}(X_1, \ldots, X_n)$

(iii). $\overline{X}$

3.33. (i). $\overline{X}/(1 - \overline{X})$

(ii). $-n/\sum_{i=1}^{n} \log X_i$

(iii). posterior distribution $\Gamma(n+1, 1 - \sum_{i=1}^{n} \log X_i)$, Bayes estimator $(n+1)/(1 - \sum_{i=1}^{n} \log X_i)$

3.34.

(ii). posterior distribution $\Gamma(n+r, \lambda + \sum_{i=1}^{n} |X_i|)$, Bayes estimator $(n+r)/(\lambda + \sum_{i=1}^{n} |X_i|)$

3.35. $\mathrm{beta}(\alpha + r, \beta + x - r)$

363

3.36. $(n-1)/(n-2)(X_{(n)}^{-(n-2)} - M^{-(n-2)})/(X_{(n)}^{-(n-1)} - M^{-(n-1)})$

3.37. (i). posterior distribution $\Gamma(X+1, \lambda+1)$, Bayes estimator $(X+1)/(\lambda+1)$
(ii). posterior distribution $\Gamma(X+\alpha, \lambda+1)$, Bayes estimator $(X+\alpha)/(\lambda+1)$

3.38. posterior distribution $\Gamma(\alpha+n, \lambda + \sum_{i=1}^{n} X_i^2)$, Bayes estimator $(\alpha+n)/(\lambda + \sum_{i=1}^{n} X_i^2)$

3.39. normal posterior distribution, Bayes estimator $(\tau^2 \sum_{i=1}^{n} X_i)/(n\tau^2 + 1)$

3.40. (i). beta$(\alpha + \sum_{i=1}^{n} X_i, n + \beta - \sum_{i=1}^{n} X_i)$
(ii). $(\alpha + \sum_{i=1}^{n} X_i)/(\alpha+\beta+n)$ and $\left((\alpha + \sum_{i=1}^{n} X_i)(n+\beta-\sum_{i=1}^{n} X_i)\right)/\left((\alpha+\beta+n)(\alpha+\beta+n+1)\right)$

4.1. $X_1, \ldots, X_n \sim N(\mu, \sigma^2)$, $H_0: \mu = 125$ versus $H_1: \mu \neq 125$ (Another possible hypothesis is: $H_0: \mu \geq 125$ versus $H_1: \mu < 125$)

4.3. (i). $X \sim \text{bin}(n_j, p_j)$ with n_j the size of the sample of boys, p_j the proportion of boys that choose mathematics, $Y \sim \text{bin}(n_m, p_m)$ with n_m the size of the sample of girls, p_m the proportion of girls that choose mathematics. $H_0: p_j \leq p_m$ versus $H_1: p_j > p_m$

4.4. $Y \sim N(\alpha + \beta x_1 + \gamma x_2, \sigma^2)$. $H_0: \beta = 0$ versus $H_1: \beta \neq 0$

4.6. (i). $H_0: \mu \leq 200$ versus $H_1: \mu > 200$ (a): do not reject H_0, (b): reject H_0
(ii). $H_0: \mu \leq 220$ versus $H_1: \mu > 220$ (a): do not reject H_0, (b): reject H_0

4.7. $X_1, \ldots, X_{100} \sim N(\mu, 1)$, $H_0: \mu \geq 50$ versus $H_1: \mu < 50$, H_0 is not rejected

4.8. (i). $K = \{x: \sqrt{n}(\bar{x} - 0)/\sigma \geq 1.64\} = \{x: \bar{x} \geq 0.66\}$
(ii). no, H_0 is not rejected
(iii). $\pi(0.5; K) = 0.34$
(iv). 0.058

4.9. (i). reject H_0
(ii). 0.0008

4.11. (i). $X \sim \text{bin}(25, p)$, $p \in [0, 1]$, $H_0: p \geq 0.6$ versus $H_1: p < 0.6$, $T = X$, $K = \{17, \ldots, 25\}$
(ii). 0.055
(iii). 0.27
(iv). 0.12
(v). no, do not reject H_0 for either value of α_0

4.12. (i). $T = X$, $K = \{20, \ldots, 25\}$
(ii). $\pi(0.6; K) \approx 0.034$, $\pi(0.7; K) \approx 0.19$, $\pi(0.8; K) \approx 0.60$, $\pi(0.9; K) \approx 0.98$ with normal approximation
(iii). 0.034 with normal approximation

4.13. (i). $e = 59$, $K = \{59, 60, \ldots, 100\}$
(ii). equal

4.14. (i). $X \sim \text{bin}(n, p)$ with $n = 250$
(ii). $H_0: p \geq 0.035$ versus $H_1: p < 0.035$
(iii). $K = \{0, 1, 2, 3\}$
(iv). 0.13
(v). more observations

4.15. $n \geq 213$

4.16. $n \geq 35$

4.17. $n \geq 263$

4.18. (i). $\hat{p} = \sum_{i=1}^{30} Y_i/1800$
(ii). the p-value equals 0.30, the null hypothesis is not rejected

4.19. $\alpha \geq 0.215$

4.20. (i). $X \sim \text{bin}(n, p)$ with $n = 1000$, $H_0: p \leq 0.9$ versus $H_1: p > 0.9$

4.21. (i). $T = \sum_{i=1}^{n} W_i \sim \text{bin}(n, p)$, two-sided binomial test
(ii). $T = \sqrt{n}(\bar{Z} - 0)/S_Z$, two-sided t-test with $K_T = (-\infty, -\xi_{\alpha_0/2}] \cup [\xi_{\alpha_0/2}, \infty)$
(iv). there need not be a mistake, because the p-values can differ; under the assumption of part (iii) we go with the second statistician, whose test is more powerful

4.22. $T = X_{(1)}$ with $K_T = (-\infty, -(\log 0.9)/n]$

4.27. $\mathrm{MSE}(\sigma^2; T_c) = \sigma^4(2c^2/(n-1) + (c-1)^2)$, minimal for $c = (n-1)/(n+1)$

4.28. $X \sim \chi_k^2$ and $Y \sim \chi_l^2$, so $X + Y \sim \chi_{k+l}^2$

4.29. $T = S_X^2/S_Y^2$, $K_T = [F_{m-1,n-1;1-\alpha_0}, \infty)$

4.30. (i). $T = S_X^2/(2S_Y^2)$, $K_T = [0, F_{24,15;0.01}] = [0, 0.346]$. do not reject H_0

   (ii). 0.0728

4.34. $T = \sqrt{5}(\overline{X} - 800)/S_X$, $K_T = (-\infty, t_{4,0.05}] = (-\infty, -2.13]$, do not reject H_0

4.35. (i). $X_1, \ldots X_{20} \sim N(\mu, \sigma^2)$ independent, identically distributed. $H_0: \mu = 3585$ versus $H_1: \mu \neq$ 3585. test: $T = \sqrt{20}(\overline{X} - 3585)/S_X$, $K_T = (-\infty, t_{19,0.025}] \cup [t_{19,0.975}, \infty) = (-\infty, -2.09] \cup [2.09, \infty)$, do not reject H_0

   (ii). $p \approx 2 \times (1 - 0.6368) = 0.73 > 0.05$ based on the normal table. Do not reject

4.36. $T = (\overline{X} - \overline{Y})/S_{X,Y} \times (1/20 + 1/32)^{-1/2}$, $K_T = (-\infty, t_{50,0.025}] \cup [t_{50,0.975}, \infty) = (-\infty, -2.01] \cup [2.01, \infty)$, reject H_0, drug B is better

4.37. $T = \sqrt{10}(\overline{Z} - 0)/S_Z$, $K_T = (-\infty, t_{9,0.05}] = (-\infty, -1.83]$, do not reject H_0

4.38. method (2)

4.39. (i). $T = \sqrt{3}(\overline{X} - \overline{Y})/S_{X,Y}$, $K_T = (-\infty, t_{10,0.05}] \cup [t_{10,0.95}, \infty) = (-\infty, -1.81] \cup [1.81, \infty)$, do not reject H_0

   (ii). $T = \sqrt{6}(\overline{Z} - 0)/S_Z$, $K_T = (-\infty, t_{5,0.05}] \cup [t_{5,0.95}, \infty) = (-\infty, -2.02] \cup [2.02, \infty)$, reject H_0, brand B lasts longer

4.40. (i). $T = \sqrt{12}(\overline{X} - 150)/S_X$, $K_T = (-\infty, t_{11,0.05}] = (-\infty, -1.80]$, reject H_0

   (ii). two-sample t-test with $T = (\overline{X} - \overline{Y})/S_{X,Y}(1/12 + 1/10)^{-1/2}$, $K_T = (-\infty, t_{20,0.05}] = (-\infty, -1.72]$, do not reject H_0

4.41. (i). $H_0: \mu \leq 10$ versus $H_1: \mu > 10$, test: $T = \sqrt{10}(\overline{X} - 10)/\sigma$, $K_T = [\xi_{0.95}, \infty) = [1.64, \infty)$, do not reject H_0

   (ii). $n \geq 39$

4.42. paired t-test, $T = \sqrt{20}(\overline{Z} - 0)/S_Z$, $K_T = (-\infty, t_{19,0.005}] \cup [t_{19,0.995}, \infty) = (-\infty, -2.86] \cup [2.86, \infty)$, reject H_0, p-value between $2 \times 0.005 = 0.01$ and $2 \times 0.001 = 0.002$

4.45. $1/3$

4.47. (i). $\lambda_n = 1$ if $X_{(1)} \leq 0$ and $\lambda_n = \exp(nX_{(1)})$ if $X_{(1)} > 0$

   (ii). $2 \log \lambda = 2n \max(X_{(1)}, 0)$. If $\theta < 0$, the limit distribution is degenerate at 0. If $\theta = 0$, the limit distribution is $\exp(1/2)$. If $\theta > 0$, the sequence tends to ∞.

4.48. (i). $\lambda_n = 1$ if $\theta_0 \geq X_{(n)}$ and $\lambda_n = \infty$ if $\theta_0 < X_{(n)}$

   (ii). $\lambda_n = \theta_0^n/X_{(n)}^n$ if $\theta_0 \geq X_{(n)}$ and $\lambda_n = \infty$ if $\theta_0 < X_{(n)}$

4.49. (i). $\lambda_n = (\overline{X}/\theta_0)^{n\overline{X}} \exp(-n\overline{X} + n\theta_0)$

   (ii). χ_1^2

4.50. (i). $\lambda_n = (\theta_0 \sum_{i=1}^n X_i^2/n)^{-n} \exp(-n + \theta_0 \sum_{i=1}^n X_i^2)$

   (ii). $K = \{(X_1, \ldots, X_n): 2 \log \lambda_n \geq \chi_{1,1-\alpha_0}^2\}$

5.1. $[17.42, 19.80]$

5.2. $[17.32, 19.90]$

5.3. (i). $\mu - \nu = \overline{X} - \overline{Y} \pm \sigma\sqrt{1/m + 1/n}\,\xi_{1-\alpha/2}$

   (ii). $\mu - \nu = \overline{X} - \overline{Y} \pm S_X\sqrt{1/m + 1/n}\,t_{m+n-2,1-\alpha/2}$

5.4. pivot $T = (n-1)S_X^2/\sigma^2 \sim \chi_{n-1}^2$, interval $[(n-1)S_X^2/\chi_{n-1,1-\alpha/2}^2, (n-1)S_X^2/\chi_{n-1,\alpha/2}^2]$

5.5. $[0.27, 0.46]$

5.6. pivot $T = (S_Y^2/\tau^2)/(S_X^2/\sigma^2) \sim F_{n-1,m-1}$, interval $[S_X^2/S_Y^2 F_{n-1,m-1;\alpha/2}, S_X^2/S_Y^2 F_{n-1,m-1;1-\alpha/2}]$

5.7. (i). pivot $T = \sum_{i=1}^n \lambda X_i \sim \text{Gamma}(n, 1)$, interval $[\Gamma_{n,1;\alpha/2}/\sum_{i=1}^n X_i, \Gamma_{n,1;1-\alpha/2}/\sum_{i=1}^n X_i]$

   (ii). $\lambda = 1/\overline{X} \pm \xi_{1-\alpha/2}/(\sqrt{n}\overline{X})$

5.8. (i). exact interval based on test: $[0.40, 1.44]$

   (ii). approximate interval: $\overline{X} \pm \sqrt{\overline{X}/10}\xi_{1-\alpha/2} = [0.33, 1.27]$

5.9.
   (ii). $[(2\sum_{i=1}^{n} X_i^2)^{-1}\chi_{2n,\alpha/2}^2, (2\sum_{i=1}^{n} X_i^2)^{-1}\chi_{2n,1-\alpha/2}^2]$

5.11. (i). $(p^2(1-p))^{-1}$
   (ii). $\overline{X}^3/(\overline{X}-1)$
   (iii). $p = 1/\overline{X} \pm (\overline{X}-1)^{1/2}n^{-1/2}\overline{X}^{-3/2}\xi_{1-\alpha/2}$
   (iv). $[0.30, 0.50]$

5.12. (i). $(p(1-p))^{-1}$
   (ii). $(\overline{X}(1-\overline{X}))^{-1}$
   (iii). $p = \overline{X} \pm \sqrt{\overline{X}(1-\overline{X})/n}\xi_{1-\alpha/2}$
   (iv). $[0.23, 0.41]$

5.13. (i). $2/\overline{X}$
   (ii). $i_\theta = 2/\theta^2$, $i_{\hat{\theta}} = \overline{X}^2/2$
   (iii). $\widehat{i_\theta} = \overline{X}^2/2$
   (iv). $\theta = 2/\overline{X} \pm \sqrt{2/n}/\overline{X}\,\xi_{1-\alpha/2}$

5.14. (i). $\hat{\lambda} = \overline{X}/2$
   (ii). $\lambda = \overline{X}/2 \pm \overline{X}/(2\sqrt{2n})\xi_{1-\alpha/2}$

5.15. (i). exact interval based on test: $[0.51, 0.87]$
   (ii). approximate interval: $\overline{X} \pm (\overline{X}(1-\overline{X})/n)^{1/2}\xi_{1-\alpha/2} = [0.54, 0.9]$

5.16. (i). $\hat{\theta} = 2/\overline{X}$ and $\lambda_n = (\theta_0\overline{X}/2)^{-2n}\exp(-2n+\theta_o\sum_{i=1}^{n} X_i)$
   (ii). $\{\theta: -2n\log 2 + 2n\log(\theta\overline{X}) + 2n - \theta\sum_{i=1}^{n} X_i \geq -\chi_{1,1-\alpha}^2/2\}$

5.17. (i). $\hat{\theta} = n/\sum_{i=1}^{n} \sqrt{X_i}$
   (ii). $\lambda_n = (\theta_0\sqrt{X})^{-n}\exp(-n+\theta_0\sum_{i=1}^{n} \sqrt{X_i})$
   (iii). $\{\theta: n\log(\theta\overline{\sqrt{X}}) + n - \theta\sum_{i=1}^{n} \sqrt{X_i} \geq -\chi_{1,1-\alpha}^2/2\}$

5.18. $\hat{\theta} = \overline{X}/2 + (\overline{X}^2+4\overline{X^2})^{1/2}/(2n)$, $\lambda_n = (\theta_0/\hat{\theta})^n\exp(-\sum_{i=1}^{n}(X_i-\hat{\theta})^2/(2\hat{\theta}^2) + \sum_{i=1}^{n}(X_i-\theta_0)^2/(2\theta_0^2))$,
   c.i. $\{\theta: -n\log\theta - \sum_{i=1}^{n}(X_i-\theta)^2/(2\theta^2) + n\log\hat{\theta} + \sum_{i=1}^{n}(X_i-\hat{\theta})^2/(2\hat{\theta}^2) \geq -\chi_{1,1-\alpha}^2/2\}$

5.19. (i). pivot $\sum_{i=1}^{n}(X_i-\mu)^2/\sigma^2 \sim \chi_{50}^2$, c.i. $\left[\sum_{i=1}^{n}(X_i-\mu)^2/\chi_{50,0.975}^2, \sum_{i=1}^{n}(X_i-\mu)^2/\chi_{50,0.025}^2\right] =$ $[3.36, 7.42]$
   (ii). pivot $\sum_{i=1}^{n}(X_i-\overline{X})^2/\sigma^2 \sim \chi_{49}^2$, c.i. $\left[\sum_{i=1}^{n}(X_i-\overline{X})^2/\chi_{49,0.975}^2, \sum_{i=1}^{n}(X_i-\overline{X})^2/\chi_{49,0.025}^2\right] =$ $[3.35, 7.45]$

5.20. (i). near-pivot $(X/200 - Y/725 - (p_1-p_2))/\hat{\sigma} \sim N(0,1)$, for $\hat{\sigma}^2 = \hat{p}_1(1-\hat{p}_1)/200 + \hat{p}_2(1-\hat{p}_2)/725$, c.i. $\hat{p}_1 - \hat{p}_2 \pm \hat{\sigma}\xi_{0.975}$
   (ii). realized interval: $[-0.011, 0.142]$, do not reject H_0, because $0 \in [-0.011, 0.142]$

6.1. $\sum_{i=1}^{n} X_i$

6.2. $\sum_{i=1}^{n} X_i$

6.3. $\sum_{i=1}^{n} \log X_i$

6.5. $(\sum_{i=1}^{n} X_i, \sum_{i=1}^{n} X_i^2)$

6.6. $(\sum_{i=1}^{n} X_i, X_{(1)})$

6.11. yes

6.13. $\overline{X}$

6.14. $X(X-1)/(n(n-1))$

6.15. $(\sum_{i=1}^{n} X_i, \sum_{i=1}^{n} X_i^2)$ is sufficient and complete, UMVU for σ^2 is $\overline{X}^2 - S_X^2/n$

6.16. $\overline{X}^2 - \overline{X}/n$

6.17. (i). $X_{(1)}$
   (ii). $(n-1)/n\,X_{(1)}$

6.18. $(n+2)/n\,X_{(n)}^2$

6.19. (i). $(X_{(1)}, \sum_{i=1}^{n} X_i)$
   (ii). $\sum_{i=1}^{n}(X_i-\mu)/n$

6.20. (i). yes

(ii). $V(X_1, \ldots, X_n) = (\sum_{i=1}^{n} \log X_i, \sum_{i=1}^{n} \log(1 - X_i))$

(iii). $n^{-1} \sum_{i=1}^{n} \log X_i$

6.21. $\overline{X} - \overline{Y}$

6.24. (i). $(X_{(1)}, \sum_{i=1}^{n} X_i)$

6.25.

(ii). no

(iii). $\alpha = m\tau^2/(n\sigma^2 + m\tau^2)$

(iv). $(\sum_{i=1}^{m} X_i + \sum_{j=1}^{n} Y_j)/(m + n)$ is then UMVU for μ

6.26. $\text{var}_\theta(T_n) \geq \theta/n$ is sharp

6.27. (i). x_1^2

(ii). $\sum_{i=1}^{n} x_i^2$

(iii). $\text{var}_\theta(T_n) \geq (\sum_{i=1}^{n} x_i^2)^{-1}$

(iv). yes

6.28. $i_\theta = (1 + 2\theta)/(2\theta^2)$, $\text{var}_\theta(T_n) \geq 2\theta^2/(n(1 + 2\theta))$

6.29. (i). $\text{var}_\theta(T_n) \geq 1/(n\lambda^2)$

6.30. (i). $\text{var}_\lambda(T_n) \geq 1/(nk\lambda^2)$

6.32. (i). $T = \sum_{i=1}^{n} e^{X_i} - n$ with $K_T = [0, \Gamma_{n,1;\alpha_0}]$

(ii). $T = \sum_{i=1}^{n} e^{X_i} - n$ with $K_T = [0, \Gamma_{n,1;\alpha_0}]$

6.33. (i). $\psi(x) = 1_{x_{(1)} \leq 2} + 4^n(4^n - 1)^{-1}(0.05 - 4^{-n})1_{x_{(1)} > 2}$ if n is sufficiently large that $0.05 > 4^{-n}$

(ii). $\psi(x) = 1_{1/2 < x_{(1)} \leq 1} + 0.05 \times 1_{x_{(1)} > 1}$

6.34. (i). $\psi(x) = 1_{x_{(1)} > 2} + 0.05 \times 1_{x_{(1)} \leq 2}$

(ii). $\psi(x) = 1_{x_{(1)} > 2} + 0.05 \times 1_{x_{(1)} \leq 2}$

7.1. (the duration is the response variable Y) (i). $Y = \alpha + \beta x + e$, $\hat{\alpha} = 109$, $\hat{\beta} = -5.6$

(ii). 0.14

7.3. $\hat{\alpha} + \hat{\beta}x$

7.6. $\hat{\beta} = (\sum_{i=1}^{n} x_i Y_i z_i^{-1} - \sum_{i=1}^{n} \sum_{j=1}^{n} x_i Y_j (z_i z_j Z)^{-1})/(\sum_{i=1}^{n} x_i^2 z_i^{-1} - \sum_{i=1}^{n} \sum_{j=1}^{n} x_i x_j (z_i z_j Z)^{-1})$, $\hat{\alpha} = (\sum_{i=1}^{n} Y_i z_i^{-1} - \hat{\beta} \sum_{i=1}^{n} x_i z_i^{-1})/Z$, $\hat{\sigma}^2 = n^{-1} \sum_{i=1}^{n} (Y_i - \hat{\alpha} - \hat{\beta}x_i)^2 z_i^{-1}$, with $Z = \sum_{i=1}^{n} z_i^{-1}$

7.7. the vector $(\sum_{i=1}^{n} Y_i, \sum_{i=1}^{n} Y_i^2, \sum_{i=1}^{n} Y_i x_i)$ is sufficient

7.8. (i). $Y_i = \beta x_i + e_i$ with $x_i = \sqrt{l_i}$ and $\beta = 2\pi/\sqrt{g}$

(ii). $\hat{\beta} = \sum_{i=1}^{n} Y_i x_i / \sum_{i=1}^{n} x_i^2$

(iii). $\text{var}(\hat{\beta}) = \sigma^2 / \sum_{i=1}^{n} x_i^2$

(iv). long

7.10. Additive: $Y_{ijkl} = \mu + \alpha_i + \beta_j + \gamma_k + e_{ijkl}$, possibly extended with interactions

7.11. (i). $P(Y = 1 | X_1, X_2, X_3) = (x_1, x_2, x_3)) = (1 + e^{-\beta_0 - \beta_1 x_1 - \beta_2 x_2 - \beta_3 x_3})^{-1}$, with X_1 the blood pressure, $X_2 = 0$ if a man and $X_2 = 1$ if a woman, $X_3 = 0$ if genotype (A_1, A_2) or (A_2, A_2) and $X_3 = 0$ if genotype (A_1, A_1).

(ii). as in (i), but with X_3 equal to the number of alleles A_2

7.13. $\lambda(t) = \lambda_0(t)e^{\beta_0 + \beta_1 x_1 + \beta_2 x_2 + \beta_3 x_3 + \beta_4 x_4}$ with X_1 the age, X_2 the weight, $X_3 = 0$ if man and $X_3 = 1$ if woman, $X_4 = 0$ if mechanical and $X_3 = 1$ if biological.

A.1. $EX^2 = \theta + \theta^2$

A.2. For $X \sim \text{Poisson}(\theta)$, we have $E(X(X - 1)(X - 2)) = \theta^3$, which equals 1 when $\theta = 1$.

A.3. $\sqrt{e}$

A.4. If $X, Y \sim \exp(\theta)$, then $Z = \max(X, Y)$ has density $2\theta e^{-\theta z}(1 - e^{-\theta z})$ for $z > 0$ and $EZ = 2/\theta - 1/(2\theta)$. If $\theta = 1$, then $EZ = 3/2$.

A.6. (i). $P(X + Y \leq 2) = \Phi(1/\sqrt{3}) = 0.718$

(ii). $\xi = 1 + \sqrt{3}\Phi^{-1}(0.95) = 3.849$

A.7. If $Z = X^2 + Y^2$, then $P(Z \leq z) = 1 - e^{-z/2}$.

A.8.

(iii). e^{x^2-y} for $y > x^2$

(iv). $E(Y|X = x) = x^2 + 1$

(v). $EY = 3/2$

A.10. $EY = n/(n + 1)$, var $Y = n/(n + 2) + (n/(n + 1))^2$

A.11. $E\overline{X}_n = \mu$, var $\overline{X}_n = \sigma^2/n$, $E(\overline{X}_n)^2 = \sigma^2/n + \mu^2$, $\text{cov}(X_i - \overline{X}_n, \overline{X}_n) = 0$

A.13. $P(X \leq 30) \approx 0.898$ (with continuity correction)

A.14. $P(\overline{X}_n \geq 4.5) \approx 0.868$ (without continuity correction)

Index